LINCOLN IN AMERICAN MEMORY

Lincoln in American Memory

MERRILL D. PETERSON

New York Oxford
OXFORD UNIVERSITY PRESS
1994

Oxford University Press

Oxford New York Toronto
Delhi Bombay Calcutta Madras Karachi
Kuala Lumpur Singapore Hong Kong Tokyo
Nairobi Dar es Salaam Cape Town
Melbourne Auckland Madrid

and associated companies in
Berlin Ibadan

Published by Oxford University Press, Inc.
200 Madison Avenue, New York, New York 10016

Oxford is a registered trademark of Oxford University Press

Library of Congress Cataloging-in-Publication Data
Peterson, Merrill D.
Lincoln in American memory / Merrill D. Peterson.
p. cm. Includes bibliographical references and index.
ISBN 0-19-506570-0
1. Lincoln, Abraham, 1809–1865—Influence. I. Title.
E457.2.N4 1993
973.7'092—dc20
[B] 93-1675

3 5 7 9 8 6 4 2

Printed in the United States of America
on acid-free paper

To the Memory of Three Teachers

W. R. PURKAPLE
Manhattan [Kansas] High School

HILDEN GIBSON
University of Kansas

PERRY MILLER
Harvard University

In Gratitude

CONTENTS

LINCOLN IN AMERICAN MEMORY

I

Apotheosis

ABRAHAM LINCOLN, PRESIDENT OF the United States of America, died at 7:22 a.m. on Saturday, April 15, 1865. He had been shot just over nine hours before at Ford's Theater, in Washington, where he had gone with his wife and two guests for a performance of the English comedy *Our American Cousin*. The day had been Good Friday on the Christian calendar.

Amid the uproar and tumult of a terrified audience, the wounded President, unconscious from a shot in the head, was carried to a lodging house across the street from the theater. The bullet had struck the most vulnerable part of the brain. Charles Sumner, the Massachusetts senator, who rushed to the scene, found Mrs. Lincoln delirious in an adjoining room and the President, stretched diagonally on a bed too small for him, breathing heavily. Sitting at the head of the bed, he took Lincoln's hand and spoke to him. "It is no use, Mr. Sumner," said one of the doctors. "He can't hear you. He is dead." But that cannot be, the Senator protested; he is breathing. "It will never be anything more than this," came the reply. Sumner sat there all night, listening to the breathing that became almost a melody. When in the cold, wet dawn it stopped, the Surgeon General, Joseph K. Barnes, crossed the dead man's hands over his breast. Captain Robert Todd Lincoln, the President's elder son, sobbed uncontrollably on Sumner's shoulder, while the few friends and strangers still in the room knelt around the bed as the Reverend Phineas D. Gurley, the Lincolns' Washington pastor, uttered a prayer. At its close, Secretary of War Edwin M. Stanton, from his vigil at the foot of the bed, raised a hand and, tears streaming down his cheeks, declared, "Now he belongs to the ages."[1]

The continent shook as the news of Lincoln's assassination sped over it. The first telegraphic dispatches after the shooting were brief and tentative, but the story quickly unfolded. How the assailant had leapt from the President's box to the stage, brandishing a dagger, and shouted "Sic semper tyrannis" before making his exit. How the actor John Wilkes Booth had been identified as the assassin. How William H. Seward, Secretary of State, had also been brutally attacked that night, and the capital went wild with fear and rumor not knowing the extent of the conspiracy against the government. As word of these events spread, people gravitated to newspaper offices for the latest bulletin. In Worcester, Massachusetts, an older form of communication abetted the newer: church bells tolled at three o'clock in the morning to call the citizenry to prayers for their stricken leader. In Philadelphia men and women wept openly in the crowded thoroughfares, and Chestnut Street was draped in black within an hour of the final bulletin.[2]

News of the assassination traveled more slowly to armies in the field, but that scarcely lessened its impact. Union soldiers had formed an affection for "Father Abraham" attributable in part to his periodic visits to camps and battlefields. "I never before or since have been with a large body of men overwhelmed by one single emotion," an officer on duty in North Carolina would recall in his memoirs. "The whole division was sobbing together." A soldier in Virginia recorded in his journal how the courier's dispatch was read at evening parade on April 17: "A silent gloom fell upon us like a pall. No one spoke or moved. . . . Quietly, quietly we went to our rest. . . . No drill. No dress parade. No nothing, all quiet, flags at half mast, lonesome was no word for us. . . . What a hold Old Honest Abe Lincoln had on the hearts of the soldiers of the army could only be told by the way they showed their mourning for him." Feelings of anger and revenge, although present, were repressed. General John A. Logan of Illinois liked to recall how he mounted his steed and flew from one regiment to another beseeching the men not to stain their heroic deeds by revenge upon the innocent, but to abide by the murdered President's teaching of magnaminity.[3]

Sorrow—indescribable sorrow—was the first emotion of most persons. Sidney George Fisher, a Philadelphia lawyer and author, discounted the report of assassination when he first heard it from his young son, but his wife went out and bought a newspaper from which she read the story "half crying and in a tremulous voice." "I felt as tho I had lost a personal friend," Fisher wrote in his diary, "for indeed I have and so has every

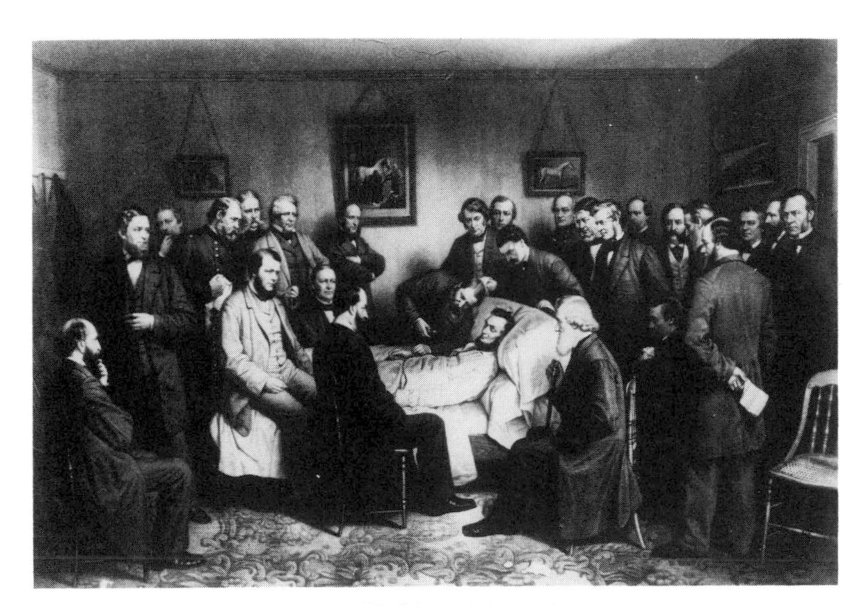

Death of Lincoln, Engraving by Alexander H. Ritchie after his painting, 1867.

honest man in the country." For many children the shock of Lincoln's death was among their earliest memories and would never leave them. Jane Addams would remember when as a child of four and one-half she found American flags draped in black on the two white gate posts in front of her Illinois home and rushed inside to inquire what they were doing there. "To my amazement I found my father in tears, something I had never seen before. . . . The two flags, my father's tears and his impassioned statement that the greatest man in the world had died, constituted my initiation, my baptism, as it were, into the thrilling and solemn interests of a world lying quite outside the two white gate posts." Samuel Gompers, already initiated into that world, "a factory boy" on New York's East Side, would write in his autobiography: "I cried and cried that day and for days I was so depressed I could scarcely force myself to work." For workingmen, he said, Lincoln represented "the spirit of humanity." In a poem, "When Lincoln Died," Katharine Lee Bates would remember herself, a five-year-old in a Cape Cod village, clutching a fold of her mother's dress as she shook the breakfast cloth from the porch "with a flourish of her own gay gallantness," when a venomous neighbor of Copperhead sympathies shouted across the yard, "They've shot Abe Lincoln. He's dead and I'm glad he's dead." The mother blanched and reeled back into the kitchen; stanching tears on the tablecloth, she declared, "I will not believe it, I'll not believe it."

But then came an old sea captain off the stagecoach with a Boston newspaper, which he carried from house to house. The poet would write:

> I heard it and hid me under the lilacs
> The mystery to prod.
> Lincoln! Lincoln! Abraham Lincoln!
> And not one angel to catch the bullet!
> What had become of God?

Lincoln's death sank into the hearts and captivated the minds of the generation that grew to maturity after the Civil War. Many of them could say, with Elbert Hubbard, "The story of Lincoln's life had been ingrained into me long before I ever read a book." So, too, the portrait image of Lincoln. It hung from a million walls like a family icon. "The child was brought up to look upon that face of sorrow and strength, to listen to that story of his wonderful patience and humanity, and to remember him as the personification of humility, love of country and sacrifice."[4]

In cities and towns across the country, ordinary business came to a halt on Saturday as people gathered in meetings to express their grief in all the mingled feelings of sorrow and rage. Sometimes the rage erupted into violence. In San Francisco two Copperhead presses were destroyed, and several "treasonable brawlers" barely escaped lynching. In military-ruled Memphis eight citizens charged with rejoicing over the assassination were sentenced to from forty days to one year at hard labor. There was no shortage of Mark Antonys to stir the crowds to revenge against the Confederacy and all who sympathized with it. For was not the assassin, Booth, an agent of the Confederacy, and was not the President the last victim of the rebellion? In Detroit the role of Antony was performed by United States Senator Zachariah Chandler, who cried *Vae victus* before a mass meeting at Campus Martius, the local parade ground. The next speaker succeeded in calming the crowd.[5]

The most notable of these public meetings took place before the Custom House in New York City. The temper of the crowd was ugly. A man who ventured to defend Jefferson Davis and the South from complicity in the crime had to run for cover under police protection. Cries of "Bring him out! Hang him!" filled the air. Hope of peace had evidently fled with the martyr's blood. General James A. Garfield, who happened to be in the city on business, sensed the mood of the crowd and sympathized with it. "It would really seem . . . ," he said from the balcony, "that this tragedy almost parallels that of the Son of God, who died saying, 'Father, forgive them; they know not what they do.' So here the rebellion has removed the barrier between themselves and justice implacable; and he [Lincoln] dying with words of tenderness and magnanimity upon his lips, forgave them all they have done. In taking away this life, the rebels have left the iron hand of the people to fall upon them." To great applause he went on, with biblical eloquence, to demand that the house of the Union "be swept and garnished . . . purged of this accursed thing." Yet, having won his audience, Garfield ended with a plea for calm. Early that morning Andrew Johnson had taken the presidential oath of office. In words that resounded from the nearby steeple of Trinity Church, Garfield declared, "Fellow citizens, God reigns and the government at Washington still lives." Although the speaker himself later became President, these words may be the most famous he ever uttered.[6]

The people who had most to lose from Lincoln's death were the freedmen, so they might have been expected to be the most vengeful. As a poor black woman cried in Washington, she had lost more than her God, for God had made her a slave, but Lincoln had made her a free woman. In Memphis, Colonel John Eaton, responding to fears of violence from the freedmen, went into their churches, and although he found a whirl of emotions he heard not one word of vengeance. "They were in despair over what might become of them now that their best friend was gone, but . . . there was no whisper against those who had sympathized with all that he opposed." And, astonishingly, for all the rhetoric of hatred and revenge in the hours and days after Lincoln's assassination, no one other than the accused murderers seems to have lost his life because of it.[7]

Easter dawned bright, but the joy and gladness of this high holiday gave way to mourning for the martyred President. It would be remembered as Black Easter. Churches were dressed in black, and ministers either laid aside their sermons celebrating the risen Christ or blended them with thoughts on the nation's tragedy. From every pulpit came the questions: What was God's purpose in allowing Lincoln to be killed? What lessons are we the living to draw from it? Two very different pairs of answers were returned. According to the first, Lincoln's work was done and so, lest he jeopardize the fruits of victory by misguided ideas of mercy and forgiveness, God had taken him to heaven and installed in his place a man of sterner stuff. "God saw this as necessary in finishing up this great rebellion. He saw that the good and kind-hearted man would not be the one for this work; for he has taken him away." It was

God's will, therefore, that traitors be punished and that the rebel states be made to feel the iron fist of justice. Yet no tears need be shed for Lincoln. Martyrdom had established his fame forever.[8]

The second set of answers, which proved fundamental to Lincoln's sanctification, conceived of his murder as an atonement for the sins of the nation and a promise of its redemption. In the words of a poet, "His blood is freedom's eucharist." That the assassination had occurred on Good Friday underscored the parallel to Jesus. Some preachers, like Edward Everett Hale in Boston, extended the parallel backward in time to the President's triumphant return from Virginia on April 9, Palm Sunday and the day of General Lee's surrender at Appomattox. "Heaven rejoices," said the Reverend Henry W. Bellows in New York, "this Easter morning in the resurrection of our lost leader . . . , dying on the anniversary of our Lord's great sacrifice, a mighty sacrifice himself for the sins of a whole people." There was no irreverence in the association of Lincoln's victory over the grave with that of Jesus Christ, Phillips Brooks assured his Holy Trinity congregation in Philadelphia. "If there were one day on which one could rejoice to echo the martyrdom of Christ, it would be that on which the martyrdom was perfected." The conception called upon the people to emulate Lincoln's virtues—the Christian virtues—of peace, love, and mercy so eloquently set forth in his Second Inaugural Address. No retribution! "I tell you," exhorted one minister, "the highest, noblest tribute you can make to Lincoln's memory is to forget how he died, in the fact that he is dead." Obviously, it made a difference whether the people of the North buried the trauma of the war with Lincoln or, instead, found motive in his assassination to wreak vengeance on the South.[9]

THE ASSASSINATION OF the President at any time would have caused an avalanche of emotion in the country; but because it came with dramatic, indeed theatrical, suddenness—a bolt in the clear sky—at the hour of victory, its force was multiplied a hundred times. The pathos was only deepened by the bolt's falling on a leader of tender and merciful heart. Walt Whitman captured the sadness in "O Captain! My Captain!" The ship of state had weathered the storm, and the captain had brought her safely to port.

> But O heart! heart! heart!
> O the bleeding drops of red,
> Where on the deck my Captain lies,
> Fallen cold and dead.

The obsequies did not end with Black Easter. They lasted for twenty days through Lincoln's funeral in Washington and the passage of his funeral train across the country to his final resting place in Springfield, Illinois. He was thus transported to immortality. But whether the assassination and the twenty days were necessary to achieve that, or whether he had created his own monument by saving the Union and emancipating the slaves—this was a question for posterity to ponder. A backward glance over Lincoln's presidency with an eye to what it foretold of his fame is requisite to understanding the emerging image of Lincoln.[10]

The man elected President in 1860 was a fifty-one-year-old lawyer-politician from Illinois. He first gained national attention in 1858 because of his celebrated debates with Stephen A. Douglas in a crucial campaign for election to the United States Senate; and he was introduced to eastern audiences by a masterly address delivered at Cooper Union in New York, and repeated in other cities, early in 1860. Several months later he was nominated for President by the Republican National Convention in Chicago. Lincoln became known to the people, as far as he became known at all, chiefly through campaign biographies. Reticent about his early life, he dismissed it as unworthy of notice, summing it up in a line from Gray's "Elegy": "the short and simple annals of the poor." He could trace his ancestry no further than his Lincoln grandfather, a Kentucky pioneer killed by Indians, for whom he was named. The grandson's life was western through and through. He had little schooling—in aggregate less than a year—but thanks to native talent, a loving mother, ambition, and a handful of books he attained the eminence of "a self-made man." He was a railsplitter, flatboatman, and country storekeeper before he became a lawyer. Tall and wiry, with a homely, sad, and thoughtful face, and a streak of humor to match his intelligence, Lincoln made himself popular with all sorts of people and acquired the affectionate sobriquet "Old Abe" while still a young man. A Whig disciple of Henry Clay in politics, he was elected to the Illinois legislature in 1834, serving there eight years, and was sent to Congress for a single term during the Mexican War. He turned his energies to the practice of law for several years, but again threw himself into politics upon the founding of the Republican party, committed to halting the expansion of slavery, in 1854.[11]

Lincoln won the presidency with only a plurality of the popular vote and no electoral votes from below the Mason-Dixon Line. Even in the North he had more doubters and scoffers than followers. Despite his impressive Inaugural Address, and the successful rally to arms after the Confederacy's attack on Fort Sumter, he appeared to many observers timid, unsure, and unequal to the task before him. "Oh, for an hour of Jackson!" was a common refrain among Republicans, while Democrats called him tyrant, and fiery abolitionists denounced him for holding emancipation hostage to the cause of the Union. Lincoln's rustic figure and manners invited ridicule, and ridicule, of course, undermined public confidence. Eastern gentility disapproved of him. He was a clodhopper and a boor—"the Illinois Baboon." Charles Francis Adams was mortified by the interview he had with the President, in the company of Secretary of State Seward, before sailing for England as United States Minister. Lincoln was shabbily dressed in ill-fitting clothes and carpet slippers; he slouched in his chair, stretched out his legs and swung his arms to his head, and, with nary a word about Adams's mission, remarked coarsely to Seward, "Well, governor, I've this morning decided the Chicago post-office appointment!" Adams never recovered from the shock. There were distinguished men—John Lothrop Motley, for instance, Lincoln's envoy to Vienna—who worshiped him from the start, but they were few. Nothing about him impressed, another departing emissary remarked, "except a certain lack of sovereignty."[12]

But Lincoln grew in public estimation as the war progressed. Obviously his place in history depended on the outcome of the war and the resolution of its tangled issues of freedom and union. Debating these matters in 1862, two congressional leaders entreated Lincoln to seek his fame under one star or the other. Said Kentucky Congressman John Crittenden: "There is a niche in the temple of fame, a niche near Washington, which should be occupied by the statue of him who shall save his country. Mr. Lincoln has a mighty destiny. It is for him, if he will but step into the niche." To which Illinois Congressman Owen Lovejoy responded: "I, too, have a niche for Abraham Lincoln; but it is in Freedom's holy fane, and not in the blood besmeared temple of human bondage; not surrounded by slave fetters and chains, but with the symbols of freedom. . . . That is a fame worth living for; aye, more, that is a fame worth dying for." These speeches, it was said, were read to Lincoln in the White House, and he seemed particularly moved by the picture Lovejoy drew.* The Emancipation Proclamation of January 1,

---

*A poet agreed with Lovejoy:

> When Lincoln's hand, O Crittenden!
> Shall dip within his heart the pen
> That writes the nation free—
> Then, towering where angels climb,
> His starry soul shall stand sublime,
> And, throned upon all Future Time
> There shall his statue be.

1863, raised Lincoln's stature among abolitionists and their sympathizers, even if it did not go to the lengths they favored.[13]

As the tides of war shifted in 1863, marked by the victories at Gettysburg and Vicksburg, Lincoln appeared ever more resolutely in command of affairs. In January 1864, James Russell Lowell, the Cambridge Brahmin, published an article in the *North American Review* which was a small milestone in the making of the Lincoln image because it showed raw western energy, previously a reproach, being transformed into native genius. Lowell praised Lincoln's statesmanship and predicted history would rank him among the truly great leaders. He drew a parallel to Henry IV, first of the Bourdon kings of France: "Henry went over to the nation; Mr. Lincoln has steadily drawn the nation over to him. One left a united France; the other, we hope and believe, will leave a reunited America." Lincoln was rude and uncultivated, but we should rejoice, Lowell said, that our leader in this struggle for the American soul was "out of the very earth, unancestried, unprivileged, unknown."[14]

Lowell's was still a minority opinion in 1864. In that election year, the President was but grudgingly supported by the Radical Republicans in Congress and faced what he feared was a formidable challenge from the Democratic candidate, General George B. McClellan, a discredited but popular military figure. The vomit of the Copperhead presses was unceasing, as in the following parody of the Union war cry "We Are Coming, Father Abraham":

> We are coming, flatboat tyrant, in mourning goods and tears,
> To hear your stories and your jokes, we trust no more for years!
> We are coming, widow-maker, from home and glen,
> A half million widows of slowly murdered men.

Lincoln need not have worried over the outcome of the election. He lost but twenty-one electoral votes and enjoyed a popular vote margin of four hundred thousand in a poll of four million. One of the Radicals, Ignatius Donnelly, acknowledged Lincoln's almost miraculous power with the people and his true greatness even as he opposed his measures:

> He is a great man. Great not after the models of the world, but with homely and original greatness. He will stand out to future ages in the history of these crowded and confused times with wonderful distinctness. He has carried a vast and discordant population safely and peacefully through the greatest of political revolutions. . . . If to adapt, persistently and courageously, just and righteous principles to all the perplexed endings and changes of human events, and to secure in the end the complete triumph of those

> principles, be statesmanship, then Abraham Lincoln is the first of statesmen.[15]

Lincoln's second term began with an auspicious inauguration. For the first time the sculptured Genius of Liberty, crowning the dome of the Capitol, looked down upon the ceremony. At the moment Lincoln rose to take the oath of office, the sun burst through the dark clouds of the day and bathed his head in a halo, sending thrills of awe through the audience, then just as suddenly disappeared, shrouding the proceedings in gloom—prophetically, some thought in retrospect. Minutes before, Lincoln had concluded his address, a speech marked by political weight, moral dignity, and unaffected solemnity, on a high spiritual note which, although it fell flatly on many ears, struck those attuned to the moral dimensions of the conflict as the most profound of all the President's utterances. "What do you think of the inaugural?" young Charles Francis Adams wrote to his father. The railsplitting lawyer was a wonder, after all. The address, Adams said, was "for all time the historical keynote" of the Civil War. "Not a prince or a minister in all Europe could have risen to such an equality with the occasion." This man, so much scorned as a "mere village lawyer," seemed destined to become one of " 'the foolish things of the world' " who confound the wise.[16]

Just after Richmond fell before General Grant's army, Lincoln made an unannounced visit to the abandoned Confederate capital. He came up the river in Admiral David Porter's flagship, landing from a barge more than a mile south of the city. Negro laborers at once recognized "Massa Linkum" in his stovepipe hat and gathered round him with cries of joy and exultation. "Bress de Lord, dere is de great Messiah!" one exclaimed. "I know'd him as soon as I seed him. He's been in my heart fo' long yeahs, an is cum at las' to free his childen from deir bondage. Glory hallelujah!" And he fell upon his knees before the President and kissed his feet, Porter later recalled. Amid an ever-growing and jubilant throng, the President, with his twelve-year-old son Tad in hand and the Admiral at his side, marched through the burned-out streets to the capitol. Men, women, and children danced and clapped hands, whooped and hollered, and sang "Glory to God! glory, glory!" Many reached out to touch the President. "I know that I am free," said one, "for I have seen Father Abraham and felt him." All the while the Richmond citizenry looked on in stony silence. After a time, Lincoln stopped to catch his breath. An old black man came to him, removed his hat, bowed, and exclaimed, with tears rolling down his cheeks, "May de good Lord bless you, President Linkum!" Lincoln lifted his hat and bowed in return. "It was a death shock to chivalry," observed a reporter.

The march completed, Lincoln entered the executive mansion recently vacated by Jefferson Davis and now the headquarters of the Union army. Whether or not he actually sat in Davis's chair, the whole country was informed that he had by Petroleum V. Nasby, the caricatured Copperhead letter-writer created by the satirist David R. Locke: "Lincoln rides into Richmond! A Illinois rail-splitter, a buffoon, a ape, a goriller, a smutty-joker, sets himself down in President Davis's cheer and rites dispatches!" The Richmond proceedings were hardly calculated to endear Lincoln to the defeated Confederacy; but they fell in with the sequence of events since the Emancipation Proclamation to consolidate the affections of the Negro freedmen for their deliverer.[17]

On his return up Chesapeake Bay aboard the *River Queen*, newspapers disclosed the day after he died, the President entertained his guests with readings from his favorite author, Shakespeare. He loved *Macbeth* above all the other plays and from it spoke the pensive lines:

> Duncan is in his grave.
> After life's fitful fever he sleeps well;
> Treason has done his worst; nor steel, nor poison,
> Malice domestic, foreign levy, nothing,
> Can touch him further.

Did the shadow of death pass across his brow as he uttered these words? Poets and philosophers might wonder. It was one of many moments in Lincoln's last days that, in retrospect, rhymed with the mythopoeic resonances of his assassination. But these last days were filled with happiness. A close friend detected a marvelous change in Lincoln's whole bearing and appearance upon his return from Richmond. "That indescribable sadness which had previously seemed to be an adamantine element of his very being, had been suddenly exchanged for an equally indescribable expression of serene joy as if conscious that the great purpose of his life had been achieved. . . . Yet there was no manifestation of exultation, or ecstasy. He seemed the very personification of supreme satisfaction." On Monday, April 10, the whole country celebrated the surrender at Appomattox. The next evening Lincoln made his last speech, a plea for speedy restoration of the rebel states to the Union, to a throng before the White House. On the fourteenth, the very day of his assassination, the Stars and Stripes was raised at Fort Sumter, exactly four years after the ensign had been struck under the barrage that commenced the war.[18]

LINCOLN'S FUNERAL TOOK place on Wednesday, April 19. It was a beautiful spring day across the country. There were obsequies everywhere. Sentiment in Washington had favored burial there. The vault under the dome of the Capitol originally intended to receive Washington's remains seemed the appropriate place. However, it was decided—by the grieving widow, Mary Todd Lincoln, twenty-one-year-old Robert, and David Davis, Associate Justice of the United States Supreme Court, who was the family's counselor—that Lincoln should rest in his hometown on the Illinois prairie. Within hours of the President's death, a public meeting of the citizens of Springfield adopted resolutions offered by the City Council urging that he be buried and memorialized in the place where he had lived and risen to fame; and a committee under Governor Richard Oglesby was named to attend to it. With the family's concurrence, the matter was settled.[19]

It was a military funeral, organized by Secretary of War Stanton with the assistance of President Johnson and several senators and representatives who happened to be in Washington. It would only begin there, for the body of Lincoln was to be placed on board a funeral train for a 1700-mile journey to Springfield, retracing, with minor variances, the route of the President-elect to the capital in 1861. So compelling was the people's wish to share in the mourning for Lincoln that precedents were pointless, but had one been wanted for this farewell journey, it might have been found in the funeral train that bore Henry Clay from Washington to Lexington, Kentucky, in 1852.[20]

The Martyr President lay in state in the East Room of the White House the day before the funeral rites. During eight hours an estimated twenty-five thousand people passed by the open casket mounted on a canopied catafalque in the center of the black-draped room. Laid out in the same black suit, with white collar and black cravat, he had worn at his inauguration in 1861, Lincoln appeared natural, with a benign smile on his face "Death has fastened upon his frozen face all the character and idiosyncracy of life," a reporter remarked. Nearly six hundred attended the noon service on the nineteenth, which was by invitation only. The widow, still distraught, remained confined to her room upstairs. The Reverend Mr. Gurley's sermon dwelt on Lincoln as a man of God who, moreover, believed himself an instrument of God. In closing, he found Lincoln's epitaph in Tacitus's *Life of Agricola:* "Never went thou in the midst of life's triumph as now in the luster death has shed upon thee."[21]

At two o'clock bells tolled and the long—two-hour long—funeral procession down Pennsylvania Avenue to the Capitol commenced. A Negro regiment just arrived from the front inadvertently placed itself at the head of the procession. It was followed by other military units—infantry, cavalry, artillery, marine—with the bands beating dirges. Marshal, clergy, and pallbearers followed. The hearse, drawn by six gray horses, afforded spectators a view of the coffin under its canopy. Just behind was the slain man's mount, followed by a carriage bearing his sons, then by delegations from the several states, the President, the cabinet, the diplomatic corps, the multitude of government employees, convalescents, Union Leagues, fire companies, workingmen's societies and lodges, and at the rear thousands of Negroes in many groups. The dead President lay in state in the Rotunda all the next day. Rain was no deterrent to mourners. From dawn to dusk two endless lines of black-clad figures filed past the coffin. For some of them the memory of Lincoln's face was fixed in the mind forever, as if carved in marble.[22]

In the cities, governing bodies proclaimed the funeral day; business was generally suspended at noon, and sermons were preached in countless churches. In Philadelphia, Phillips Brooks, the young Episcopal divine, who spoke so often about Lincoln during the season of mourning that he seemed to become "an old and close friend," thanked God for allowing the President to live until every black face in the land was radiant with freedom. In Concord, Massachusetts, Ralph Waldo Emerson saluted Lincoln as an "aboriginal man" unspoiled by European culture and tradition. The manners of the living Lincoln had troubled Emerson; with many others, however, he had struggled to overcome that prejudice, and he now embraced him as the inspired leader of the nation. "Rarely was man so fitted to the event." The philosopher thought Lincoln fortunate in his death, not only its timeliness but its manner as well. In his journal he recalled Zoroaster saying "that violent deaths are friendliest to the health of the soul."

Union soldiers in the field halted to pay last respects to their Commander in Chief. General Joshua Chamberlain, of the Twenty-first Maine, ordered the chaplain to prepare a sermon and had the troops drawn up on an open square. "The old flags were brought to the front—battle-dimmed and torn. . . . The troops stacked arms and stood calmly behind them." The chaplain took John the Baptist's beheading as his text and unfolded the analogy in thrilling and vengeful words. Some soldiers, deeply moved, went for their arms. Chamberlain took out his own. But the chaplain abruptly closed with a prayer that quieted the troops. In some other cities, as in Washington, processions followed the solemnities. Twenty thousand marched in Memphis, and the procession in San Francisco was the largest ever seen on the Pacific coast.[23]

Early Friday morning the cortege bore Lincoln's bier from the Capitol

PROCESSION THROUGH WASHINGTON WITH THE BODY OF PRESIDENT LINCOLN

From *Illustrated London News*, 1865

From *Illustrated London News*, 1865

ARRIVAL OF THE BODY OF PRESIDENT LINCOLN AT THE CITY HALL, NEW YORK

to the Baltimore and Ohio railroad depot, where it passed before two regiments of Negro troops. "They stood with arms reversed, heads bowed, all weeping like children at the loss of a father." The funeral train consisted of eight coaches, six of them carrying mourners, one for the military guard of honor, and one—a sixteen-wheel marvel that had been built to transport the Commander in Chief to the fields of battle—for the martyr, who was accompanied on this last journey by the remains of his son Willie, over three years in his grave. The train arrived in Baltimore under rain-filled skies at 10:00 a.m. The citizens' wartime quarrels with Lincoln were seemingly forgotten in their grief for him. The coffin was borne to the Exchange Building and opened to the view of an estimated ten thousand people during three hours.

The next stop was Harrisburg. A pilot engine chugging ten minutes ahead of the funeral train announced its coming, and crowds lined the platforms of the stations along the route. "Battle-flags are displayed everywhere. Banners are at half-mast. Badges of black are seen everywhere. All elevated points, balconies, windows, housetops, are occupied by people of both sexes and all conditions." In the Pennsylvania capital, the coffin was carried through drenching rain to the State House, and under its dimly lit dome thousands of mourners paid their respects that night. "At midnight they were stopped, but armies of farmers, outlying pilgrims who had driven miles with their families, sat up all night to be there when the doors opened at seven the next morning."[24]

Later that day, in Philadelphia, Lincoln's festooned hearse was drawn through thick walls of humanity lining Broad Street to Independence Hall. There the martyr lay in state Saturday evening and all day Sunday. Many citizens recalled the address Lincoln had delivered four years and two months ago when raising a new American flag, with the star of Kansas upon it, above Independence Hall. Significantly, he had chosen to remember, not the Constitution framed in this hallowed hall, but the Declaration of Independence. After offering its egalitarian principle as the basis upon which the country must be saved, he had said, "But, if this country cannot be saved without giving up that principle"—and he paused with what seemed like ominous foresight, then continued, "I was about to say I would rather be assassinated on the spot than to surrender it." When the doors of Independence Hall closed at one o'clock Sunday morning, hundreds of mourners remained in ranks until they swung open four hours later. By ten o'clock the column wound for three miles, from the Delaware to the Schuylkill, and many waited up to five hours to see the coffined President. At Holy Trinity that Sabbath morning, Brooks delivered another eulogistic sermon, in which he portrayed Lincoln as the ideal American character.[25]

The procession rolled across New Jersey and over the Hudson River into New York City on Monday morning. The nation's most famous preacher, Henry Ward Beecher, of Brooklyn's Plymouth Church, delivered from his Sunday pulpit an eloquent prologue to the big event.

> And now the martyr is moving in triumphal march, mightier than when alive. The nation rises up at every stage of his coming. Cities and States are his pallbearers, and the cannon beats the hours with solemn progression. Dead—dead—dead—he yet speaketh. . . . Four years ago, O Illinois, we took from your midst an untried man . . . , we return him to you a might conqueror. Not thine any more, but the Nation's; not ours, but the world's. Give him place, ye prairies.

With no love of Lincoln in life, Beecher became his celebrant in death. And so did Tammany-dominated New York City. The hearse bearing the martyr was drawn by sixteen white horses through crowded streets darkened with long streamers of crepe. People looked on in silence; only the sounds of bells and minute-guns were heard. At City Hall mourners queued up to look upon Lincoln's face. It was no longer natural; indeed, some thought it grotesque, the skin discolored, the eyes and cheeks sunken. At least 120,000 people passed by the bier during twenty-four hours. Thousands more were disappointed when the casket was closed at noon Tuesday. A spectacular parade—the grandest the city had ever seen—followed. It was marred, unfortunately, by the refusal of the City Council to permit Negroes to march. At the last minute an incensed Police Commissioner Thomas C. Acton, who had previously been responsible for opening the streetcars to Negroes, allowed some two hundred who had shown up to form the rear of the procession. Four freedmen led this brave troop under a banner inscribed "To Millions of Bondsmen He Liberty Gave." At Union Square, after the parade passed, George Bancroft, the historian, delivered an oration; prayers were said, hymns were sung; and the crowd heard a poem from the pen of William Cullen Bryant.

> In sorrow by the bier we stand.
> Amid the awe that hushes all,
> And speak the anguish of a land
> That shook with horror at thy fall.[26]

The cortege moved up the Hudson to Albany, thence along the Erie trunkline to Buffalo. As it progressed, every city en route sought to surpass the pomp and pageantry of prior tributes. More magnificent arches over the tracks, more splendid catafalques, more beautiful garlands for the coffin, more mournful dirges. As Lloyd Lewis would write, "The thing had become half circus, half heartbreak." And not only in the cities. Whole villages lighted the night with bonfires and torches and lined the tracks to bid goodbye to Abraham Lincoln. Nothing was more impressive, said one of the mourners on the train, than the frequent sight of a farmer in the field dropping his hoe or halting his team, removing his hat, and bowing his head as the cortege passed. Cleveland, Columbus, Indianapolis. The caravan chugged into Chicago on Monday, the first of May. The city bulged with visitors from miles around. The procession followed Michigan Avenue to the Court House. "Under a tremendous arch surmounted by Lincoln's bust, the casket was carried, thirty-six high school girls strewing flowers upon it and walking beside it." Lincoln lay in state for twenty-six hours. Mourners numbering approximately 125,000 entered the rotunda through a door bearing the inscription "Illinois clasps to her bosom her slain but glorified son" and existed under the legend 'The beauty of Israel is slain upon the high places."[27]

The funeral train reached its destination on May 3, twenty days after the assassination in Washington. (The assassin, Booth, had preceded Lincoln to the grave.) Emotions ran high in Springfield. Everyone remembered Lincoln's affectionate farewell in February 1861: "Here I have lived a quarter of a century, and have passed from a young to an old man. . . . I now leave, not knowing when, or whether ever, I shall return, with a task before me greater than that which rested upon Washington." The town fathers, with the Governor's support, had made plans to bury Lincoln, initially in a temporary vault, on grounds acquired near the city's center, upon which a fitting tomb and monument would be erected. However, Mrs. Lincoln objected to the separation of her husband's remains from those of other members of the family, beginning with the lamented Willie, and insisted upon burial in a newly opened wooded cemetery, Oak Ridge, on the outskirts of Springfield. She prevailed. Lincoln's open casket rested in Representatives Hall in the State House, whose walls had once resounded to the "House Divided" speech. The last tears shed over his body were the tears of his friends and neighbors of central Illinois. On Thursday afternoon, the fourth of May, a huge, sweating throng marched to Oak Ridge, where Methodist Bishop Matthew Simpson delivered the last ponderous eulogy over the deceased. The services closed with a hymn sung to hallow the grave of the immortal:

The Funeral Procession in Chicago. (Illinois State Historical Library)

Mourners before the Lincoln Home in Springfield. (Illinois State Historical Library)

This consecrated spot shall be
 To Freedom ever dear
And Freedom's son of every race
 Shall weep and worship here.

O God! before whom we, in tears,
 Our fallen Chief deplore;
Grant that the cause, for which he died,
 May live forever more.

The Reverend Mr. Gurley, the author of the hymn, then delivered the final benediction.[28]

Thus ended the grandest funeral spectacle in the history of the world. An estimated one million people had looked upon the martyr's face. This was the stuff of myth and legend. Prairie farmers would say the brown thrush did not sing for a year after Lincoln's death. Watchmakers' clocks, according to a persistent legend, were henceforth set at 7:22, forever to mark the time of that death. After such a spectacle few men cared to dispute Lincoln's heroic epitaph. The worst effect was yet unrealized, a journalist suggested: "It has made it impossible to speak the truth of Abraham Lincoln hereafter." Observers remarked on the pervasive unity and depth of feeling among the bereaved, and wondered whether the cause lay with the events Lincoln made or with the character of the man himself. The people, rather the white people, of the states of the defeated Confederacy were doubtless an exception—a significant exception—but in the North even Copperhead Democrats, like Clement Vallandingham and the editors of the *New York World*, repented and acknowledged Lincoln's greatness.[29]

Among the poets, Walt Whitman summoned the inspiration to engrave the funeral permanently in American memory.* Whitman, oddly, had never met Lincoln, though in his comings and goings as a "wound dresser" in Washington he had observed him so often, and so admiringly, that he seemed a friend. "When Lilacs Last in the Dooryard

---

*Of other poems of the apotheosis, Richard Henry Stoddard's *Abraham Lincoln. An Horatian Ode* (New York, 1865) most effectively incorporates the processional theme:

Go darkly borne, from State to State,
Whose loyal, sorrowing cities wait
 To honor all they can
 The dust of that good man.

Go, grandly borne, with such a train
As greatest kings might die to gain
 Thus the just, the wise, the brave,
 Attend thee to the grave.

Bloom'd" is a deeply felt elegy in which the theme of love and death and mourning merges into a hymn of praise for the nation. The poem employs three symbols from nature. The lilac outside the door, "with delicate-color'd blossoms and heart-shaped leaves of rich green," will be forever associated with the poet's mourning. The evening star, which in that spring hung low in the western sky and burned unusually bright before falling, represents his departed comrade. The hermit thrush, described to Whitman by a naturalist and friend as "the most sweet and solemn of all our songsters, heard at twilight in the dusky cedars," warbles "the carol of death."

> Ever-returning spring, trinity sure to me you bring,
> Lilac blooming perennial and drooping star in the west,
> And thought of him I love.

Having formed these symbols into a sustained threnody, the poet breaks off a sprig of lilac and lays it on Lincoln's coffin.

> Coffin that passes through lanes and streets,
> Through day and night with the great cloud darkening the land,
> With the pomp of the inloop'd flags with the cities draped in black,
> With the show of the States themselves as of crepe-veil'd women, standing,
> With processions long and winding and the flambeaus of the night,
> With countless torches lit, with the silent sea of faces and the unbared
> heads,
> With the waiting depot, the arriving coffin, and the sombre faces,
> With dirges through the night, with the thousand voices rising strong and
> solemn,
> With all the mournful voices of the dirges pour'd around the coffin,
> The dim-lit churches and the shuddering organs—where amid these you
> journey,
> With the tolling tolling bells' perpetual clang,
> Here, coffin that slowly passes.
> I give you my sprig of lilac.

The poet asks what pictures he should hang on the walls of the burial-house of him he loves, and breaks into a love chant for America reminiscent of *Leaves of Grass*. The pictures are of "fresh sweet herbage under foot," "Manhattan with spires," "Ohio's shores and flashing Missouri." The person of Lincoln has no presence in the poem, nor is it about his services and ideals. It is a love song

For the sweetest, wisest soul of all my days—and this for his dear sake,
Lilac and star and bird twined with the chant of my soul,
There in the fragrant pines and cedars dusk and dim.[30]

The nation's show of mourning did not end when Lincoln's body was laid to rest in Springfield. His sacrifice could not be forgotten amid events celebrating the victory of Union arms. The greatest of these was the last review of the Army of the Potomac in the nation's capital on May 23. General Chamberlain, as he marched at the head of his soldiers before the reviewing stand occupied by President Johnson, General Grant, members of the cabinet, foreign ambassadors, and other dignitaries, felt the absence of the hero of it all. "But we miss the deep, sad eyes of Lincoln coming to review us after each sore trial. Something is lacking in our hearts now—even in this supreme hour." In perhaps the most imaginative lyric of the apotheosis save Whitman's, Henry Howard Brownell invoked the Grand Review:

Grandest of mortal sights
The sun-browned ranks to view—
The colors ragg'd in a hundred fights
And the dusty Frocks of Blue!

Unhappily, the soldiers on marching by did not meet Father Abraham's eye, but the poet conjures up a grander review of all the lost battalions in heaven. The colors ripple, the drums roll, as all the faithful dead salute their President.

For thee, O Father we died!
And we did not die in vain,

. . .

Close round him, hearts of pride!
Press near him, side by side,—
Our Father is not alone!
For the Holy Right he died,
And Christ, the Crucified
Waits to welcome his own.[31]

President Johnson, in official acknowledgment of Lincoln's death, proclaimed a national day of fasting and prayer on June 1. The Fast Day sermons generally turned into eulogies of Lincoln and praise of his moral and spiritual virtues. Boston chose this day for its tribute. Church services in the forenoon were followed by a grand parade in eight divisions, which wended its way through the crooked streets and stopped before the

Music Hall at 4:00 p.m. There the observances included a choral hymn written by the city's poet laureate, Oliver Wendell Holmes. Sung to the music of Luther's "Judgment Hymn," it implored God to be "thy orphaned Israel's friend" and with the blood of the father wash away the sins of the children.

The program was chiefly memorable for Charles Sumner's eulogy, "Promises of the Declaration of Independence and Abraham Lincoln." The Senator pointed out that this was only the second time the American people had been summoned by the President to unite on an appointed day to commemorate the death of a great leader, the first having been George Washington. There was, moreover, a hidden link between these events, for Lincoln completed the work began by Washington. His turning in the 1850s to the promise of liberty and equality in the Declaration of Independence, then clinging to it "with all the grasp of his soul"—"this grand pertinacity"—was the most interesting incident in his biography and his chief title to glory. Sumner seemed to have a clearer view of this matter than anyone else. In the end, he thought, "it was the sacrificial consecration of those primal truths embodied in the birthday Declaration of the Republic, which he had so often vindicated, and for which he had announced his willingness to die," that caused Lincoln's death. Sumner, of course, had thought Lincoln slow to act on these truths, yet he had acted, and for the prodigious achievement of emancipation commanded a great place in history. His example, moreover, was "authority for the future." The Senator finally turned his eulogy into a stump speech for Negro suffrage.[32]

The voice of Europe, of the world, on the assassination was heard in the United States in May and June. English opinion had been overwhelmingly unfriendly to President Lincoln, a mixture of calumny, contempt, and ridicule. It had begun to change for the better, however, after the Emancipation Proclamation and the Gettysburg Address. Goldwin Smith, professor of modern history at Oxford, wrote an appreciative article in *Macmillan's* magazine. The President's reelection exhibited his strength with the people, and with the noble Second Inaugural Address his reputation soared. The assassination struck everywhere, William Gladstone said, "with a thrill of horror." John Bright, always Lincoln's friend and advocate, thought no event had created such a sensation in fifty years. "The whole people positively mourn," he wrote to Sumner, "and it would seem as if again we were one nation with you, so universal is the grief." Even the *London Times* made amends. But no publication had had so much fun at Lincoln's expense as *Punch*. Tom Taylor, who, curiously, was the author of *Our American Cousin*, penned the magazine's apology under a cartoon, "Britannia Sympathizes with Columbia," in which Britannia tearfully lays a wreath at the martyr's bier. What! the poet asks, you who with mocking pencil traced his "length of shambling limb . . . his unkempt, bristling hair . . . his garb uncouth . . . his lack of all we prize as debonair"?

> Yes, he had lived to shame me for my sneer,
> To lame my pencil, and confute my pen—
> To make me own this kind of prince's peer,
> The rail-splitter a true-born king of men.* [33]

The verdict in France, where there had always been more sympathy for the United States, was even more emphatic. The Comte de Montalembert, a prominent writer, in a leading review, said the United States had astonished the world by placing a common man at the head of affairs, yet under his leadership had vindicated not only its Constitution but liberty, democracy, and humanity. Lincoln's eulogy was everywhere, and it behooved Frenchmen "to engrave in our souls with our lives this pure and noble memory." John Bigelow, the American minister in Paris, was amazed by the outpouring of grief. The French Academy offered a prize for the best poem on "La Mort du President Lincoln." Forty thousand French citizens subscribed to a memorial medal for presentation to Mrs. Lincoln. "Tell Mrs. Lincoln," Bigelow was instructed when he received the medal, "that in this little box is the heart of France."

That the nations of the world should send condolences upon the death of a foreign head of state was to be expected, but all ranks and conditions of men grieved for Abraham Lincoln, as if to say that in homage to him all men were brothers. The address of the citizens of an Italian village declared, "Abraham Lincoln was not yours only—he was also ours . . . a brother whose great mind and fearless conscience guided a people to union, and courageously uprooted slavery." The legation at Bern sent to the Secretary of State two bound volumes of three hundred addresses from twenty-one Swiss cantons, municipal governments, and associations, plus twenty thousand signatures, "the aggregate and congregate voice of all Switzerland." The government of Argentina proclaimed three days of mourning. In Santiago and Valparaiso, Chile, men wept in the streets, passed resolutions, and marched in processions to honor the man who was "the incarnation of modern democracy."

---

*To which Alice Cary made a devastating retort:

> What need hath he now of a tardy crown,
> His name from mocking jest and sneer to save
> When every ploughman turns his furrow down
> As soft as though it fell upon his grave.

Cartoon by John Tenniel in *Punch*, 1865.

The full extent of international tribute would become manifest upon the government's publication early in the new year of a 717-page volume, with the cheerless title *Appendix to the Diplomatic Correspondence of 1865*, in which the condolences were gathered. As a historian has said, it is "one of the most interesting and deeply affecting books in the English language." The assassination of Lincoln touched the world's heart because Lincoln was the human being he was, but it also impressed itself on the mind of the world because the nation's survival and triumph over this catastrophe proved the strength and resiliency of American democracy.[34]

THE APOTHEOSIS OF Abraham Lincoln had to do with myth as well as history, but remembered history is always penetrated with myth. The martyr was instantaneously deified both because of the dramatic structure of the events surrounding his death and because of public esteem for him as a man and statesman. The image people had of Lincoln—the way they perceived, thought, and felt about him—was inseparable from the Civil War. Yet in his magnificent humanity Lincoln transcended the war. Indeed, for many in the North, his person and his character constituted the chief glory of the war; and as its colors faded into the past, the memory of Lincoln became the most treasured legacy of that conflict. The immediate aftermath of his death was hardly the time to form a just estimate of Lincoln's place in history; nevertheless, editors, politicians, poets, portraitists, and preachers essayed that task. Five main themes run through this work, and like the movements of a symphony they interpret and reinterpret each other. These themes present Lincoln as Savior of the Union, Great Emancipator, Man of the People, the First American, and the Self-made Man. Nationality, Humanity, Democracy, Americanism, and the individual opportunity which is its essence: these are the building blocks of the Lincoln image.[35]

Lincoln's great object in the war was to preserve the Union, and he died in the knowledge he had succeeded. Again and again, to the Confederate leaders who would destroy the Union and to the abolitionists whose primary purpose was emancipation of the slaves, Lincoln underscored his resolve. As he wrote in a famous letter to Horace Greeley in 1862, "My paramount object in this struggle *is* to save the Union . . . If I could save the Union without freeing *any* slave I would do it, and if I could save it freeing *all* the slaves I would do it, and if I could save if by freeing some and leaving others alone I would also do that." The victory of northern arms crushed state sovereignty and established the permanency and supremacy of the Union. "Beginning in darkness and doubt," a writer in the *North American Review* observed, "national interests and policy have insensibly conquered the first place in the estimation of the people." The fiery furnace of war had forged a new nation, a new people, for whom nationality was almost a religion, declared Yale theologian Horace Bushnell. "We have gotten a pitch of the grand, new, Abrahamic statesmanship, unsophisticated, honest and real." In the political catechism of Union Leagues, Loyal Legion commanderies, and the Grand Army of the Republic formed in the war's aftermath, the first lesson taught by Lincoln's life was the inviolability of the American Union. And the lesson taught by his death was the heroic sacrifice needed to sanctify the nation. "The blood of the martyrs was the seed of the church," editorialized a Philadelphia newspaper. "So the blood of the noble martyr to the cause of freedom will be the seed that will fructify to the great blessing of this nation."[36]

Lincoln's name was constantly coupled with Washington's. In fact, that began before the President's death, but the apotheosis firmly enforced the pairing in the public mind. Funeral banners proclaimed, "Washington the Father, Lincoln the Savior." Their portraits, with vignettes of their lives and deeds, were joined in lithographs by Currier and Ives and other printmakers. An old engraving of Washington's apotheosis was adapted to serve Lincoln's, while John Sartain inserted Washington among the angels ushering Lincoln into heaven. The six-

Abraham Lincoln the Martyr Victorious. Engraving by John Sartain, 1865.

teenth President, like the first, was that historical rarity, a great and powerful man who was also good. Not surprisingly, some people urged that he be entombed beside Washington at Mount Vernon. Of course, there was more to contrast than to compare between Washington and Lincoln. Even if it could be argued that the latter completed the work begun by the former, Washington, as Sumner pointed out, had been quite oblivious to the conception of a nation founded on Jefferson's proposition "that all men are created equal." Washington, too, was a god in the South as well as the North. He remained, as he had always been, a truly national hero, while Lincoln was, and for some time would remain, a partial one. Lincoln might be a symbol of nationality, reborn and reinvigorated, but he had almost no place in the hearts of a large portion of his countrymen. Mainly for this reason Lincoln did not displace Washington's paramouncy in the nation's pantheon. Yet there were those, like Rutherford B. Hayes, the Ohio Republican, who dissented from the usual qualification of Lincoln's greatness: "second only to Washington." Said Hayes, "The truth is, if it were not sacrilege, I should say Lincoln is overshadowing Washington. Washington is formal, statue-like—a figure for exhibition. But both men were necessary to complete our history. Neither could have done the other's work."[37]

Lincoln's fame as the Great Emancipator flowed from the Emancipation Proclamation, announced September 22, 1862, and decreed January 1, 1863. "Now who had done the greatest deed / Which History has ever known?" wrote a Union journalist of these events.

> And who in Freedom's direst need
> Became her bravest champion?
> Who a whole continent set free?
> Who killed the curse and broke the ban
> Which made a lie of liberty?
> You, Father Abraham—you're the man!

The fact that the Emancipation Proclamation freed virtually no slaves, since it extended only to areas still in rebellion against the United States, did not deter the enthusiasm of abolitionists, white and black. The war had been "invested with sanctity," Frederick Douglass said. The President who "began by playing Pharaoh . . . ended by playing Moses." Driven by his "grand pertinacity," Lincoln became convinced that for the Union to live slavery must die. The Thirteenth Amendment added the necessary constitutional sanction. Among the many icons generated by the war, perhaps none left a more vivid impression than Francis B. Carpenter's painting *The First Reading of the Emancipation Proclamation Before the Cabinet*. It was exhibited in the Rotunda of the Capitol before Lincoln's second inauguration and then toured the northern cities. Engravings of it were hung in homes, public offices, and schoolrooms.[38]

The image of Emancipator belonged particularly to American Negroes. They had come to honor, love, and revere Lincoln during his lifetime. The demonstration in the streets of Richmond was only the most dramatic instance of that worship. The New Year's Day reception at the White House in 1865 was remarkable for the attendance of Negroes for the first time. The President welcomed them, it was reported, "with a heartiness that made them wild with exceeding joy. They laughed and wept, and wept and laughed, exclaiming through their blinding tears, 'God bless you!' 'God bless Abraham Lincoln!' 'God bless Massa Linkum!' " Already Negroes were beginning to celebrate the anniversary of the Emancipation Proclamation as their "independence day." No people had more cause to mourn the President's death. Like the Christ of "The Battle Hymn of the Republic," he had died to make them free. Every Negro church was clad in black for his funeral; everybody wore bands of black crepe, or made do with strips cut from bonnets or flour sacks dipped in chimney soot. Soon Lincoln's bust might be seen on church altars. Negroes were among the very first to propose raising a monument to him.

But the image of Lincoln as Emancipator was not the exclusive possession of Negroes, nor was it limited to the historical act of the emancipation of the slaves. For the act was taken as the harbinger of a new era of freedom for oppressed humanity everywhere. The foreign response to Lincoln's death showed that. William Lloyd Garrison, the Massachusetts abolitionist leader who, with many of his followers, had begun by damning Lincoln but ended by praising him, understood it. "And, after all," he asked when addressing the annual meeting of the American Anti-Slavery Society three weeks after the assassination, "what was it that endeared Lincoln to mankind so universally? It was not simply that he came from among the people, a 'rail-splitter,' reaching the highest position in our country; it was not simply that he was an honest and amiable man; it was the consciousness that he incarnated in his position, and in his heart, the great cause of universal freedom."[39]

The image of Lincoln as Man of the People had two main aspects. First, it represented him as one of the plain people, a child of poverty and toil, rough-grained like the country he inhabited. As Richard H. Stoddard visualized him in his Horatian ode:

> A laboring man, with horny hands,
> Who swung the axe, who tilled the lands—
> One of the people, born to be
> Their conscious epitome.

He was not a remote and splendid figure, like Washington or Napoleon. He seemed more like Cromwell, warts and all, some writers suggested. All agreed, fundamentally, that he was unlike anyone who had before appeared on the world stage. Lincoln was a new kind of hero, the democratic hero, "elevated from the people, without affluence, without position, either social or political, with nothing to commend him but his own heart and sagacious mind." Editor Greeley, of the *New York Tribune*, whose political bouts with Lincoln were notorious, observed just after his death that, Lincoln being essentially a commoner, "the masses thought of him as one with whom they had been splitting rails on a pleasant Spring day or making a prosperous voyage down the Mississippi on an Illinois flat-boat, and found him a downright good fellow." Other Presidents may have sprung from the common people, but none was so absolutely one with them, instinct with their humanity, Greeley thought. There was a paradox here, however, for nobody believed that Lincoln was merely a popular type. He was rather, an uncommon common man: the peerless democrat.[40]

The other facet of the image celebrated Lincoln's faith in democracy, his shrewdness as a popular leader, his uncanny reading of the public mind, and his ability to inspire trust in people. The fact that he succeeded in his perilous task, that he trusted the people and, no less important, that the people had the wit to trust him, was the ultimate test, in European eyes, of the fledgling democratic experiment in America. It was axiomatic in Europe that great power corrupted the state and its rulers. Lincoln disproved that. His record was not spotless, but he had used extraordinary power responsibly for virtuous ends. Greeley still felt that Lincoln had suffered as a wartime leader from excesses of patience, constancy, and leniency. In the general opinion, however, these attributes contributed to his democratic genius. "It was the surest proof of Mr. Lincoln's sagacity and the deliberate reach of his understanding," Lowell wrote, "that he never thought time wasted while he waited for the wagon that brought his supplies. The very immovability of his purpose, fixed always on what was attainable, laid him open to the shallow criticism of having none—for a shooting star draws more eyes . . . than a planet—but it gained him at last such a following as made him irresistible." The writer George H. Boker voiced the same sentiment in his verse epitaph:

> Great in his goodness, humble in his state,
> Firm in his purpose, yet not passionate,
> He led the people with a tender hand,
> And won by love a sway beyond command.

The model American democrat had also, at Gettysburg, offered a definition of democracy, "government of the people, by the people, for the people," that was as deep as it was succinct—one that would capture the imagination of popular leaders the world over.[41]

The idea of Lincoln as the First American received its classic statement in Lowell's "Commemoration Ode," written for the Harvard Commencement in July 1865. In this long meditation on the heroic dead, on freedom, and on the nation's promise, the poet laid his wreath at the urn of the Martyr Chief:

> Nature, they say, doth dote,
> And cannot make a man
> Save on some worn-out plan,
> Repeating as by rote:
> For him her Old-World moulds aside she threw,
> And, choosing sweet clay from the breast
> Of the unexhausted West,
> With stuff untainted shaped a hero new,
> Wise, steadfast in the strength of God, and true.

Lowell went on to evoke a man of "clear-grained human worth," for whom "outward grace is dust," and concluded the tribute:

> Great captains, with their guns and drums,
> Disturb our judgment for the hour,
> But at last silence comes;
> These are all gone, and, standing like a tower,
> Our children shall behold his fame,
> The kindly, earnest, brave, foreseeing man
> Sagacious, patient, dreading praise, not blame,
> New birth of our soil, the first American.

Lowell never met Lincoln, and he appears in the poem as an idea rather than a flesh-and-blood human being. Nor, in fact, did the poet recite this encomium at Harvard; it was added prior to publication, and would be remembered long after the rest of the ode was forgotten. There were, of course, many intimations of Lincoln's Americanism. Lowell himself had hinted at it in his 1864 essay. His literary friend Nathaniel Hawthorne had described the President as "the pattern American" after an interview in 1862, though Hawthorne's Boston editor thought the portrait too indelicate for publication while Lincoln lived. Brooks, in his sermon the day Lincoln lay in state in Independence Hall, called him "the best and most American of all Americans," while the editorial writer for the *New York Herald* said Lincoln was "as indigenous to our soil as the cranberry crop."[42]

In Lincoln, it was sometimes said, were the characteristic lineaments of each of the great sections of the country. He was the veritable specimen, physically, of the Yankee, as Hawthorne noted. By birth and parentage he was a southerner, whence he may have derived his keen understanding of the South. Above all, he was a westerner, his mind "genial, level-lined, fruitful and friendly," like the prairie, Lowell said. The first President whose entire life was western, he brought to his work habits of mind and character bred by a new country. Much was said of his plainness and simplicity. "His simple presence," Sumner remarked, "was like a Proclamation of Human Equality." Many marveled at the entire absence in him of vanity and affectation, malice and guile. He was open, candid, and honest. "Honest Abe" was more than a political slogan; it was God's truth. An engraver published a memento of Lincoln's death in which Diogenes with his lantern appears above a medallion portrait of Lincoln. The inscription reads:

> Diogenes his lantern needs no more;
> An honest man is found; the work is o'er.

Certain traits of character, such as kindness and charity, added to the wonderful individuality of Abraham Lincoln, though there was nothing "western" about them. One defining trait, humor, with a pungent western accent, was scarcely noticed amid the solemnities of Lincoln's death. In years to come, however, it would add a bold and salty strain to Lincoln's Americanness, indeed his folksiness, as the people went about the work of discovering themselves in the mirror held up to Lincoln.[43]

The final image, in a sense the central myth of the Lincoln epic, was that of the Self-made Man. Although by no means the first of the type, Lincoln, unancestried, unprivileged, uneducated, unknown, incarnated the ideal. "No man ever lived, who was more a self-made man than President Lincoln," wrote a biographer just after his death. "Self-made or never made" was the homey adage, and Lincoln himself took pride in the success story he had become. "I happen, temporarily, to occupy the White House," he remarked in a little address in 1864. "I am a living witness that any of your children may look to come here as my father's child has."

The campaign biographies of 1860 hinted at the conception, but it took flight in William M. Thayer's *The Pioneer Boy and How He Became President* in 1863. Twenty-six thousand copies were sold within two years, when fire destroyed the plates. Although intended especially for young readers, the book appealed to all ages. Viewing the phenomenon of Lincoln's rise, Thayer asked, "How was it done?" And he answered that

Diogenes His Lantern Needs No More. Engraving by M. B. Hall, 1865. (The Lincoln Museum, Fort Wayne, a part of Lincoln National Corporation)

Lincoln's success followed from practicing the proven virtues of honesty, kindness, temperance, industry, and pluck. The moral and educational value of the story was inestimable. It centered, of course, not on what Lincoln did or said, but on what Lincoln was, in short on his personal character. Parson Weems, in his famous *Life of George Washington*, which the "pioneer boy" had so much admired, summing up the greatness of Washington had put little stock in his public attainments. "Oh no! give us his private virtues! In these, every youth may become a Washington—a Washington in piety and patriotism—in industry and honor—and consequently a Washington in what alone deserves the name, SELF ESTEEM AND UNIVERSAL RESPECT." What most deserved the name in Lincoln's life was the triumph of the virtues of American democratic character.[44]

The five themes here set forth would be developed, entwined, embellished, revised, and recast, through several generations. While not all-inclusive, they are the guiding themes in the nation's memory of Abraham Lincoln, and, although changeable, they are remarkably resilient. The public remembrance of the past, as differentiated from the historical scholars', is concerned less with establishing its truth than with appropriating it for the present. It relies on written records, but it draws as well upon oral tradition and personal reminiscence, upon stories endlessly repeated, and upon the meanings conveyed by pictorial images, patriotic rituals, and sanctified sites. While heightening consciousness of the nation's heritage, it restages the past and manipulates it for ongoing public purposes. History, as a critical science, is only a later phase of the enterprise. The magnitude of Lincoln's achievement combined with the drama of his death made his memory especially important to the American people. He was a masterpiece, a national treasure to be preserved, loved, revered, and emulated. "It may be," said one memorialist, "that one purpose of that strange manner of summoning him to the skies was to . . . engrave his traits all the deeper into the memories of coming generations." To which Phillips Brooks, in his great sermon, added the ultimate benediction: "May God make us worthy of the memory of Abraham Lincoln."[45]

2

Shapings in the Postwar Years

THE APOTHEOSIS OCCURRED against a background of stirring national events: the final capitulation of Confederate forces, the trial of Booth's co-conspirators, and President Johnson's measures to restore the Union before the Thirty-ninth Congress convened in December. Congress had yet to pay its respects to Lincoln. Among its first acts was a joint resolution inviting Secretary of War Stanton to deliver an oration on the martyr's fifty-seventh birthday, February 12, 1866. Stanton declined, and the invitation went to George Bancroft, the notable historian of the United States. Whether this was prompted by his identification with the Johnson administration, his eulogy in Union Square in April, or other reasons—he had pronounced Andrew Jackson's eulogy before Congress in 1845—Bancroft was an unhappy choice. Until the war all his political affiliations were Democratic, and he had but slight personal acquaintance with President Lincoln.

Addressing a distinguished audience, Bancroft, spare of frame, with a clear voice but devoid of oratorical graces, first surveyed a broad field of history, then came to the Civil War and epitomized Lincoln's fame in the Emancipation Proclamation. This was the act that aroused the sympathies of people throughout the world. Bancroft had little to say about Lincoln the man, yet coolly suggested that he may have been lacking in executive ability, courage, and imagination. The oration was chiefly memorable for its attack on the British government as friend and abettor of the Confederacy. Uttered before the entire diplomatic corps seated in the gallery of the House of Representatives, the criticism was particularly offensive to the British minister, and certainly would not have been countenanced by Lincoln himself. After almost two and one-half hours of grave and embellished discourse, the orator closed with a comparison between Lincoln and Viscount Palmerston, the British Prime Minister who had died not long after Lincoln and been buried in Westminister Abbey. The comparison struck many as inept, indeed demeaning to a statesman of Lincoln's humanity. Among literary men the oration found admirers, but to Lincoln's friends, especially to his longtime Illinois friends, it was cold, ill-informed, and presumptuous. "I believe that Lincoln is well understood by the people," remarked John Hay, formerly the President's private secretary, "but there is a patent-leather, kid-glove set who know no more of him than an owl does of a comet blazing into his blinking eyes. . . . Their effeminate natures shrink instinctively from the contact of a great reality like Lincoln's character."[1] This biting criticism not only implied a higher estimate of Lincoln's character and fame than Bancroft's but also pointed up the importance which some persons were already placing on a full and truthful understanding of his life.

Who should write Lincoln's biography? Who should paint his portrait and carve his features in marble? Who should build his grand monument? And where? Such questions begged for answers in the aftermath of Lincoln's death and funeral. The record, the sermons and essays, the verse and oratory of the apotheosis added immensely to the stock of Lincolniana. New biographies, hastily got up to satisfy public demand, appeared within a year of the President's death. But, of course, whatever their value, none was the ample, rich, and authoritative life that was wanted. With one voice a host of eastern litterateurs asked William Cullen Bryant, the New York poet and journalist, to undertake the work. He had met Lincoln in Illinois, introduced him to an eastern audience at Cooper Union in February 1860, and as editor of the *New York Evening Post* warmly supported his administration. But Bryant was seventy-one years of age and declined anything so laborious. Charles Eliot Norton, a Boston Brahmin with impressive literary credentials, applied to Robert Todd Lincoln for use of his father's papers and support for an authorized biography. With others of his class he had come to revere the man he had first reviled. "I conceive Lincoln's character," Norton said, "to be on the whole the greatest net gain from the war." But Robert Lincoln thought an authorized biography premature.[2] The man who set himself up as Lincoln's chief memorialist—in a manner of speaking, his Boswell—was William H. Herndon, his former law partner in Springfield. In Springfield, too, would rise the first great public monument to Lincoln.

The time was unpropitious for building a fitting monument to Lincoln in the nation's capital. Preoccupied with the deep and divisive problems of Reconstruction, Congress was content with tokens of veneration. Lincoln, of course, could not escape the history of Reconstruction. It was a testing ground for his principles and policies, and in this regard he could not avoid being measured against his successor. The southern image of Lincoln, without changing fundamentally, was subtly reshaped and redefined during the course of Reconstruction.

Lincoln, Reconstruction, and the South

The great perplexity of Lincoln and Reconstruction was simply this: Would it have been different had he lived? And there are the related questions of how it would have been different and how this would have affected his fame. Lincoln's first tentative policy on Reconstruction was developed toward Louisiana after it came under federal control in 1863. The plan granted amnesty, with certain exceptions, to rebels who would take an oath of loyalty to the United States, and it offered recognition of new state governments founded on the will of a loyal 10 percent of the 1860 electorate, contingent upon acceptance of emancipation of the slaves. With the backing of a military governor, Louisianans—whites and free blacks—proceeded to abolish slavery, adopt a new constitution, and elect a government. Congress balked at seating its senators and representatives, however, and neither Louisiana nor any rebel state was restored to the Union before Lincoln's death.

The impasse grew out of a breach between "Radicals" and "Moderates" in the Republican party. The division ran all through Lincoln's administration. Fundamentally, the factions disagreed on the relative priority assigned to winning the war and saving the Union, on the one hand, and emancipating the slaves and remaking southern society, on the other hand. The split also involved, increasingly, a conflict over presidential versus congressional control of Reconstruction. Lincoln had managed by masterful leadership to maintain the balance between these forces, alternately siding with one or the other to achieve his purposes, but without permanently alienating either. In 1864 the Radicals, generally, would have preferred another presidential candidate but finally supported Lincoln. They were dissatisfied with his Reconstruction plan, and in the Wade-Davis Bill put forward their own, one which required an oath of loyalty from a majority of white males as a prerequisite for restoration. The President used his pocket veto to defeat the bill in July. He still conceived of Reconstruction as a strategy for hastening an end to the war and, of course, remained committed to the new regime in Louisiana.[3]

The issue of Lincoln and Reconstruction turned, in a sense, on one's reading of the Second Inaugural Address. Speaking of the magnitude of the war, so far beyond the expectations or fears of either side in 1861, Lincoln sought a religious perspective on the conflict. "Both read the same Bible, and pray to the same God; and each invokes His aid against the other." After observing how strange it seemed that men could ask God's assistance "in wringing their bread from the sweat of other men's faces," Lincoln declared emphatically, "but let us judge not that we be not judged." This admonition against the dangers of self-righteousness and fanaticism was followed by reflection on God's purpose in inflicting this terrible war on the nation and culminated in a powerful passage:

> Fondly do we hope—fervently do we pray—that this mighty scourge of war may pass away. Yet, if God wills that it continue, until all the wealth piled by the bond-man's two hundred and fifty years of unrequited toil shall be sunk, and until every drop of blood drawn from the lash, shall be paid by another drawn with the sword, as was said three thousand years ago, so still it must be said "the judgments of the Lord, are true and righteous altogether."

Having thus avowed his determination to carry this moral struggle to conclusion, Lincoln ended with a no less powerful "coda of healing":

> With malice toward none; with charity for all; with firmness in the right, as God gives us to see the right, let us strive on to finish the work we are in; to bind up the nation's wounds, to care for him who shall have borne the battle, and for his widow, and his orphan—to do all which may achieve and cherish a just, and a lasting peace, among ourselves, and with all nations.

Lincoln did not contradict himself. With a chastening sense of the limits of his own moral position, he called for charity and magnanimity toward the enemy; but "with firmness in the right," he also called for finishing the work. Unfortunately, these two parts of a complex truth were easily separated, and too many men read the address as commanding one thing or the other—magnanimity or righteousness.[4]

Lincoln did not live to adapt his own plan of Reconstruction to peacetime conditions. His last public address, delivered at the White House two days after General Lee's surrender at Appomattox, was basically an appeal for the Louisiana plan. He warned that Reconstruction was "fraught with great difficulty," partly because of "the disorganized and discordant elements" with which the government must work in the South, and partly because of "deep divisions among ourselves" on how to proceed. He begged northern men to put aside pernicious abstractions on whether or not the seceded states had committed suicide or were in or out of the Union. With those states safely at home, it was immaterial whether they had ever been abroad. Pleading for the Louisiana government, he conceded it was "only what it should be as the egg is to the fowl," but averred, "We shall sooner have the fowl by hatching the egg than by smashing it." The Radicals in Congress seemed determined to smash it.[5]

Some of the Radicals secretly rejoiced in Andrew Johnson's succession to the presidency. He had never disguised his hatred of the planter aristocracy and had personally reconstructed Tennessee as its military governor. Indiana Congressman George W. Julian, disgusted with Lincoln's "false magnanimity," called his death "a godsend," and Ohio Representative Ben Wade warmly congratulated the new President, saying, "Johnson, we have faith in you. By the gods, we'll have no trouble now in running the government." Johnson's early actions seemed to justify this confidence. On May 2 he issued a proclamation for the arrest of Jefferson Davis and several other "rebels and traitors" for inciting, concerting, and procuring the assassination of Lincoln and the attempt on Seward's life. A reward of $100,000 was offered for Davis; and he would be arrested, imprisoned, and indicted, though never brought to trial. On May 29 Johnson proclaimed a general amnesty to which, however, he made numerous exceptions of persons who must apply for pardon to him, among them any with taxable property of $20,000 or more. In none of his proclamations or speeches did Johnson tie his policies to Lincoln's or represent himself as following in the martyr's footsteps. Horace Greeley, who was no Radical but an advocate of speedy reconciliation, had been unable to suppress the thought that assassination had deprived Lincoln of his true calling and, ironically, that Johnson should have been the war leader with Lincoln the healer and peacemaker at its close. For some time, certainly, Greeley found no reason to change his opinion.[6]

But the Radicals were soon disenchanted with Johnson. They dominated the Thirty-ninth Congress, and when he presented a restored Union as a fait accompli, they spurned it and proceeded to enact legislation and to impose governments designed to transform southern society and secure equal civil and political rights under the national authority. The fight in Washington culminated in the President's impeachment, while in the South Negroes rejoiced as the bottom rail came to the top, and many of the white folks wondered if the pains and penalties of Reconstruction were not worse than those of the war.

Might it all have been different had Lincoln lived? When the question was put to Norman B. Judd, an Illinois Republican who had been close to Lincoln politically, he replied: "If Lincoln had lived he would have tried charity to the rebels till all his radical friends were outraged and until he discovered the real animus of the rebels; then he would have tried severity and would have become the most popular man in the nation and would have been re-elected for a third term." The answer reflected admiration for Lincoln's superior political skills, for his pragmatism and flexibility, always responsive to the unfolding logic of events, for his hold on the affections of the people and his commitment to equal rights. Surely he would never have abandoned the freedmen to the mercies of a restored master class. As to Congress: "It is scarcely conceivable," wrote James G. Blaine, who knew that body as well as anyone, "that had Lincoln lived any serious difficulty would have arisen between himself and Congress respecting the policy of reconstruction." If he was not himself a Radical Republican, he had always sympathized with the Radicals' aims and held the upper hand with them. "Oh, if only Lincoln had lived!" Carl Schurz exclaimed at Philadelphia in 1866. "Alas, that the good President is dead! We have learned to measure the greatness of our loss by what he left behind him."[7]

The opposite opinion held that President Johnson faithfully adhered to Lincoln's policy; therefore the fate that befell one would have befallen the other. Or as an early historian of Reconstruction put it: "If Lincoln was right so was Johnson, and vice versa." Senator James R. Doolittle of Wisconsin, who in 1864 had declared that next to God Almighty he believed in Abraham Lincoln, forcefully spelled out the singular identity of "the Lincoln-Johnson policy of reconstruction" in 1866. When the unfortunate President was impeached, his friends recalled Lincoln's veto of the Wade-Davis Bill and surmised that the Radicals would have impeached him, too, given the opportunity. There was still another view of this matter, one especially favored among southerners. Had Lincoln lived, he would have pursued his policy of magnanimity, but unlike the inept Johnson he would have succeeded. "O, mysterious Fate!" moaned a southern congressman, "to quench the warm pulses of that kind heart, and turn us over to the tender mercies of bitter, implacable tyrants!"[8]

While most of the players on the political stage professed to be true to Lincoln's legacy, they were unable to agree on what it was. In part, this was because his policy on Reconstruction was not cast in stone and would undoubtedly have been modified after hostilities ended, though in just what way it was impossible to say with authority. The fact that Lincoln's own close associates divided on Reconstruction contributed to the difficulty. In the cabinet, which Johnson inherited, Edwin Stanton was on one side, Gideon Welles on the other; and in the Senate, with Sumner at one pole, Doolittle at the other, and Lyman Trumbull of Illinois in between, the whole spectrum of opinion was represented. Before long most men realized that the argument from Lincoln's authority was quite futile. For that reason, Lincoln's name did not appear prominently in debate. However, public discussion of several incidents touching on Lincoln's attitude offered illuminating, if inconclusive, testimony of another kind.

There was, for instance, the matter of Jefferson Davis's escape. Johnson, as noted, issued a ukase for the fleeing president's arrest. But on the testimony of General William T. Sherman, Lincoln had wanted Davis to escape. Not long after the assassination Sherman was repudiated and reprimanded for offering generous peace terms to General Joseph E. Johnston upon the surrender of his army in North Carolina. On the road to Washington to defend himself, the unhappy Sherman told a reporter of his meeting with Lincoln at City Point, on the James River, in March. Sherman asked if he wanted him to capture Davis, and the President replied, as was his habit, by telling a story. An old temperance preacher, traveling the dusty roads of Illinois on a hot summer day, stopped at a house to rest, and upon being offered a glass of lemonade was asked if he wouldn't like it sweetened with something stronger. He replied, "No, no," for he was a total abstainer, but after a pause remarked, "If you could manage to put in a drop unbeknownst to me, I guess it wouldn't hurt too much." Lincoln drew the moral: "Now, General, I'm bound to oppose the escape of Jeff Davis, but if you could manage to let him slip out unbeknownst like, I guess it wouldn't hurt too much." And that was all he heard on the subject, Sherman said, until Stanton berated him for letting Davis escape. (The story was a favorite with Lincoln. In another version an Irish soldier took the place of the preacher; and the President was also reported telling it to commend the escape of Jacob Thompson, the notorious Confederate agent.) Alexander H. Stephens, vice-president of the Confederacy, heard the same anecdote from Sherman; and General Grant confirmed that Lincoln wanted Davis to escape. In his *Memoirs*, Grant reflected that a generous peace in the spirit of this fable was the wise course and would have succeeded had Lincoln lived.[9]

Lincoln's position toward the South at the abortive Hampton Roads Conference in February 1865 became a subject of controversy two months after his death and continued as such well into the twentieth century. The President had agreed to talk peace with a Confederate commission, although he suspected it was under instructions from Davis to insist upon southern independence, so the talks would come to nought. No official record was kept, but all three of the Confederate commissioners—Stephens, R. M. T. Hunter, and Judge John A. Campbell—later wrote accounts of the conference.

Stephens, in his *Constitutional View of the Late War Between the*

States, represented Lincoln as holding that the future status of the slaves, except for those already freed under the Emancipation Proclamation, was a question to be decided by the courts, and that the Confederate states should be speedily restored to the Union. Stephens may have embroidered this account in his private conversation. Two statements were attributed to him. First, Stephens said that Lincoln offered $400 million in compensation for the slaves if the rebels would lay down their arms and return to the Union. Back in Washington Lincoln did, in fact, lay such a proposal before the cabinet, where it was unanimously rejected, as he knew it would be. He would not, in any event, have made an offer of his kind on his own responsibility. Hunter, in his account, dismissed it as only an offhand expression of opinion. Second, Lincoln supposedly said to the Georgian, whom he had known and respected since they were Whig congressmen together in the 1840s, "Stephens, let me write 'union' at the top of this page, and you may write below it whatever you please." Neither of his fellow commissioners, nor any authentic record, substantiated this statement, either Lincoln's making it or Stephens reporting it. But if such an overture was made, and relayed to Jefferson Davis, and he rejected it, southerners might wonder if it he had not let a golden opportunity slip through his fingers. The effect of the story was to present Lincoln as big-souled and yielding and Davis, by implication, as stubborn and heartless. Henry Watterson, the redoubtable editor of the *Louisville Courier-Journal*, often told the story of Lincoln's spurned generosity in his lectures and writings. Prominent Georgians independently confirmed it seventy-five years after the event. It was further proof that if Lincoln had lived, the long drawn-out horrors of Reconstruction would have been avoided.[10]

Judge Campbell added that Lincoln advocated convening the rebel legislature of Virginia to expedite peace. Several weeks later at Richmond he ordered General Godfrey Weitzel, in command of the occupation army, to issue a proclamation calling the legislature into session. When he was asked by another if he did not mean, rather, to convene the so-called "restored" government which was maintained at Alexandria under a loyal governor, Francis B. Pierpont, Lincoln replied, "The government that took Virginia out [of the Union] is the government that should bring her back." Regrettably, it was said in the South, Secretary of War Stanton protested Weitzel's order, and Lincoln countermanded it before his death. According to Pierpont, Lincoln told him the plan was to convene the rebel assembly to do a single act, that is, take Virginia out of the war, after which Pierpont and his government would be installed at Richmond. He recalled Lincoln's exact words: "But if I had known General Lee would surrender so soon, I would not have issued the proclamation." And so it did not imply a generous and forgiving plan of Reconstruction.[11]

Because Negro suffrage became a leading objective of the Radicals, anything Lincoln may have said on the issue was important. On March 13, 1864, he wrote a letter congratulating Michael Hahn upon his election as the first governor of liberated Louisiana.

> Now you are about to have a Convention which, among other things, will probably define the elective franchise. I barely suggest for your private consideration, whether some of the colored people may not be let in—as, for instance, the very intelligent, and especially those who have fought gallantly in our ranks. They would probably help, in some trying time to come, to keep the jewel of liberty within the family of freedom.

The new constitution made no provision for Negro suffrage. The President's recommendation had been made in confidence; it was well known to black leaders in Louisiana, however. Lincoln himself alluded to it in his last address. Not long after the assassination, Hahn, who had been elected to the United States Senate, consented to publication of the letter favoring a limited Negro franchise. As a modern historian of Reconstruction has observed, it was "hardly a ringing endorsement of black suffrage." Nevertheless, the proposal to enfranchise black soldiers who had shot at southern white men and who were sufficiently enlightened to lead the Negro community was quite advanced for the time, and its direction was unmistakable. Looking back, years after the adoption of the Fifteenth Amendment, Blaine said, "Lincoln's meaning was one of deep and almost prophetic significance. It was perhaps the earliest proposition from any authentic source to endow the Negro with the right of suffrage."

The letter to Hahn did not stand alone. There was Lincoln's regret expressed, in the last speech, that the Louisiana constitution omitted to award the franchise to black soldiers. There was his letter to General James Wadsworth in January 1864, in which he coupled universal amnesty with universal suffrage, "or, at least, suffrage on the basis of intelligence and military service." This letter like the one to Hahn, was widely published after Lincoln's death. Finally, there was the light of Lincoln's guiding principle: "No man is good enough to govern another without the other man's consent." Although he never envisioned a reconstruction on the scale of the Radicals' program—one that looked to a virtual social revolution in the South—he seemed to share their commitment to the civil and political rights of the Negroes.[12]

THE CONFEDERATE IMAGE of Lincoln took form in the election of 1860 and developed in the early years of the war. In it he was not only a "Black Republican" but a figure of vulgar satire, half-buffoon and half-gorilla. He was mercilessly caricatured as a harem-dancer lifting the veil to reveal a Negro face, as a Don Quixote astride a weary horse in pursuit of racial equality, as a coarse demagogue retailing his politics in a country store. On the stage he was broadly satirized in such plays as *The Royal Ape* and *King Linkum the First*. The satire was heavy-handed, but it conveyed the prevalent opinion that President Lincoln was no gentleman, indeed that he was the very antithesis of the Cavalier. By 1864 the ridicule was giving way to grudging respect. As Mrs. Chesnut wrote in her remarkable diary, "They have ceased to carp at him as a royal clown. . . . You never hear of Lincoln's nasty fun; only of his wisdom."[13]

Rebel reaction to Lincoln's murder ran the gamut from rejoicing to regret. "Hurrah! Old Abe Lincoln has been assassinated!" exclaimed a seventeen-year-old South Carolina girl. Lincoln had been the hatred enemy for four painful years, the incarnation of oppression and conquest, and so his death was hailed by masses of people as just retribution. "We had seen his face over the coffins of our brothers and relatives and friends, in the flames of Richmond, in the disaster of Appomattox . . . ," wrote the Virginian John S. Wise. "We greeted his death in a spirit of reckless hate, and hailed it as bringing agony and bitterness to those who were the cause of our agony and bitterness. To us Lincoln was an inhuman monster." If God Almighty ordered the event, then John Wilkes Booth must have been his avenging angel. The depiction of Booth in some places as "Our Brutus"—the last protagonist of Southern Chivalry—contributed to northern belief that the assassination was a plot of the Confederate government.

Of course, Lincoln's canonization galled the feelings of Confederate Hotspurs. Edmund Ruffin was so "utterly disgusted by the servile sycophancy, the man-worship, of a low-bred and vulgar and illiterate buffoon" that on June 18, with one last execration of the "vile Yankee race," he put a bullet in his head. Reflecting on the "*monomanic Apotheosis*," the Charleston-born poet Paul Hayne wrote, "That a man like Lowell . . . should designate the Illinois pettifogger as "the *first American*,' confounds one's sense of right and wrong, and produces a chaotic condition of the brain!" These emissions of southern loathing were echoed in the scurrilous last gasps of northern Copperheadism. Lincoln's legacy was a half-million graves of murdered men, a dissevered Union, a monstrous public debt, and the "mongrelization" of America. Three years after Lincoln's death, the New York journalist "Brick" Pomeroy was still denouncing "the shameless tyrant, justly felled by an avenging hand,

[who] rots in his grave, while his soul is consumed by eternal fires at the bottom of the blackest pit in hell." When Radical Republicans in Atlanta had the audacity to campaign for a monument to Lincoln in that city, the reaction of the natives was epitomized by the newspaper editor who suggested that it be made of the bleaching bones of Confederate soldiers and cemented with their blood: "a fitting monument of Southern devotion to the memory of him who will pass into history as the greatest and most wicked murderer of his or any other time."[14]

Yet in the wake of the assassination, editors, generals, and public officials all across the South voiced the opinion that the region had lost its best friend. Indignation meetings, so-called, were convened in many places. Lincoln stood for peace, mercy, and forgiveness. His loss, therefore, was a calamity for the defeated states. This opinion was sometimes ascribed to Jefferson Davis, even though he stood accused of complicity in the assassination. At the trial of the conspirators in Washington, a witness from Charlotte, North Carolina, testified that the fleeing Davis, who had stopped there, read the dispatch about Lincoln's death from the steps of his, the witness's, house, then said to the surrounding troops, "If it were done, it were better it were well done."* Davis himself, backed by sympathetic witnesses, flatly denied this, however. He read the telegram as sad news, and when it was greeted with an exultant shout raised his hand to check the demonstration. "For an enemy so relentless . . . we could not be expected to mourn," the Confederate president remarked. But next to the day of Lee's surrender, the darkest of all days for the South was the day of Lincoln's assassination. In the work of healing the nation's wounds he was infinitely preferable to Johnson. "He had power over the Northern people," Davis wrote in his memoir of the war, "and was without malignity towards the people of the South; his successor was without power in the North, and the embodiment of malignity to the Southern people."[15]

Respect for Lincoln and regret over his loss had nothing to do with liking and approval, however. In the literature of the Lost Cause, the two presidents were often compared, as in a Plutarchian parallel, always to Davis's advantage. "The one was an unsettled, shifting, vulgar, rollicking man—the other serious, grave, dignified, and determined. The one was plebian by nature—the other a nobleman." One met the death of a tyrant, while the other "may still stand erect over the grave of his dead foe." By the time of Davis's death in 1889, the similarities, more

* If Davis said this, he was paraphrasing *Macbeth:*

> If it were done when 'tis done, then 'twere well
> If were done quickly.

than the differences, in the lives of the two men were coming to symbolize the tragedy of the Civil War. Both were born in a Kentucky log cabin but eight months and a few miles apart. Both served in the Black Hawk War. It was even said that Davis, an army lieutenant, swore in Lincoln as captain of a company of recruits. (In fact, they did not meet.) The resemblance extended to their physical appearance. Yet, as fate would have it, one became an Illinois lawyer, the other a Mississippi planter, and both the chief protagonists of the two great sections in the Civil War.[16]

In the South's defense and vindication at the bar of history, Lincoln was inevitably the primary culprit. Had he lived, he might have redeemed himself; the war's breach might have been quickly healed, and southern feelings, "disentwined of much of the passionate clinging to the past" that made up the Lost Cause, might have been redirected into the currents of national life. But this happy outcome, as a thoughtful Virginian observed, was crossed by Lincoln's death. Many white southerners accepted the verdict of the war, and a few reveled in it, but for some of the most articulate and influential the Lost Cause was not only to be remembered but to be regained. Just as it had been the duty of every southern son to fight for the Confederacy, Wade Hampton exhorted a Richmond audience during Radical Reconstruction, "so, now, when the country is prostrate in the dust . . . every patriotic impulse should urge her surviving children to vindicate the great principles for which she fought."[17]

Edward A. Pollard's *The Lost Cause*, in 1866, started the southern apologia. A fiery Richmond editor during the war, he imbibed all the South's prejudices against Lincoln and held him ultimately responsible for provoking the war. He "revolutionized" the government by his "despotic" rule, making it good for nothing forever after. Trivial though it might seem, the President's request for the band to play "Dixie" at the celebration of Appomattox before the White House was his "last joke" on the South. Two able constitutional defenses of the Confederate cause were authored by Albert T. Bledsoe and Alexander H. Stephens. The former's *Is Davis a Traitor; or, Was Secession a Constitutional Right Previous to the War of 1861?* has been called "the classic Southern apologia." A tightly constructed legal brief, it was free of personal aspersion of Lincoln. He and the author had been acquainted in the 1840s when Bledsoe, a struggling young lawyer in Springfield, "suffered the disgrace," as he put it, of living in the same community with Lincoln. Bledsoe vented his feelings in the pages of the *Southern Review*, which he founded in Baltimore in 1867.

Stephens's ponderous treatise, *A Constitutional View of the Late War*

Between the States, appeared in two volumes in 1868 and 1870. Stephens, too, had known Lincoln; indeed, he had corresponded with him on the eve of the war and, of course, talked to him about peace at Hampton Roads four years later. Alone of the southern apologists, Stephens held Lincoln in high regard. "The Union with him in sentiment," said the Georgian, "rose to the sublimity of religious mysticism." And so he misunderstood the Constitution and inaugurated the war at Fort Sumter. In this decision he may have been misled by his advisers, as Davis also conceded; still, the responsibility was his. Having started the war, he proceeded upon a series of usurpations unparalleled outside imperial Russia that turned the Union of the fathers into a "Centralized Despotism." Greeley and other nationalists made the mistake of supposing the war was over slavery when in fact, said Stephens, along with the most die-hard defenders of the Confederacy, it was about civil liberty and the sovereignty of the states upon which that liberty depended. The causes of the war traced back to the original conflict between Jefferson and Hamilton over the nature of the Union. Slavery was but an incident. Yet Stephens, in his celebrated address after his election as vice-president of the Confederate States of America, had declared that slavery was "the immediate cause" of the war and that the cornerstone of the new government was "the great truth that the Negro is not equal to the white man; that Slavery . . . is his natural and normal condition." The abstract legalism of *A Constitutional View* helped to keep the southern mind from recognizing the true cause of the war, certainly as Lincoln understood it, and nurtured the illusion that what was surrendered in 1865 was not the state-rights principles that justified secession but only the defense of those principles by arms.[18]

In all the pathos and patriotism of the Lost Cause, whether expressed in monuments to the Confederate dead, in veterans' reunions, in the writing of history (the Southern Historical Society was founded in 1869), or in the cult of General Lee, Lincoln was little more than a dim, dark shadow. Just as he became the incarnation of the national ideal, Lee, rapidly eclipsing Davis in the southern pantheon, became the incarnation of the Lost Cause. Lee was a gallant Virginia gentleman, with an ancestry the equal of Washington's, who had known instinctively in 1861 where his primary loyalty lay. Apparently he never said a harsh word of Lincoln, or Lincoln of him, though his wife Mary was a good hater and a force in the glorification of Lee after his death in 1870. The South required its own heroes. Many southerners felt diminished and threatened by the apotheosis of Lincoln. They might utter regrets over his loss, but after that found no good reason to remember him. He was neither a symbol of reconciliation nor a historical character worthy of emulation. Southern children grew up knowing almost nothing about him. A South Carolinian recalled, "I went through school without having heard of the Gettysburg Address or First Inaugural." Yet southern patriots continued to complain, as before the war, of the subversive influence of textbooks imported from the North. Stephens wrote a history of the United States for use in the schools, and a Copperhead work manufactured in New York, Rushmore D. Horton's *Youth's History of the Great Civil War*, was long popular in the South. But the big battles over the portrayal of Lincoln and the war in textbooks awaited the twentieth century.[19]

As Reconstruction passed into history, and northern opinion largely conceded its injustice and failure, there were foreshadowings of a more appreciative view of Lincoln in the South. In 1878 "Little Ellick" Stephens, who again represented his Georgia district in Congress, praised Lincoln for his wisdom, kindness, and generosity in a well-publicized speech seconding the acceptance of the gift of Francis B. Carpenter's famous painting of Lincoln and the Emancipation Proclamation. Stephens, his wasted frame bound to a wheelchair, without in any way modifying his opinion about slavery or the Constitution, added his voice to a rising chorus calling for the dissolution of old prejudices and hatreds and supposed that Lincoln's fame was an appropriate vehicle for this work. The editor Henry Watterson, a repentant rebel, dared to pay tribute to "The Nation's Dead" on Memorial Day at the National Cemetery in Nashville in 1877. He yearned for the time when "the South" would be at last simply a geographical expression. "The day of the sectionalist is over. The day of the nationalist has come." Recalling a line from the Gettysburg Address, "that . . . these dead shall not have died in vain," Watterson affirmed, "They did not die in vain. The power, the divine power which made us a garden of swords . . . will reap thence for us, and for the ages, a nation truly divine."

Three years later a young law student at the University of Virginia, Thomas Woodrow Wilson, speaking for the southern generation that grew to maturity after the war, declared, "I yield to no one precedence in love for the South. But because I love the South, I rejoice in the failure of the Confederacy." Wilson absorbed the nationalism of the New South creed in which Lincoln loomed as a symbol of reconciliation.

The leading proponent of that creed was Henry W. Grady, editor of the *Atlanta Constitution*. In 1886 Grady, thirty-six years old, was invited to address the New England Society of New York on the 266th anniversary of the landing of the Pilgrims at Plymouth. General Sherman, seated on the platform, was an honored guest, and the band played "Marching Through Georgia" before Grady was introduced. Pronouncing the death of the Old South, he lauded the New South of union and freedom and progress. And he offered Lincoln as the vibrant symbol not alone of reconciliation but of American character. Lincoln, he said, comprehended within himself all the strength and gentleness, all the majesty and grace of the republic." He was, indeed, the first American, "the sum of Puritan and Cavalier, in whose ardent nature were fused the virtues of both, and in whose great soul the faults of both were lost."[20]

Books, Portraits, and Monuments

Mary Todd Lincoln, the disconsolate widow of the martyred President, was not an endearing woman. Coming from a prominent Kentucky family, she had the pride of her class and a domineering temperament. Her husband's background and disposition were entirely different, of course, yet their marriage was loving and fruitful; they were partners politically, and totally devoted to each other. In Washington she was a highly visible but never popular First Lady. She put on "the airs of an Empress," according to the man who looked after her White House receptions, and her spendthrift ways proved embarrassing to the President. Shattered by his murder, she went into permanent mourning. Always willful, she became more so, to the point where her son Robert, with others, feared for her sanity. She quarrelled with the town fathers of Springfield, first over her husband's burial place and then over the site of his monument. Five or six acres near the center of the city, the Mather Block, had been acquired for the proposed monument. But Mrs. Lincoln demanded that the monument be erected over Lincoln's remains. On June 14 the Monument Association submitted. Mrs. Lincoln won the fight but at the cost of Springfield's good will. Against Robert's advice, she chose to live in Chicago.

Having left Washington in debt, she pleaded poverty. Horace Greeley started a subscription fund for her, but efforts in that direction collapsed after it was reported on good authority in the press that Lincoln had left an ample fortune. The estate was in the hands of the family lawyer, Justice David Davis, and although payments were made from it to the widow and her two sons, it was not yet settled. Meanwhile, Mrs. Lincoln carried her campaign to Congress. She asked that the full four year's presidential salary, amounting to $100,000, be paid to her. Congress adhered to the precedent set in the case of President William Henry Harrison a quarter-century before, under which the widow received the balance of the annual salary of $25,000. Mrs. Lincoln protested against the comparison of her husband's service with Harrison's lackluster month in office, but Congress declined to reconsider. Mrs. Lincoln had exhausted the patience of the few friends she had in the capital by forcing President Johnson to cool his heels at the door of the White House for five weeks before vacating it.[21]

In September 1867, Mrs. Lincoln traveled to New York City and, joined by a female companion, checked into a dingy hotel room under an assumed name. Her identity along with her business was soon disclosed: to sell through commission agents dresses, furs, and jewelry from the former First Lady's wardrobe, together with certain gifts she had received. "Mrs. Lincoln's Second-Hand Clothing Sale," the New York press screamed. Articles valued at $24,000 were exhibited on Broadway. What was behind this bizarre scheme? Mrs. Lincoln, on her part, continued to plead poverty. Her meager income, $1,700 a year, she said, was beneath her station. Some observers thought there was method in her madness, for it put pressure on Congress to make adequate provision for her. The publicity backfired, however. Thurlow Weed, the New York Republican chief, attacked her for ingratitude and said that if she had conducted herself becomingly as First Lady, Congress would have felt more generous toward her. Many congressmen suspected that, in addition to over-furnishing the White House while she lived there, she "unfurnished" it when she left.[22]

The sale grossed $824, barely enough to cover expenses, and Mrs. Lincoln returned to Chicago under a cloud of humiliation. It did not go away. In the new year her companion in this misadventure, Elizabeth Keckley, published several of Mrs. Lincoln's letters to her in a highly personal account of their relationship, *Behind the Scenes*. "Lizzy" Keckley, a strikingly handsome light mulatto, had purchased freedom for herself and her son some years before securing employment as dressmaker, or modiste, to the First Lady. The two women became friends. Lizzy started one of the first contraband relief societies for former slaves who flocked to Washington, and Mrs. Lincoln contributed generously to this work. Mrs. Keckley's book offered unusual glimpses and anecdotes of domestic life in the White House. It was written, she said, with the intent of placing Mrs. Lincoln in a better light before the world. Its effect, however, was to draw attention to her faults and foibles—for instance, her prodigality, her jealousy of any woman who paid attention to the President, her distrust of his cabinet officers. "If I listened to you, I should soon be without a Cabinet" was Lincoln's intemperate response on one occasion. "I never in my life saw a more peculiarly constituted women," Mrs. Keckley confessed. "Search the world over, and you will not find her counterpart." Mrs. Lincoln might well pray to be saved from her friends. She felt betrayed, in particular, by the publication of the anguish and despair she had expressed in letters meant to be private. Angrily, she terminated the friendship—one of the few she had.[23]

President Lincoln's estate was settled in 1868. It came to about $110,000, of which the widow received one-third. Two years later Congress grudgingly voted her an annual pension of $3,000. With this addition her income could be deemed almost princely. She went abroad to live for several years, disappearing from public view, doubtless to the country's relief. Through all of this, she had very little influence on the posthumous fame of her husband. In the main, her negative image tended to increase sympathy and admiration for him. The opinion formed that his domestic life had been an ordeal, that his wife was such a termagant he must have been a saint to survive it. Among the acid portraits etched of her was that of Mary Logan, wife of the Union general from Illinois. She called Mary Lincoln a spoiled child with a violent temper and boundless ambition. After the blow of her son Willie's death in 1862, all the world began to realize what her husband had known for years, that Mrs. Lincoln was "semi-insane." In the country's hour of need, she was engrossed in household trivialities, and, of course, she did nothing to earn the country's gratitude after her husband's death. Unfortunately, Mary Lincoln's tragedy only deepened in the years to come; and many decades passed before a sympathetic portrait of her emerged.[24]

THE APOTHEOSIS SET AFOOT many movements to build grand monuments to Lincoln. Had all of them succeeded the country would have bristled with shafts and columns to the martyr-hero. They proved too ambitious, however, and with one exception all the plans were either abandoned or radically reduced in scale. Three of these movements are particularly significant for what they did, and did not, accomplish.

Washington was the most appropriate place for a major monument. And one might have been early built had Lincoln's remains been left there. His political friends and associates in the capital formed the National Lincoln Monument Association. With a design already in hand, it was incorporated by Congress in 1867. James Harlan, the Iowa senator who had been a political intimate and whose daughter, Mary, would wed Robert Todd Lincoln, was the association's president. Congress, while making no appropriation, put its stamp of approval on the association, authorized it to raise funds, and offered a site east of the Capitol. The design was by Clark Mills, best known for his equestrian statue of Andrew Jackson in Lafayette Park. Mills, with his two sons, had his studio in the city and a foundry for casting in bronze nearby. His plan was really a war memorial with Lincoln the dominant figure. Six equestrian statues of Union generals would fill the triangular granite base; on a second tier would be twenty-one statues of statesmen, philanthropists, and other civilian leaders; above that would be sculptured representations of soldiers, nurses, freed slaves, and so on; above that, allegorical figures of Justice, Liberty, and Equality; finally, crowning the whole, seventy feet in the air, a colossal statute of President Lincoln, seated at a desk, in the act of signing the Emancipation Proclamation.

The association claimed that Lincoln himself had approved of the design, or an earlier version of it, before his death. But this surpasses belief. He could not have approved of anything so grandiose. Nothing to equal it had been built in modern times. The halting progress, after a quarter-century, of the Washington Monument did not augur well for anything on the scale of Mills's project. The association set out to raise $400,000, three-fourths of it by subscriptions, one-fourth from the particular friends of the personages honored in the monument. Fundraisers were employed and twenty-two thousand subscription books put in circulation. But the movement collapsed amid ensuing political and economic turmoil and fell into oblivion.[25]

Meanwhile, the National Lincoln Monument Association organized under the laws of Illinois in May 1865 met with a better fate. Its roots were in Springfield, its principal support in the state of Illinois, but it appealed to the nation and aimed from the first to build *the* national monument to Lincoln. Simultaneous with its birth, an association of soldiers at nearby Camp Butler published a broadside appeal for a "Soldiers and Sailors Monument" to their martyred chief. In June, however, it withdrew in favor of the powerful Springfield association headed by Governor Richard J. Oglesby. It was this group, of course, which came to terms with Mrs. Lincoln on the Oak Ridge site. The city conveyed to the association a deed to six acres of land adjacent to the cemetery.

A campaign was launched to raise $250,000. In every county of Illinois agents solicited one-dollar donations; army regiments, Sunday schools, fraternal societies contributed. Children who gave fifty cents were promised a picture of the proposed monument. One later remembered drilling corn all day to earn his half-dollar and growing impatient when the picture failed to materialize. Governors of other states were asked to set aside plans they might have to memorialize Lincoln and get behind the Springfield monument. The association tried, without success, to bring the Negro-initiated Freedman's Monument (discussed below) under its wing. Some of the funds raised elsewhere for other projects, or for the relief of Mrs. Lincoln, came into the association's treasury. Still it was a long way from the goal as the year 1867 began. Oglesby, therefore, under his governor's hat, appealed to the legislature for a $50,000 appropriation. The infusion raised the fund to $125,000.[26]

The association had already solicited designs for the monument, offering a prize of $1,000 to the winner. Thirty-one artists responded, six of them submitting two designs. Rotundas and towering columns were the favored motifs, the former covering the sculptured figure, commonly Lincoln the Emancipator, the latter elevating it. During the first ten days of September 1868 the designs were on public view in the capitol. The association then awarded the prize and a huge $70,000 contract to Larkin G. Mead, Jr. Mead was a thirty-three-year-old Vermonter working in Italy. He had been the first in the field of competition and mobilized influential support. Almost from the hour of the assassination, Mead said, a fitting monument to Lincoln became the great ambition of his life; he had actually worked up his design before learning of the Springfield project. In 1866 he shipped a plaster and wood model to New York, then came home to preside over its exhibit. This plan, with some minor changes, prefigured the monument. Entrance at ground level gave access to a spacious crypt and memorial hall. Exterior stairs led to a terrace fourteen feet above. From the center of the massive base rose a towering granite obelisk, before which stood the bronze figure of Lincoln. Just below, on rounded pedestals at the four corners of the base, were military groups, representing infantry, cavalry, artillery, and marine.[27]

The monument was dedicated with much ceremony on October 15, 1874, before any of the military groups were in place. (Responsibility for raising money to complete them was assigned to four cities: Chicago, Philadelphia, New York and Boston.) President Ulysses S. Grant was among the dignitaries in attendance. He offered a few remarks, but the orator of the day was Richard Oglesby, now United States senator from Illinois, who had headed the Monument Association from its beginning. As he concluded, looking out over the crowd of twenty-five thousand people, Oglesby declared, "I dedicate this monument to the memory of the obscure boy, the honest man, the illustrious statesman, and great Liberator, and the martyr President, Abraham Lincoln, and to the keeping of Time. 'Behold the image of the man.' " The red-and-white veil fell, the choir broke into song, and the crowd applauded as it gazed upon Mead's statue.

It is a faithful and dignified likeness. Lincoln is standing at rest, his right hand on a monolith, his left extending a scroll, and there is an air of gravity on his face. Some contemporary critics thought that the monument was too grand for a man of plain and simple character; that the proportions between the statue and the obelisk were wrong (a defect later corrected); and that while it might do justice to the emancipator and martyred chief, the statue did not reach the loftier ideal of the prophetic statesman. When the monument was finally completed with the instal-

Lincoln National Monument, Springfield, 1874, by Larkin G. Mead. (Illinois State Historical Library)

lation of the military groups, it took on the appearance of a soldiers' and sailors' monument. These excited figures, in marked contrast to the impassive Lincoln, are the triumph of the monument. What are they doing there? As Lorado Taft observed, "The tomb of Lincoln is forgotten in the ill-timed vehemence of these superfluous performers."[28]

Hopes that the Springfield monument would become a mecca were not soon realized. It was plagued with problems. The foundations collapsed. Often in disrepair, the monument would be twice rebuilt during sixty years. Custody was repeatedly criticized in the press. In November 1876 the custodians suffered the indignity of an attempted robbery of Lincoln's remains by a gang of crooks. The record of troubles kept alive the conviction of many Lincoln worshipers that he deserved a better monument than the shrine in Springfield.[29]

The Freedman's Monument had its inception in the heart of a newly emancipated woman, Charlotte Scott, in Marietta, Ohio. On April 15, immediately upon hearing of the President's death, she said to her employer, "The colored people have lost their best friend on earth. Mr. Lincoln was our best friend, and I will give five dollars of my wages toward erecting a monument to his memory." The gift went by way of a local minister to James E. Yeatman, president of the Western Sanitary Commission, in St. Louis, which now made itself the agent for the freedmen's memorial fund. Although founded to care for wounded soldiers, the commission had turned to helping the freedmen along the Mississippi River. The biggest contributors to the fund were the black soldiers in that region. Colonel W. C. Earles of the Seventieth U.S. Colored Infantry, in Mississippi, for instance, forwarded almost $3,000 collected from 683 enlisted men. They want the monument built only with the coloreds' money, he said. And if it is to be built "proportionate to the veneration with which the black people hold his memory, then its summit will be among the clouds—the first to catch the gleam and herald the approach of coming day, even as President Lincoln himself proclaimed the first gleam as well as glorious light of universal freedom." By the year's end the fund had grown to $16,000. The momentum flagged, however, and the fund increased hardly at all for several years.[30]

The impulse to praise and honor Lincoln was widely felt among the freedmen. He was their Moses. To some he was a father figure—"Father Abe"—in the special sense that he was the nearest approximation to a male parent they had known. At Fourth of July celebrations and in the several black state conventions held during the summer of 1865, Lincoln was exalted as their friend and liberator. Some thought the advancement of education the best memorial to him. Two black regiments in Missouri contributed $5,000 toward the founding of Lincoln Institute in Jefferson City. In 1866 nine-year-old Ashman Institute, near Philadelphia, the first Negro college, changed its name to Lincoln University. But Charlotte Scott was not alone in calling for a monument. Major Martin Delaney, whom Lincoln and Stanton had sent to Charleston to raise a black regiment, proposed to levy one cent on every Negro man, woman, and child, and envisioned a monument which would have as its main feature a kneeling African female shedding tears into an urn—emblematic of the four million American Negroes—with arms stretching forth to God.

In Washington, July Fourth was celebrated under the auspices of the Colored People's National Lincoln Monument Association, whose president was Henry H. Garnet, a Presbyterian minister long active in the cause of his race. Part of the long program was a poem on slavery and emancipation, which concluded with the lines:

Our Lincoln Memorial of One shall speak,
Like Moses faithful, and like Moses meek;
Who led the people through a redder sea
Than Israel passed, to light and liberty.
Of him, who, humbly trusting in the Lord,
Moved by the Holy Spirit, spoke thy word;
And, as that word was plainly, *firmly*, spoken,
The bonds-man's chains fell off, the tyrant's rod was broken.

Nothing more was heard of this association, however. Very likely it merged its efforts with the freedmen's memorial under Yeatman's direction.[31]

Early in 1867 the Western Sanitary Commission tentatively accepted a design for the monument submitted by Harriet Hosmer, a young American sculptor in Europe whose patron, Wayland Crow, recommended her to his St. Louis friend, Yeatman. Her design was on the same grand scale as those of Mead and Mills. On a sixty-foot-square base of granite, the monument rose through several levels to a marble "Temple of Fame" enclosing the recumbent figure of Lincoln on a sarcophagus. There were many figures, bas-reliefs, and inscriptions, all treating the twin ideas of Savior of the Union and Great Emancipator. It was all terribly elaborate. The estimated cost, $250,000, placed it out of reach. When this became obvious the following year, Crow, on Hosmer's behalf, shipped her design to Springfield, but it apparently found little favor there.[32]

Luckily, in 1869 a member of Yeatman's commission, William G. Eliot, happened to visit the studio of the Boston-born sculptor Thomas Ball in Florence, where his astonished eyes fell upon a small marble statue, half life-size, of Lincoln in the act of emancipating a slave. Ball, famous for his equestrian statue of Washington in Boston, had been inspired to carve this "Emancipation Group," as he called it, upon learning of Lincoln's death. He now offered the statue to the commission for nothing more than the cost of casting it in bronze. Suddenly the prospect of realizing Charlotte Scott's dream revived. Eliot returned to St. Louis with photographs, and the commission promptly ordered the statue in bronze with one condition: that the figure of the slave, who appeared passive and lacking Negroid features in the original marble, be remodeled. Aware of the defect, since he had modeled the slave on a looking-glass image of himself, Ball gladly accepted the condition. He remodeled the figure upon photographs of a freedman, Archer Alexander, sent from St. Louis. The location of the monument, if ever in doubt, was settled in 1874 when Congress appropriated $3,000 for the base and pedestal of the statue and offered a fine Washington site, Lincoln Park, so named in 1866, on the main axis east of the Capitol.[33]

Freedman's Monument, Washington, D.C., 1876, by Thomas Ball. (The Lincoln Museum, Fort Wayne, a part of Lincoln National Corporation)

The Freedman's Monument was dedicated on April 14, 1876. Congress had declared a holiday in the city. The Negro community was out in force; benevolent and fraternal societies, delegations from afar, marching bands, paraded down Pennsylvania Avenue to the Capitol, thence one mile to Lincoln Park, where a large assemblage, black and white, had gathered. Yeatman presented the monument on behalf of the Western

Sanitary Commission. President Grant pulled the cord unveiling the statue to gasps of awe, followed by loud applause accompanied by booming cannon and the strains of the Marine Band. The twelve-foot-high-statue, mounted on a ten-foot pedestal, showed a benign, half-smiling Lincoln, his right hand holding the Emancipation Proclamation, his left extended graciously over the slave, who is rising on one knee, his wrist shackles broken, his face turned upward as if ready to race for freedom. He is receiving the gift of emancipation, but not submissively. Most Negroes warmly approved of the statue; some, however, disliked the kneeling posture of the freedman and wished for a more manly attitude. With respect to Lincoln, Ball froze in bronze the weary frame with the care-lined face, performing an act both noble and ideal, yet without loss of that homely kindness that was peculiar to him. The group is, as Taft said, "one of the inspired works of American sculpture."[34]

The dedication was made more memorable by Frederick Douglass's oration. Now almost sixty years of age, the bearded, white-haired Douglass, with an imposing manner and a rich voice—a consummate orator—was the leader of his people. Early in the war, with other abolitionists, he had criticized Lincoln for subordinating the cause of freedom to the cause of union, but with the Emancipation Proclamation he became a firm supporter. He went to Washington in 1863 to confer with the President on the treatment of Negro soldiers. In later accounts of that interview Douglass described the trepidation he felt upon entering the White House and how the President at once put him at ease. Douglass grasped what he called "the educating tendency of the conflict" upon Lincoln. After that he talked to him on several occasions, becoming convinced that Lincoln was "emphatically the black man's President." "He treated me as a man," Douglass told a friend, "he did not let me feel for a moment that there was any difference in the color of our skins! The President is a most remarkable man!"[35]

Yet, for Douglass, there was always the gnawing doubt that kept him from a full embrace, and he bared his complicated feelings toward the Great Emancipator in this oration. They were gathered, black and white, at the nation's capital to perform a national act, he began. It was first and fundamentally an American act. From this day, he continued, "let it be known everywhere, and by everybody . . . , we the colored people, newly emancipated and rejoicing in our blood-bought freedom . . . have now and here unveiled, set apart, and dedicated a monument of enduring granite and bronze, in every line, feature, and figure of which the men of this generation may read, and those of after-coming generations may read, something of the exalted character and great work of Abraham Lincoln." All acknowledged his greatness. Nevertheless, "truth compels me to admit," even at this moment consecrated to his memory, Lincoln was not in the fullest sense "either our man or our model." He was preeminently the white man's President with the white man's prejudices toward the black. And looking upon the white faces in the crowd, the orator declared, "You are the children of Abraham Lincoln. We are at best his step-children; children by adoption, children by force of circumstances and necessity." That Father Abraham was only a stepfather to the colored people was a startling statement, the more so on this occasion; still it was a truth that pierced clouds of sentimental rhetoric, one which thoughtful Negroes could never wholly evade. Its recognition did not, Douglass went on to say, diminish one iota the Negroes' debt of gratitude to Lincoln. They understood why he had been slow and hesitant, why saving the white man's Union came before their rescue from bondage, yet they never doubted that he was destined to be their liberator.[36]

As cities in the North and West—San Francisco, Washington, Brooklyn, New York, Philadelphia, Springfield, Boston, Chicago—dedicated statutes to the Martyr President in the twenty years or so following his death, the question of Lincoln as a subject of heroic portraiture came into discussion. The general impression was that he was an ugly man. Lincoln's humor mocked the trait. His long, lank frame and craggy features could not be depicted with the dignity and beauty required in monumental art, it was often said. There was nothing of the Roman about him. "Tall as he was," the journalist Donn Piatt wrote, "his hands and feet looked out of proportion, so long and clumsy were they. Every movement was awkward in the extreme. . . . He had a face that defied artistic skill to soften or idealize." That face—high forehead, bushy eyebrows, sorrowful grey eyes, dark complexion, sunken cheeks, prominent mole, long nose, thick lower lip, incisive bearded chin, all set between big flopping ears—could not be ennobled or prettified. And that was only the beginning of the problem, for what was one to do with the long frame and its loose-fitting clothes? The ridicule heaped upon Horatio Greenough's monumental statue of Washington—he looked like a Roman just from the bath, it was said—was a warning against idealization.

But could Lincoln be portrayed realistically? Artists sent to Springfield to paint or sculpt his portrait in 1860 were instructed "to make it good-looking whether the original would justify it or not." This proved less a problem than expected; on the whole, the first portraitists were enchanted with Lincoln as a person and as a subject. One of them, Charles A. Barry from Boston, said the Illinoisan's head and jaw were Jacksonian, signifying Old Hickory's power and determination, while the eyes in that extraordinary face, when animated, were "like coals of living fire," which recalled Daniel Webster. Unfortunately, in the opinion of the artist, Lincoln decided at the time of the presidential election to grow a beard, thereby disguising the best features of his face and adding to the melancholy of his countenance in repose. The bearded image of President Lincoln became the public image. It took some time for artists to rediscover the face under the beard.[37]

Actually, few painters or sculptors ever observed Lincoln at first hand. For the most part, they worked from photographs, of which there were a great number. But they could be misleading. Some had been retouched to prettify the subject. And they emphasized Lincoln's *gravitas* almost to the exclusion of his humor. In part, this was a defect of the medium, which demanded the subject maintain a single pose and expression for one minute. To smile was impossible. Mrs. Lincoln thus spoke despairingly of his "photographer's face"—not his real face—and his private secretary, John G. Nicolay, referred to "that serious far-away look" in these camera portraits. It "petrified" one mood, one look, when he was a person of many. The Marquis de Chambrun, who observed the President closely during the last months of his life, said that no portrait captured both the sadness and the fun of the man or exhibited psychological depth. This was a hasty judgment to pass upon such masters as Mathew Brady and Alexander Gardner, but it reflected the prevalent opinion summed up by one who did paint a life portrait, Edward D. Marchant, that President Lincoln was "the most difficult subject ever to task the skills of an artist."[38]

Lincoln, for better or worse, had no Gilbert Stuart to turn him into an icon. Of the few life portraits in oil, none made a distinct public impression. G. P. A. Healy, the foremost painter to whom Lincoln sat in 1860–61, did not complete his likeness until after the President's death, using a model and photographs, and only then as part of his large historic group *The Peacemakers;* but both Robert Lincoln and General Sherman—one of the group—thought it the best portrait of him. The painter who reached the most people was Francis B. Carpenter, a second-rate artist at best. He had been accorded the special privilege of living in the White House for six months in order to paint his historical canvas *First Reading of the Emancipation Proclamation.* While he did not make a great painting, Carpenter employed a realistic brush to fix permanently in the mind of the American people the epochal achievement of the Emancipation Proclamation. After its exhibition days were over, Carpenter sold the painting, and in 1878 it was given to Congress for installation in the Capitol. Separate portraits of Lincoln were by-products of the big painting. One of these, a life-size standing Lincoln, was acquired for the New York state capitol. Carpenter also painted *The Lincoln Fam-*

First Reading of the Emancipation Proclamation. Engraving by Alexander H. Ritchie after the painting by Francis B. Carpenter, 1867. (Illinois State Historical Library)

ily for the express purpose of facilitating an engraving, which when published became very popular. A year of so after Lincoln's death Carpenter retained an engraver, Frederick W. Halpin, to reproduce one of his portrait studies of the President for the middle-class market. At the same time William E. Marshall, who had never seen Lincoln and had just returned from Europe, produced a fine engraving, and a fine likeness, for the Boston publishing firm of Ticknor and Fields. The two artists, Carpenter and Marshall, vied with each other for endorsements and patronage. Their works were only the most prominent of the many prints that conveyed the physical image of Lincoln to posterity.[39]

Although the Union Lincoln preserved built no soaring monument to him, Congress, in 1866, voted $10,000 for a marble sculpture to be placed in the Capitol, and it directed that the commission be given to a nineteen-year-old girl, Vinnie Ream. Senator Sumner objected vociferously. The artist was unworthy of such a trust. "She might as well contract to publish an epic poem, or the draft of a bankrupt bill." The Capitol ought not be turned into an artists' junkyard, he said. But Vinnie Ream, an accomplished, high-spirited young woman from Missouri who learned sculpture from Clark Mills, had won the hearts of many congressmen, and Sumner's protest was dismissed as ungallant. She had carved several busts, including Lincoln's, from life, and was hailed as a young genius from the western prairies. Vinnie Ream was given a makeshift studio in the Capitol, so that congressmen might observe the progress of her work. The plaster model was completed in 1869. Orville H. Browning, of Illinois, said he felt upon viewing it the "actual presence" of his departed friend. The full-length marble statue was unveiled in the Rotunda two years later. Lincoln is standing, a cloak draped loosely from his shoulders, the right hand grasping the Emancipation Proclamation, his head bowed and his eyes bathed in melancholy. In this latter aspect the statue moved the viewer; for the rest; it was damned with faint praise. Of it Mrs. Logan remarked, "The first glance . . . is the most satisfactory that you will ever have." It stands today in the Capitol's Statuary Hall.[40]

The supreme work of this first phase of Lincoln sculpture was Augustus Saint-Gaudens's statue in Chicago. The will of a local businessman, Eli Bates, left $40,000 for the erection of a statue of President Lincoln in the park named for him in 1865. Saint-Gaudens, thirty-five years old, won the prized commission in 1883. He had attained fame several years before for his dramatic bronze of Admiral Farragut in New York City. As a youth he had caught a fleeting glimpse of Lincoln, and when his body lay in state at City Hall, he passed by the coffin, not once, but twice. Still he knew little about him. The artist's friends supplied him with books to extend his understanding to Lincoln the poet and prophet as well as the benevolent statesman. Richard Watson Gilder, a minor poet and editor of *Century* magazine, brought to his attention the newly rediscovered life mask of a clean-shaven Lincoln made by Leonard W. Volk, of Chicago, in 1860. (Volk, as it happened, was runner-up for the Bates commission; and he later executed a monumental bronze for Rochester, New York.) The plaster mask had been lost from sight until Gilder came upon it in the home of Wyatt Eaton, a New York painter, who had secured it from Volk's son. Better than any photograph, it was, said Gilder in a sonnet inspired by the mask, "the very form and mold of our great martyr's face."

> Yes, this is he;
> That brow all wisdom, all benignity;
> That human, humorous mouth; those cheeks that hold
> Like some harsh landscape all the summer's gold;
> That spirit fit for sorrow, as the sea
> For the storms to beat on; the lone agony
> Those silent, patient lips too well foretold.*

*Helen Keller, visiting the home of a Princeton collector where a copy of the mask hung from the wall, felt every feature and said without any prompting, "Why, it looks like Lincoln!" And, she added, it was the first time she had *seen* him without a beard.

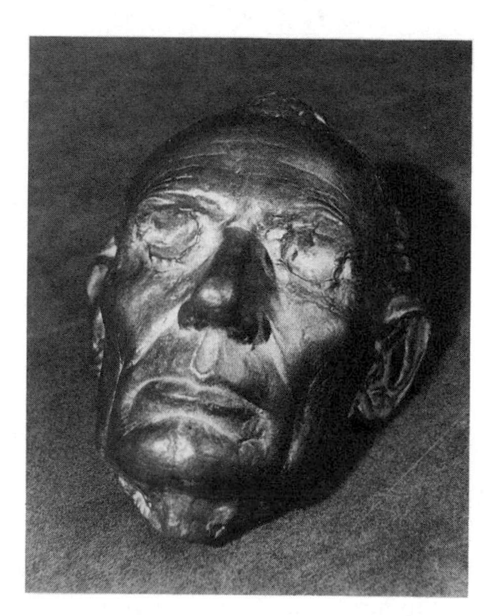

Life Mask of Lincoln. Bronze from plaster, 1860, by Leonard Volk. (Illinois State Historical Library)

Saint-Gaudens made the first bronze of the life mask, and it exerted a direct influence on his modeling of Lincoln's head. No less important was the fact that the Lincoln statue and the long relief monument known as Shaw Memorial progressed together in Saint-Gaudens's New York studio. Visitors observed them in intimacy with each other, although the latter, destined for Boston, was completed much later. Something of the nobility and the tragedy of Captain Robert Gould Shaw and his Negro soldiers got into the face of Saint-Gaudens's *Lincoln*.[41]

The statue was dedicated on a blustery October day in 1887. Leonard Swett, a survivor among Lincoln's lawyer friends, delivered the address, and the fifteen-year-old grandson of the President pulled the cord to unveil the gigantic bronze. "The truthfulness of the whole was instantaneously realized," it was said. Lincoln is represented standing before an august chair of state, as if he had just risen and was about to speak, his left hand grasping the lapel of his coat, his right hand behind his back. Of the early statues, Saint-Gaudens's was only the second to drop the allusion to the Emancipation Proclamation, thereby freeing the mind for other thoughts. Lincoln's head is thrown slightly downward, giving "the contemplative look" so characteristic of the man, as Gilder said. The expression is thoughtful, reserved, weighty, yet with a glimmer in the eyes of the humor that is close to sorrow. Lincoln is dressed in clothes such as he wore; plain and unadorned, they are creased and rumpled like the face. The sculpture gains much from the spacious and elegant architectural setting which Stanford White gave it. But the sculpture itself brilliantly laid to rest any doubts about the realistic portrayal of Lincoln.[42]

There was nothing to rival Saint-Gaudens's *Lincoln* for a quarter-century or longer. For many, seeing the statue for the first time was an electrifying experience; some returned to it again and again with feelings akin

Lincoln, Chicago, 1887. By Augustus Saint-Gaudens. (The Lincoln Museum, Fort Wayne, a part of Lincoln National Corporation)

to reverence. To Franklin K. Lane, a naturalized American, it expressed "all that America is, physically and spiritually." The long, lanky body contained all the youthful awkwardness of the country, while the eyes shone with idealism, courage, and compassion. To David Graham Phillips, the statue was "the apotheosis of Democracy." Nothing about it suggested rank or aristocracy. "What a countenance! How homely, yet how beautiful; how stern, yet how gentle; how inflexible, yet how infinitely merciful; how powerful, yet how tender; how common, yet how sublime." To the poet Carl Holliday, it was the wrinkled vest the sculptor carved, more than the noble head, that thrilled the spirit.

> That vest!—All furrowed, crumpled. Oh,
> No devotee of fashion could
> Have worn those creases to and fro
> Across his garb. No egotist
> Inquiring: What think they of me?
> Do I look prim and neat, or missed
> Some touch that adds to nicety?
> Not so this man—that wrinkled vest
> Bespeaks the toiler—one who bore
> A nation's burdens . . .

Regardless of interpretation, all admirers agreed that Saint-Gaudens had captured the soul of Lincoln.[43]

The Early Biographers: Herndon and Others

The eulogistic tone of the apotheosis colored the early biographies of Abraham Lincoln. The atmosphere was favorable to the generation of myths, pious falsehoods, and wayward traditions. The best of the books rushed into print in 1865 were by men who had known Lincoln as President and had already written a campaign life of him. Joseph H. Barrett, an Ohio Republican named by Lincoln Commissioner of Pensions, had authored one of the original lives, which he brought up to date in 1864, and again after the President's death. The 842-page book, like a number of others, was long on speeches and state papers, short on portrayal and interpretation. But Barrett presented Lincoln as a "true example of Christian character" as well as a great statesman; and in a concluding panegyric, he drew a parallel to Robert Burns, for like the Scottish poet, Lincoln, too, had risen from the poor and humble, kept his identity with them, and conducted himself always as a true democrat. Henry J. Raymond, editor of the *New York Times*, became the President's ardent champion in 1864 when, with his subject's approval, he wrote his *History of the Administration of Abraham Lincoln.* The martyr was scarcely in his grave before the publisher of that work, Derby & Miller, prevailed upon Raymond to write a "definitive life." With two assistants he worked at breathless speed in order to catch the flood tide of emotion from the apotheosis. The result was substantially a new book, not simply an expansion of its political predecessor. Yet, if somewhat fuller and more skillful, Raymond's biography differed hardly at all from Barrett's. Both authors understood that Lincoln's magnificent achievement arose from extraordinary personal character, but neither understood the sources of that character. Amazingly little had been added during five years to the sparse record of his personal life before 1860. In order to cast some light on "the elements of originality" in Lincoln's character, Raymond appended the White House reminiscences Francis B. Carpenter had published during the summer in the *Independent,* a New York weekly.[44]

This, of course, was Carpenter the painter. Three days after Lincoln's death, he communicated to the *New York Evening Post* the poem he called the President's favorite and had copied from his own lips. The poem, "Mortality" was a long and melancholy refrain, known by its opening line, "Oh! Why should the spirit of mortal be proud?" (George C. Pearson immediately set the poem to music, published in Boston.) Lincoln had encountered it in a newspaper many years earlier and committed it to memory. Neither he nor Carpenter knew its author, but the *Post* identified him as William Knox, a Scotsman little known to fame. Carpenter kept a diary in the White House, and although he had left without intending to write anything of his experience, his articles aroused so much interest that he was led to enlarge them into a book, *Six Months at the White House with Abraham Lincoln.*

Because of the unusual liberties granted to the artist, he was able to offer a close and private view of the President. He observed him for hours at a time conducting the business of the government; he observed him with members of his family; and, utterly enchanted, he had lengthy conversations with the President on a range of subjects. Talking of Shakespeare, Lincoln not only recited long passages with dramatic emotion but showed a critic's judgment in, for instance, favoring the King's soliloquy after the murder to Hamlet's. Once, apropos of Bulwer-Lytton, Lincoln remarked, "It may seem somewhat strange to say, but I never read an entire novel in my life! . . . I once commenced *Ivanhoe,* but never finished it." Incredibly, he missed Dickens altogether and presumably knew nothing of Cooper or Hawthorne or Melville. Carpenter reported touching incidents of Lincoln's executive clemency. He offered examples of his humor and anecdote, and said, contrary to some reports, he never heard anything that would be out of place in a lady's drawing room. Underneath the humor, Carpenter said, "ran a deep undercurrent of sadness, if not melancholy," and the relation between them obviously offered a clue to unraveling the mystery of the man. Conceding that Lincoln was not religious in the accepted sense, Carpenter agreed with those who said he was, nevertheless, as true a Christian as ever lived. The book was much read, for it helped to satisfy the public craving for an intimate hero, one who, unlike Washington, was loved as well as venerated.[45]

By far the most popular of the early biographies was Josiah Holland's *Life of Abraham Lincoln*, published in 1866. Some eighty thousand copies were sold. Already a successful journalist and author, Holland was editor-in-chief of a leading Republican newspaper in Massachusetts, the *Springfield Republican*, when Lincoln died. He delivered the city's eulogy of the Martyr President. It caught the eye of a Boston publisher, and Holland, recognizing the timeliness of the literary venture, contracted to write a biography for a broad audience. He had never met or seen Lincoln. In May he traveled to Illinois to talk to the people who had lived and worked with him before he became President.

Limited as this research was, it enabled Holland to write more fully and perceptively of Lincoln's early life than previous biographers. His primary interest, as he stated in the preface to his book, was in Lincoln's character, and for it his admiration was unbounded. Lincoln's triumph over the rude conditions of his birth and upbringing was a marvel. "No man ever lived, probably, who was more a self-made man than President Lincoln." Holland amended this judgment, however, by attributing to Lincoln's " 'angel mother' " a decisive influence on his development. "His character was planted in this Christian mother's life." It owed nothing, certainly, to the half-savage environment nor to his shiftless father. With typical eastern condescension, Holland apologized for Lincoln's storytelling habit, doubtless acquired in the coarse society of circuit-riding lawyers, and regretted the "great misfortune to Mr. Lincoln that he was introduced to the nation as pre-eminently a rail-splitter." It belittled him in the eyes of cultivated people. "It took years for the country to learn that Mr. Lincoln was not a boor."[46]

Holland was a devout Christian. Obviously, whether or not Lincoln was a Christian had little bearing on the future of the faith; nevertheless, for the sake of his religion, his country, and the martyr's memory, Holland set out to place Lincoln before posterity as a Christian. That he had never joined a church or made a confession of Christian faith was well known. Yet no President had drawn more deeply from the Bible or more faithfully invoked divine aid than he. When Holland visited Springfield, he asked the local oracle, William H. Herndon, about Lincoln's religion. "The less said the better," the oracle snapped. "O never mind," said the biographer with wink and a shrug, "I'll fix that." And he did. He gave great weight to the testimony of Newton Bateman, the Illinois Superintendent of Public Instruction, who had conversed at length with Lincoln just before the presidential election of 1860. Examining a book purporting to show how every citizen of Springfield would vote, the Republican candidate noted with dismay that only three of twenty-three ministers were for him. How could this be? "Mr. Bateman, I am not a Christian—God knows I would be one—but I have carefully read the Bible, and I do not understand this book." He paused. Then after a time he rose and paced the floor, stopping finally before the stunned Bateman, and in a trembling voice, cheeks wet with tears, he broke out, "I know there is a God and that he hates injustice and slavery. . . . I know I am right because I know that liberty is right, for Christ teaches it, and Christ is God."

For Holland this remarkable conversation, given him orally and in writing by Bateman, furnished "a golden link in the chain of Mr. Lincoln's history." It revealed a hidden dimension of religious experience which was the source of all Lincoln's passion for humanity. It showed that Lincoln had lived a "double life"—a shallow, fun-loving outer life and a deep inner life of sublime piety. A contradiction in himself, he must appear differently to different men. "Men caught only separate aspects of his character—only the fragments that were called into exhibition by their own qualities." Holland went on in the biography to demonstrate that Lincoln was "eminently a Christian President." And he grew in faith as the war progressed. His Christian character, said Holland, "will become the basis of an ideal man." The black race have already made him a saint. The white people will not be far behind.[47]

Lincoln's old friends and companions in Illinois were reluctant to surrender his fame to eastern men of letters like Holland and Bancroft. As they had exercised preemption rights in the building of Lincoln's public monument, so should they claim a special place in fashioning his literary monument. These Illinois luminaries included prominent and influential men: David Davis; Richard Oglesby; Leonard Swett, a Chicago lawyer; Orville Browning, who became President Johnson's Secretary of the Interior; Jesse Dubois, the state auditor; Joseph Medill and Horace White, respectively publisher and editor-in-chief of the *Chicago Tribune*; Jesse Fell, editor and publisher of the *Bloomington Pantograph*; and Isaac N. Arnold, a former congressman appointed an auditor in the Treasury Department before Lincoln's death. Swett contemplated a biography but produced nothing more weighty than reminiscence and oratory. White made valuable contributions to the political history of Lincoln's time but wrote nothing substantial on him. Browning kept a diary that would be published sixty-five years later. Of this group, only Arnold wrote a biography. A serious and studious Chicago lawyer, he had known Lincoln for twenty years and worshiped his memory.

Arnold had been an important anti-slavery congressman, so it is not surprising that his first essay in biography focused on Lincoln the Emancipator. *Abraham Lincoln and the Overthrow of Slavery* was published in 1866, the year Holland's biography. Arnold's avowed purpose was to crown Lincoln the hero of the whole long American conflict between slavery and freedom. Evidence to the contrary was simply omitted. Thus in his account of the Lincoln-Douglas debates, Arnold passed over the hero's equivocations on Negro equality and his refusal to attack slavery in the states. A New Yorker by birth, of stately and refined bearing, the author had no natural affinity for the western elements of Lincoln's character. Nevertheless, he underscored those very elements, portraying Lincoln as a rude product of the frontier, "the child of the wilderness," solitary, awkward, self-made. Arnold knew these things more from literature than from acquaintance, as suggested in his lecture on Lincoln which compared him to James Fenimore Cooper's ideal frontiersman: rough-hewn but kind and sagacious. Arnold lectured on Lincoln both in the United States and England. He completed his *Life of Abraham Lincoln* just before his death in 1884. It was not as interesting a book as the earlier one, though it helped to fill out the Lincoln portrait.[48]

Among the Illinoisans, William H. Herndon was category unto himself. He had been a young man, only just admitted to the bar, when Lincoln asked him to become his law partner in 1844. People wondered at the choice, for Lincoln was a leading attorney in the Eighth Circuit and could have acquired a better-established partner. But for whatever reason, the partnership of Lincoln and Herndon worked, and was not officially dissolved until the death of the senior member. They had some things in common. Both had been born in Kentucky; both were lawyers; both were ardent Whigs. "Billy" Herndon, however, came from a good middle-class family, attended Illinois College, and conversed with the world of books and learning. The two men were of utterly different temperaments: Lincoln cool, slow, and cautious; Herndon hot, rash, and impulsive. Lincoln's mind ran to concrete facts, Herndon's to abstractions. The junior partner read a lot and fancied himself a philosopher. To his chagrin Lincoln read little outside the newspapers, though Herndon credited him with thinking more, and with more clarity and precision, than anyone else. In an age of addiction to strong drink, Lincoln was the most temperate of men, while Herndon waged recurrent battles with Demon Rum. Some loss of social respectability was the result. Local citizens were quick to notice that President Lincoln omitted to appoint his law partner to office. Unfortunately, between him and Mary Lincoln there were only bad feelings from the day they met, and these reverberated sadly in the Abraham Lincoln constructed by Herndon's imagination.[49]

After recovering from the shock of Lincoln's death, Herndon decided to do two things: first, to develop an evidential record of Lincoln's life, especially his early life; and second, to write his biography. One might be considered necessary to the other, but Herndon sometimes thought of them as independent projects. His first idea for a book was a study of Lincoln's inner or subjective life; gradually, it changed into a full-scale biography. Some of Herndon's friends thought him uniquely qualified for such an effort. "You were perhaps more truly *en rapport* with him than any other," Horace White wrote, and ought not to wait until strangers twisted and deformed the man they knew and loved. Herndon agreed. Boasting he knew Lincoln better than Lincoln knew himself, he set out on a mission of truth to save the *real* Lincoln, not from his enemies—his enemies scarcely mattered—but from sickly sentimentalists and hero-worshiping friends. Otherwise Lincoln's biography would likely become a series of fables and parables and homilies, similar to the Washington of Parson Weems. "Would you have Mr. Lincoln a sham, a reality or what, a symbol of an unreality?" he challenged Arnold. "Would you cheat mankind into a belief of a falsehood by defrauding their judgments? Mr. Lincoln must stand on truth or not stand at all." Unhappily, as Herndon discovered, he had neither the talent nor the discipline for writing a biography. His pen was that of a lawyer, and a passion for truth went only so far.[50]

Herndon was in his element, however, in the collection of facts, testimony, and memorabilia. Letters of inquiry were dispatched in every direction to elicit knowledge of Lincoln as he had been before history took notice of him. Advertisements for letters, records, relics, and scraps of information were placed in newspapers. Interviews were conducted and transcriptions made. Dennis Hanks, a garrulous cousin who had grown up in close proximity to Lincoln, proved a prolific source, one about whom Herndon's critics charged he was much too credulous. In September 1865, Herndon traveled to southern Indiana, to Gentryville and the Pigeon Creek farm where Lincoln grew from boyhood to manhood. He planned repeatedly to visit Hodgenville, Kentucky, and the area thereabouts, where Lincoln was born, but regrettably he never did. The richest vein of memory tapped by the researcher was among the people of Petersburg, Menard County, Illinois. It was to Petersburg that many of the inhabitants of New Salem moved when the town was abandoned not long after Lincoln himself left for Springfield. He had spent six years of his young manhood in New Salem, and dozens of old folks remembered him well. Their testimony filled huge gaps in Lincoln's biography and in at least one instance, the Ann Rutledge romance, enabled Herndon to contribute a chapter not only novel but more imaginative than anything in Holland's despised work. He became convinced that New Salem, so little known to history, was the crucible of Lincoln's life. "I have been with the People," he wrote excitedly of this discovery, "ate with them—slept with them, and thought with them—cried with them too. From such an investigation—from records—from friends—old deeds and surveys etc. etc. I am satisfied, in connection with my own knowledge of Mr. Lincoln . . . that [his] whole early life *remains to be written.*"[51]

Meanwhile, in December, as he was compiling his "Lincoln Record," lawyer Herndon delivered the first of several lectures before an appreciative Springfield audience. The second lecture two weeks later continued and completed the first, the subject of both being the mind and character of Lincoln. If anyone doubted Herndon's intimate knowledge of his former partner, the lecture set all doubts to rest. No more penetrating verbal portrait had ever been given of him. Beginning with a physical description, Herndon at once showed the acuteness of his observation. "The whole man, body and mind, worked slowly, creakingly, as if it needed oiling. . . . In walking Mr. Lincoln put the whole foot flat down on the ground at once, not landing on the heel." The lecture went on to characterize his intellect. "Mr. Lincoln's perceptions were slow, cold, precise, and exact. Everything came to him in its precise shape and color." Accordingly, his imagination was weak. So, too, was his faculty of judgment, or what the world called common sense. Reason—stern reason—was his guide in everything. The idea that he was ruled by his heart, by feelings of love and benevolence, was wrong, in Herndon's opinion. Some of Herndon's observations seemed harsh at first glance, but his friends, who had been Lincoln's friends, understood that he was trying to change the mind-set from one of apotheosis to one of truth and honesty, never doubting the Lincoln would emerge the greater because of it. Carpenter liked the lecture so much that he pirated it for his book. Oglesby, Arnold, Swett, all the Illinois Lincolnians, applauded Herndon for a keen and just appreciation. Thanking one of them, Herndon touched on the animus behind his approach. "The writers on Mr. Lincoln, in my humble judgment, make him a feeble ass—a soft fool, and this I could not stand. Lincoln was great headed."[52]

The third lecture, on Lincoln the statesman, stirred little interest. In it, however, Herndon reinforced his conception of Lincoln's tough-mindedness. Had he been all mercy and kindness, the lecturer averred, the war would not have been won, the Union saved, and slavery abolished.[53]

Herndon's fourth lecture bore the enigmatic title, "Abraham Lincoln. Miss Ann Rutledge. New Salem. Pioneering, and the Poem Called Immortality." It came after an interval of ten months, during which Holland's biography had appeared, joining the issue, as it were, between him and Herndon; yet few in the small audience at the Sangamon County Court House that November evening realized, as David Donald has said, that they were witness to "a historic occasion." Turning to New Salem, Herndon sought to show how it "fixed the man." New Salem was a pioneer village, Lincoln's family were pioneers, and to understand Lincoln it was essential to understand the influence of pioneering upon him. Holland plainly did not. And so he wrote in his book, "A good deal of what is called 'the pioneer spirit' is simply the spirit of shiftless discontent." A case in point was Thomas Lincoln, father of the President, an emigrant by virtue of weakness and inefficiency. "What!" growled Herndon, "are Grant and Jackson, Douglas and Benton, Clay and Lincoln, inefficient men, coming west from the spirit of shiftless discontent!"

But the defense of western character was the minor theme of Herndon's discourse. The major theme—the sensation of the evening—was the revelation of Lincoln's youthful romance with Ann Rutledge. Her name would not be found in Holland's book or any other; nor was it known to more than a handful in the audience at Springfield. The pretty daughter of a tavern keeper, with whom the newcomer had boarded when he first came to New Salem, Ann fell in love with Abe Lincoln in 1835. Unfortunately, she was already betrothed to another, John McNamar, who had gone back to his old home in New York to attend to some business but expected to return to marry her. Thus the young lady from no fault of her own became engaged to two men at the same time. Emotionally torn, she could neither eat nor sleep. She fell ill. Lincoln visited at her bedside. Sadly, she died of a raging fever in August.

To construct this Victorian romance, Herndon drew upon the testimony of a dozen or more former denizens of New Salem. He selected his evidence with care. John McNamar, who returned soon after Ann died, said he had never heard anyone suggest that Lincoln had addressed her in a "courtship way." Herndon passed over such statements. Had he let the story end with Ann's death, it would have had little significance for the Lincoln image; but the lecturer went on to maintain that Lincoln's heart lay buried in Ann Rutledge's grave. Overcome by grief, he suffered a fit of insanity and rambled aimlessly for days over the hills and through the woods along the Sangamon. "He slept not, he ate not, he joyed not. . . . His mind wandered from its throne." Finally, in the home of his friend Bowling Green, he recovered. In the larger sense, however, he never recovered. For Ann was the only love of his life, and the strain of melancholy that led him to embrace William Knox's poem settled into his soul forever.

> The maid upon whose cheek, on whose brow, in whose eye
> Shone beauty and pleasure—her triumphs are by;
> And the memory of those who loved her and praised,
> Are alike from the mind of the living erased.

But Lincoln did not forget. The poem was for him, a remembrance of Ann as well as a sermon on the fragility of human life.[54]

Among those in Herndon's audience that evening was a visitor from Boston, Caroline H. Dall. A forty-four-year-old reformer and essayist, she was indebted to Herndon for supplying pertinent information on Illinois statutes for her book *Women's Rights Under the Law*, published in 1862. She was now on a lecture tour in the Midwest, and while in Springfield she became intellectually engaged with Herndon and his work on Lincoln. He opened to her his records and papers, even disclosing some things he was not yet ready for the world to hear. The prairie lawyer was hungry for recognition, especially in the cultured East. "What phases of Western life and character do you, *in the East*, want brought out and developed," he asked one correspondent, "so that Mr. Lincoln's life can be understood by reflection?" Herndon, who had treasured his correspondence with Theodore Parker, found in Mrs. Dall a new conduit to the mind of the East. She became fascinated with Herndon's Lincoln and even thought to write her own biography of him.

Herndon encouraged Mrs. Dall to advertise his work within the limits of discretion, and she did so immediately upon her return to Boston. In December the *Daily Advertiser* published an article over her initials, which was copied in other eastern newspapers. She wrote of Herndon's intimacy with Lincoln and his understanding of the pioneer society from which he came. People in the East, who knew the President, knew virtually nothing of this man. Holland's *Life* was a joke in western opinion, she said. "Gammon! . . . pious cant!" The *Advertiser*, with its sensibilities, had been shocked by Herndon's November lecture. But the Ann Rutledge romance was a necessary link in Lincoln's biography, Mrs.

Dall wrote. "It could no more be left out of a true life of Lincoln than Dante's love of Beatrice, or Petrarch's love of Laura . . . could be omitted in the biographies of these poets, and still leave them intelligible. For thirty years after the period it describes, his terrible sorrow continued to move him at times out of himself, and it is the only explanation of many significant facts."

The nation owed a double debt to Herndon, Mrs. Dall said. First, it was he who had converted Lincoln to anti-slavery Republicanism in the 1850s. The recognition must have warmed Herndon's heart, for he always fancied himself Lincoln's political mentor. The second debt, of course, was the biography he was researching. "Will the world ever hear that you should tell the whole truth?" she asked Herndon. "It must have that or nothing," he replied. "And what would Lincoln have desired?" His eyes closed, a strange tenderness stole over his face, and he answered, "I seemed to see and hear him just as I used to do, and he says, 'Go ahead, Billy.' "[55]

Mrs. Dall went on to write an article, "Pioneering," published in the *Atlantic Monthly* in April 1867. Here she expatiated on the barrenness and sadness of the young Lincoln's life. She retold the story of Ann Rutledge, adding new details and, in fact, reading more into it than Herndon had. Through the ordeal of her death, said Mrs. Dall, Lincoln gained mastery over himself and over others. Meanwhile, Herndon was being discovered by other visitors to the Illinois capital in search of the *real* Lincoln. The well-known journalist George H. Townsend, who wrote under the name "Gath," stopped at the office of Herndon and Zane (Charles S. Zane became Herndon's partner in 1865) and met the man whose exterior appearance—rough dress, wind-hardened face, bushy beard, teeth discolored by tobacco—belied the philosopher and remembrancer of Abraham Lincoln. Herndon's national celebrity was accompanied by considerable local displeasure over his fourth lecture. He shelved plans for a fifth on Lincoln's childhood, in which he presumably would have disclosed terrible doubts about Lincoln's and his mother's legitimacy.

Carpenter, who had been enthusiastic about the first lecture, was disturbed by the fourth. "It seemed to me an invasion of a sacred chamber," he wrote. But Carpenter did not yet understand Herndon's theory of "counter mines." The world would discover the truth sooner or later, he reasoned; better that it be discovered, and analyzed, by Lincoln's friends than by his enemies. "Sacred lies will not protect us," he wrote to another correspondent. "Hence as Mr. Lincoln's friend I propose to sink and cut a counter mine. I propose to throw overboard in other words all things now and avoid the whale and the shark."[56]

Because no one could offer proof to refute the Ann Rutledge yarn, criticism of Herndon focused on its offensiveness to Mrs. Lincoln and her sons. Robert attempted to keep the story from her; upon learning of it, she was outraged. "This is the return for all my husband's kindness to this miserable man!" she exclaimed. The slur upon her marriage—that her husband's heart had secretly belonged to another—came with especially bad grace because of Mary Lincoln's courtesy to Herndon only two months earlier. At his request through Robert, she agreed to be interviewed and traveled to Springfield for that purpose. The substance of the interview was unimportant except as it showed her deep devotion to her husband and touched on his religion. Lincoln was "a religious man always," though he was not, she conceded, "a technical Christian." This was grist for Herndon's campaign against Bateman, Holland, and the preachers. Their testimony was "all bosh," he insisted. Lincoln was an infidel or perhaps, like Herndon himself, a "theist" on the order of Theodore Parker. Anxious to curb Herndon's pen, Robert made a hurried trip to Springfield after the November lecture. He had an obsession with privacy. He would not, years after the event, release for publication Queen Victoria's letter of condolence to his mother, despite the fact that it was meant for publication. Knowing he could not silence Herndon, Robert pleaded for consideration of his mother's feelings, and he seems to have had momentary success.[57]

The matter flared up again from a distant quarter in March. The Reverend James Smith had been pastor of Springfield's First Presbyterian Church, where Mrs. Lincoln and the children worshiped. Lincoln held Smith in high regard and appointed him American consul at Dundee, Scotland. There, in a local newspaper, he read an account of Herndon's lecture. He had scarcely regained his composure when he received a rudely phrased letter from Herndon inquiring what evidence he might have that their late friend believed "the *Bible was God's special miraculous revelation!*" And if he did, why hadn't he joined the pastor's church? Smith refused Herndon the courtesy of a private reply. Instead he sent his response, with the letter that provoked it, to the *Dundee Advertiser,* trusting that it would be picked up in the United States. The correspondence appeared in the *Chicago Tribune* on March 6, 1867. Why, said the angry preacher, Herndon had broken the widow's heart and shamed her sons! He was worse than the assassin Booth, for he had sent his friend "down to posterity with infamy branded on his forehead, as a man . . . destitute of those feelings and affections without which there can be no real excellence of character." As to religion, Lincoln most certainly did avow his faith in Christ and the authority of Scripture. But Smith did not choose to pursue the subject with the diabolical Herndon.

Mary Lincoln felt revenged. The *Tribune* rejoiced in the Reverend

Mr. Smith's rebuke. The story of the alleged romance, even if true, should never have been published; Lincoln's whole life pronounced it a lie, for "to suppose that during so many years he could hide his sorrow and conceal the anguish of a broken heart, would be to attribute to him a duplicity foreign to his frank and open-hearted nature."

Some of Herndon's friends stood by him. Robert Dale Owen, who had contracted to write a biography of Lincoln soon after his death, only to despair of it after learning of Herndon's work, wrote sympathetically, saying the minister's attack was "the meanest exhibition of the Pharisee" recently encountered and that their dear friend ought to be esteemed, rather than disgraced, for his love of Ann Rutledge.[58]

Another of Lincoln's friends who aspired to write his biography was Ward Hill Lamon. Big, bluff, and blustery, Lamon had ridden circuit with Lincoln and served all during his presidency as Marshal of the District of Columbia. In 1865 he resumed the practice of law in Washington. On a visit to Illinois in search of materials for the biography in 1869, Lamon soon discovered that Herndon was way ahead of him. He therefore approached him with a proposition to purchase the Lincoln Record, which Herndon had bound in three huge leather volumes, and use it as the basis of a biography. Herndon was receptive, in part because his own literary efforts had come to a standstill, in part because he needed the money. His law practice had dwindled; he took up farming, alas without success; and he returned to his old drinking habit. He had, as he said, "passed through several hells" since 1866–67. Lamon paid $4,000 for Herndon's records, not the originals but copies, with the full right to use and dispose of them. He returned east and entered into a contract with a Boston publisher, Osgood & Company, for a Lincoln biography. First, however, acknowledging his own literary incompetence, he drew up a contract with Chauncey F. Black to ghostwrite the biography mainly from Herndon's materials. Young Black was the son of Jeremiah S. Black, a distinguished attorney, prominent Democrat, and former Attorney General in President Buchanan's administration, with whom Lamon was now associated at the bar; and the son religiously followed his father in politics. So the first significant Lincoln biography after Holland's would be the product of a curious troika: Herndon, Black, and Lamon.[59]

Working with Herndon's records, Lamon and Black quickly realized that they raised as many questions as they answered. Explanations were wanted, and on some delicate issues Herndon was encouraged to replace doubt with certitude. "I want to possess myself of your *mind* in regard to Lincoln," Lamon wrote. Under these proddings, the Illinoisan, who was naturally loquacious, grew bolder. On the basis of Lincoln's offhand comments, the testimony of correspondents in Kentucky, and scattered bits of evidence, Herndon suspected that Lincoln was of illegitimate birth and that one Abraham Enlow was his father. Now, while admitting the evidence was inconclusive, he thought it might be stated as the truth. This was all the support Lamon and Black needed. The evidence of Lincoln's illegitimate birth is "absolutely overwhelming," the latter said. Nor was the truth something to be ashamed of. "It is to Mr. Lincoln's glory that he was a bastard, and knowing himself to be a bastard, he never hid his face like a craven, but steadily walked straight forward and upward." Nancy Hanks, his mother, did the world a favor by receiving the embraces of a gentleman rather than low-born Tom Lincoln. The lower the depths from which Lincoln came, the greater his fame. "We must point mankind to the diamond glowing on the dunghill," Black declared. Carpenter had called Herndon "a pre-Raphaelite in biography." As Benjamin Thomas later remarked, "If Herndon was a pre-Raphaelite, Black was a surrealist!"[60]

The matter of Lincoln's paternity, together with Black's prejudiced treatment of his accession to the presidency, caused alterations in the biography on the eve of publication in 1872. Stories were later told of how the Illinois guardians of Lincoln's fame forced Lamon to expunge the slander of illegitimacy, for which Judge Davis paid him three hundred dollars. Whether or not this was a bribe, the money was the more willingly paid because the besmirchment of the Republican idol might reflect badly on Davis's presidential candidacy in this election year. Black was even more outraged by the demand of the publisher to drop from the book the last chapter, actually written by his father, purporting to give a true history of the Buchanan administration during the secession crisis. This was an act of political censorship, in young Black's opinion. He was angry at Lamon for going behind his back and submitting to these and other "mutilations" of the book. "Many of the chapters were shamefully mutilated and the most creditable as well as the most interesting parts ruthlessly struck out," he complained to Herndon. His narrative was reduced to "hog-wash."[61]

The Life of Abraham Lincoln from His Birth to His Inauguration as President appeared under Lamon's name on June 1, 1872. It ran to 547 pages and was to be followed by a second volume, which never came. There was no hint of the true author, Chauncey Black, but Herndon received acknowledgment for the research, and his facts and interpretations were so often cited in the text that the discerning reader would suspect the book was fundamentally his. At first, he applauded the work. "There never was as true a biography written in this world," he declared. And there was much for him to applaud. If the assertion of illegitimacy was suppressed, it was implied by the alleged absence of proof of marriage, for instance a marriage license or family Bible record. Six of the twenty chapters were devoted to the New Salem years. Here, in the words of one of Herndon's informants, was the rustic portrait of the young Lincoln as he would be depicted in numerous books and, later, in motion pictures:

> He wore flax and tow linen pantaloons—I thought about five inches too short in the legs—and frequently he had but one suspender, no vest or coat. He wore a calico shirt. . . , coarse brogans, tan color; blue yarn socks, and straw hat, old style, and without a band.

Here were the stories of Lincoln's bouts with the Clary Grove Boys, clerking in Denton Offut's store, and the Ann Rutledge romance. Another young lady, Mary Owens, whom Lincoln briefly courted, was introduced. From her Herndon had received Lincoln's letters, and Lamon himself had obtained from Mrs. Browning the rejected suitor's amusing account of the affair, which many would read as a damning exhibit of his vulgarity.

Lamon described New Salem as a freethinking community and said Lincoln had written an "infidel book" based upon the writings of Paine and Volney; but one day, fortunately for his political future, Samuel Hill—so his son told Herndon—snatched it from his hands and threw it into the fire. Lincoln was a "trickster" at the bar and a wily politician, "never agitated by any passion more intense than his wonderful thirst for distinction." Mary Todd was also ambitious and married him to satisfy her craving for power. The marriage came about after the first engagement had been broken at the last minute and the intended groom had gone off "crazy as a loon." Of course, the marriage proved an ordeal. Lamon described Lincoln's character in terms just the opposite of those normally assigned to it. He was heartless, friendless, and uncharitable. And the root cause of this character, including his melancholy, was the rejection of Christian faith. "The fatal misfortune of his life . . . was the influence of New Salem . . . which enlisted him on the side of unbelief."[62]

To such lengths had gone Herndon's passion for biographical truth. He had set up a false dichotomy between an *ideal* Lincoln and a *real* Lincoln, and supposed that in representing the latter he would foil the fancied enemies of the man he loved and revered; but in this biography, ironically, the troika of friends seemed to do the work of the enemies. Generally, the book met with a cool to icy reception. Curiously, but not surprisingly, the longest and warmest review appeared in the pages of the *Southern Review*, written by the editor, Albert T. Bledsoe. He took the biography as the revelation of Lincoln's "innermost soul," and it was fearful to behold. All of Lincoln's "good nature," it appeared, was fake, put on to further an ambition that was, in turn, revenge upon his poor-white origins. As sick and disgusting as his career was, he proved to be the right man at the head of "the Northern Demos," and in that character the South had reason to remember him. "For if, as we believe, that was the cause of brute force, blind passion, fanatical hate, lust for power, and the greed for gain, against the cause of constitutional law and human rights, then who was better fitted to represent it than the talented but the low, ignorant and vulgar, railsplitter of Illinois?" Bledsoe spoke with authority in the South.[63]

The two most important reviews in the North came from the editors of two of the leading magazines, the *Atlantic Monthly* and *Scribner's Monthly*. William Dean Howells, in the former, thought the book in bad taste and doubted its authority. It treated young Lincoln's love affairs with "fumbling melodramatic sentimentality." It never accounted for Lincoln's eminence in history. Yes, he was ambitious, said Howells—who, it may be recalled, wrote a hasty life of the Republican candidate in 1860—"but nothing betrays less of selfish design, less of ignoble expedient, in our history, than the development of Lincoln's greatness." And after all the indignities Lamon piled upon Lincoln, nothing in the book need make his admirers ashamed. The review in *Scribner's* was by its founding editor, Josiah Holland. He, of course, knew very well where the book came from and stated flatly it was "very largely" Herndon's work. (Black's part remained a secret.) While Herndon-Lamon complained of the idealization of Lincoln, their own portrait, supposedly realistic, embodied their own preconceptions and prejudices, Holland wrote, and nothing in the philosophy of the volume allowed for the transformation of Lincoln in the presidency. As for the attack on Lincoln's religious character and the depiction of his domestic life, the book was an outrage on decency. Given such a reputation, Lamon's biography was little read. James Russell Lowell dismissed it with a genteel sneer: "The author was a vulgar man who vulgarized a noble subject."[64]

Holland opened the last chapter of this literary melodrama in July 1873 by publishing in *Scribner's* "The Later Life and Religious Sentiments of Abraham Lincoln," by the Reverend James R. Reed, who now occupied the pulpit of the First Presbyterian Church in Springfield. He had prepared this lecture, and given it before many audiences, to refute the "infamous charges" in Lamon's book. Basically, he repeated Smith's contention that, while not a Christian in the strict sense, Lincoln believed in the divine authority of the Scriptures and grew in the faith.

Reed also introduced statements obtained from two of Herndon's informants charging that their testimony had been falsified.

Black, meanwhile, disgusted with Lamon, was urging Herndon to distance himself from the book and join him in writing a true Lincoln biography purged of sophistry. Herndon agreed, more or less, and on November 24 published a "card," drafted by Black, in the *Chicago Tribune* denying that he had had any part in the writing of Lamon's book or had seen it in advance of publication. Black, at the same time, goaded Herndon to reply to Reed. "They are combining against us everywhere—Holland, Arnold, the preachers, and the press." Being a "ghost," Black's hands were tied, and Lamon was incompetent, so the defense fell to poor Herndon. On December 12 he returned to the platform in Springfield, and in a bold, intemperate address, scornful of his critics, reiterated his opinion that Lincoln lived and died an infidel: "and now let it be written in history and on Mr. Lincoln's tomb—'He died an unbeliever.' " The *Illinois State Register* printed the lecture as a broadside, and copious extracts appeared in the nation's press. The editor of the *New York Herald* named Herndon a Judas—"Judas in Springfield"—and said that no words could adequately describe the blackness of his treachery.

On that note the chapter ended. Nothing came of Black's project to collaborate with Herndon, since Lamon refused to sell his rights to the book and the Lincoln Record. When in 1873 Caroline Dall, reacting to the failure of Lamon's book, revived her interest in writing Lincoln's biography, Herndon offered friendly encouragement. "*Privately*," he wrote with emphasis, "*the Life of Lincoln has yet to be written*." Moreover, since he still possessed the originals of the Lincoln Record, he would happily sell it to Mrs. Dall! She faltered, however, and Herndon waited many years for another opportunity.[65]

3

Filling Up the Image

WILLIAM H. SEWARD, Lincoln's Secretary of State, died in October 1872. Six months later, in Albany, New York, when the legislature was in session, Charles Francis Adams delivered a memorial address on the statesman who had been a major player in state and national politics for forty years. Adams, of course, had served the President and Secretary as minister to Great Britain; but his appointment as orator on this occasion was especially gratifying because Seward had performed the same office on the death of Adams's father, John Quincy Adams, a generation before. The oration was admirable, except that in bringing into relief the virtues of Seward as a war statesman Adams denigrated Lincoln. He portrayed the new President in 1861 as a bumbling provincial wholly out of his element in Washington and far below the mark required by the crisis. "I must then affirm, without hesitation," Adams declared, "that in the history of our government, down to this hour, no experiment so rash has ever been made as that of elevating to the head of affairs a man with so little previous preparation for his task as Mr. Lincoln." Fortunately, he had enough sense to recognize his inadequacy, and so surrendered direction of affairs to his political and intellectual superior, the first cabinet officer, who became "the master mind" of the government. In foreign affairs, especially, Lincoln "absolutely knew nothing," said Adams, and owed everything to Seward's adroit management.[1]

The falsity of this portrait was instantly recognized by knowing and fair-minded men. Gideon Welles, Lincoln's Secretary of the Navy, now retired in Hartford, undertook to reply to Adams. The indefatigable diarist of both the Lincoln and Johnson administrations, Welles had already begun to publish historical papers based upon his personal record. Three long papers were now contributed to *Galaxy* magazine and subsequently published as a book, *Lincoln and Seward.* Cheered on by Montgomery Blair, another alumnus of the cabinet, Welles painstakingly reversed Adams's conception of the relationship between the President and his Secretary of State. Lincoln was the master, Seward the fumbling subordinate saved more than once from embarrassment in his own department of foreign affairs. Welles alluded to an early dispatch to Adams himself, outlining a course of conduct in diplomacy with Britain, which Lincoln "criticized, modified, and with his own hand expurgated, corrected, and improved," until, unknown to Adams, he was substantially its author. Poor Adams. Welles suspected he had been misled by Seward's alter ego, Thurlow Weed, who sat directly in front of the orator at Albany.[2]

The shock of Ward Lamon's biography undoubtedly contributed to the reaction against Adams. The Lincoln therein portrayed was scarcely a model for emulation; now, it seemed, the presidential Lincoln might be similarly diminished. Robert Todd Lincoln, who viewed Lamon and Herndon as traitors to his father's memory and thought Adams false and misleading, concluded that the time had come for a proper biography. For this purpose he put the President's papers at the disposal of John G. Nicolay and John Hay, the two young Illinoisans who had been Lincoln's secretaries during the war and, having just returned from posts abroad, were eager to undertake the work. They had only begun when Adams's eulogy of Seward appeared; before they finished seventeen years later, the country was awash in Lincoln memory and reminiscence.

The Flood of Reminiscence

"Reminiscence," it would be said, "lies warm upon the life of Lincoln." What was New Salem but a reminiscence? What was Herndon—his own direct observation notwithstanding—but a retailer of reminiscences? Reminiscence, in its nature, is the recollection of past persons and events largely without benefit of historical documents. Its authenticity depends upon the memory of the remembrancer. But memory fades and, as everyone knows, is subject to tricks: of vanity and conceit, of partiality, error, and displacement. In a literate culture, reading corrupts or displaces memory. Book knowledge enters the stream of oral tradition. "Dennis Hanks told many things he read in Thayer," a twentieth-century scholar noted, adding, "I wouldn't give a nickel for an interview with an informant who had read a book on Lincoln." Reminiscence is like storytelling; it goes on more or less continually and changes with the telling. One reminiscence triggers another, and so the process feeds upon itself. Reminiscence, as the product of memory, is not simply imprinted but constructed by the mind. In it truth and error dwell so closely together that one seems lost without the other. Reminiscence is the opposite of inquiry. One professes through memory to recover something once present in the mind; the other professes through knowledge to validate the past.[3]

Reminiscences were fruitful sources of myth and legend, yet the historical Lincoln is inconceivable without this mass of recollective data, much of it oral originally but all of it transmitted through the written word. This was especially true with respect to the pre-presidential Lincoln, so little known to fame. A popular biography of some four hundred pages in 1865 devoted less than one page to the hero's life before his twenty-first year and only forty-five pages to the half-century before his nomination for President. Reminiscences filled the yawning gaps between the documents. They might help to establish the record of persons and events in Lincoln's life; more important, however, their accumulation reflected the people's fascination with Lincoln's character. It was the most singular character in the annals of history. It defied the conventional standards of greatness. Ordinary historians could not be expected to comprehend "the germ of a character so extremely uncouth, so pathetically simple, so unfathomably penetrating, so irresolute and yet so irresistible, so bizarre, grotesque, droll, wise, and perfectly beneficient." But the common people, for whom greatness was not a historical type, and who were moved by their affections as well as by their reason, might comprehend him. Taking possession of Lincoln for their own purposes, they fashioned images, like the "everyman" of Carl Becker's famous essay, from traces of memory, from casual threads of information, and from sentiments gathered about him.[4]

Reminiscence, of course, had not waited for the martyr's apotheosis. It was coextensive with Lincoln biography, and some uniquely valuable contributions were made during his presidency. On September 1, 1864, for instance, the *Independent* published the Reverend John P. Gulliver's account of a conversation he had with Lincoln on a railroad journey in Connecticut in 1860. The pastor of a Norwich church, and also a literary scholar, Gulliver had been deeply impressed by Lincoln's speech the evening before—in substance his celebrated Cooper Union Address. Lincoln was taken aback by such high praise from a learned easterner. What was so remarkable? he asked. Gulliver answered, "The clearness of your statements, Mr. Lincoln; the unanswerable style of your reasoning, and especially your illustrations, which were romance and pathos and fun and logic all welded together." This was the first time Lincoln had been told he had a style.

Probing the source of his power, Gulliver asked about his education. Lincoln made his customary dismissal of school, but reflected on a childhood passion to wrestle an idea "till I had bounded it north and bounded it south and bounded it east and bounded it west," and also described how after reading law he had studied Euclid until he knew what was meant by demonstration beyond the possibility of doubt. The pastor thought Lincoln had got hold of political truth the same way. On parting, he said, "Be true to your principles, and we will be true to you, and God will be true to us all." Touched by this earnestness, Lincoln replied, "I say amen to that! Amen to that!" Gulliver thought, four and one-half years later, that Lincoln had proven true, and he published this reminiscence in the hope that it might sway any doubting Thomases to vote for the President in the coming election. It became a staple of Lincoln biography.[5]

The tide of reminiscence rose slowly after Lincoln's death, became a flood in the 1880s, and crested at the centennial of his birth in 1909. Nearly all this reminiscence was of northern origin, and nearly all of it was sympathetic, indeed most of it affectionate or even reverential. Several southerners made important contributions, however: Stephens, Bledsoe, Watterson. And among northerners there were distinguished men whose memories rankled with enmity and discord.

Moncure Conway, a native Virginian who went north and became a leading Unitarian minister and abolitionist, liked Lincoln personally and said he answered Frederick the Great's definition of a prince, "the first of subjects," yet was politically alienated from him. In 1863 Conway sailed for England to lecture on behalf of the North but was repudiated by his friends after he proposed to sacrifice the Union in return for Confederate emancipation. He became an expatriate. In his *Autobiography*, he blasted the war as "a great catastrophe" and held Lincoln personally responsible for the death of half a million men.

Lyman Trumbull, the three-term Republican senator from Illinois, refused to participate in the canonization of Lincoln, and sometime before his death in 1896 penned his critical estimate of the man he had known for a quarter-century. He was a shrewd and able politician, said Trumbull, but far from artless and simple. His ambition was relentless. "A more ardent seeker after office never existed." Partly for that reason, he was a follower rather than a leader in politics. "Without attempting to form or create public sentiment, he waited till he saw whither it tended, and then was astute to take advantage of it." He was given credit for ending slavery and suppressing the rebellion, but, said Trumbull, "a man of more positive character might have accomplished the same result in half the time, and with half the loss of blood and treasure." This seemed ungracious, considering that Lincoln had subordinated his ambitions to Trumbull's in the 1854 Senate race. As Welles remarked, Trumbull always acted as if he were better fitted than Lincoln to be President.

Horace Greeley, so often at odds with Lincoln, was another man who felt that he had been an inept war leader. Greeley's immensely popular two-volume history of the war, *The American Conflict*, denigrated the President by ignoring him. In 1864 he had upbraided Lincoln for failing to respond to alleged peace overtures from the Confederacy; and when in July Lincoln indulged him, sending him with Hay to Niagara Falls to talk to rebel agents, Greeley got nowhere, yet contrived to lay defeat on the President's faithlessness rather than his own folly.

In 1869 the eccentric journalist delivered a lecture on Lincoln. This was surprising, though some thought that he had a memoir or biography in view. In the lecture, published long after Greeley's death, he recalled his relations with Lincoln from their first meeting in Washington in 1848. As if trying to reach a more just estimate of the man, one unencumbered by memories of discord, Greeley emphasized his growth, his democratic character, and the power of his example for "the humblest American youth." "Looking back . . . ," he concluded, "I clearly discern that the one providential leader, the indispensable hero of the great drama—faithfully reflected even in his hesitation and seeming vacillations the sentiments of the masses—fitted by his very defects and shortcomings for the burden laid upon him, the good to be wrought through him, was Abraham Lincoln."[6]

Almost anyone who had encountered Lincoln in a significant way felt compelled sooner or later to tell about it, often less with a view to adding to the historical record than to caressing Lincoln's memory. As the supply and demand for reminiscences fed each other, they were sought out, collected, and published in large volumes. The first of these was *The Lincoln Memorial: Album Immortelles*, edited by Osborn Oldroyd in 1882. An Ohio-born Union veteran, Oldroyd may have been the first dedicated collector of Lincolniana; and about the time this book was published he moved into the old Lincoln homestead in Springfield and turned it into a museum of relics and memorabilia. The book of over five hundred pages was a hotchpotch, containing excerpts from Lincoln's letters and speeches and miscellaneous tributes as well as heretofore unpublished reminiscences.[7]

In 1885 Allen T. Rice, publisher of the *North American Review*, set out to get every notable person who had known Lincoln to contribute recollections for a small fee. These were sold to newspapers through the magazine's syndicate, then published in a permanent memorial volume.* *Reminiscences of Abraham Lincoln* contained thirty-three articles of uneven quality and was much read and consulted for many years. In 1886 Francis F. Browne, a Chicago magazine editor, published *The Every-day Life of Abraham Lincoln.* A book of over seven hundred pages, lavishly illustrated, it contained a great many glimpses of Lincoln which, collectively, were intended to admit the reader into intimacy with him. While Rice's *Reminiscences* focused on the sixteenth President, Browne's volume focused on the Lincoln of Illinois. In 1895 the *Independent* published a "Lincoln Number" consisting of forty reminiscences; these were gathered in *Abraham Lincoln: Tributes from His Associates.* Much of this was previously unpublished. The supply of reminiscences seemed inexhaustible.[8]

They touched on every episode and aspect of Lincoln's life from his birth to his death. Old Dennis Hanks told how he had held the newborn in his arms and when the baby cried uncontrollably had returned him to Nancy, saying, "Aunt, take him! He'll never amount to much." The boy Abe was not long in changing his opinion. In Indiana he acquired a reputation for book learning, became the oracle of Jones's store in Gentryville, and boasted to Mrs. Josiah Crawford, a neighbor, "I'll tell you, I intend to be President of the United States before I die." He must have been pleased to hear this confirmed by a voodoo fortune-teller on the second of his youthful voyages to New Orleans. "You will be President," she predicted, "and all the Negroes will be free." Isaac Arnold, with others after him, called this "a tradition," and offered it as evidence of Lincoln's superstition as well as his early anti-slavery bent. (Asked by Arnold what he knew about it, Herndon said, "It *seems* to me *just* now" that he had once heard it, which suggests that his own memory was

*Rice himself had been too young to know Lincoln, but with Lowell and others he regarded him as a native genius of great power. In 1886 he published in facsimile Seward's instructions to Adams, as revised by the President, of May 21, 1861. Welles had alluded to this document. During the Grant administration, Secretary of State Hamilton Fish brought it to a cabinet meeting where it aroused so much interest that photographic copies were made for each member under a seal of secrecy. The negative was destroyed, but one of these photographs later came into Rice's hands. The corrected dispatch showed, said Rice, that Lincoln was the country's "ablest diplomatist." The instructions concerned the Union's response to British recognition of the belligerent rights of the Confederacy; and Lincoln, by his careful attention to literary and diplomatic nuance, "saved the nation from a war with England," according to George S. Boutwell. (*Abraham Lincoln: Tributes from His Associates*, William Hayes Ward, ed. [New York, 1895], 74). See "A Famous Diplomatic Dispatch," *North American Review*, v. 142 (1886), 402–10.

sometimes as vague as that of his informants.) Arnold also reported, on the testimony of John Hanks, Lincoln's outraged reaction to a slave auction in New Orleans: "If I ever get a chance to hit that institution, I'll hit it hard." Herndon vouched for the truth of this assertion; yet it was later discovered Lincoln himself had stated that Hanks left the boat in St. Louis.

Anecdote and reminiscence traced his proverbial honesty to his storekeeping years in New Salem. A poet of the Lincoln legend later reduced it to rhyme:

> One night, Abe saw the intake of the store
> Had totaled six cents higher than was fair.
> He searched and found the error, closed the door,
> And walked three miles to pay the debtor's share.
> The woman owed cried, "Honest Abe!" Now, lore
> And legend still extol his noble flair.

However, T. G. Onstot, who remembered Lincoln as a boarder at his father's tavern in New Salem, and whose first knowledge of him was as "a great marbles player," said he earned the title "Honest Abe" after becoming a lawyer, part of whose business was collecting debts for merchant clients. Not only was he efficient, but he actually turned back to the clients the money he collected, thus the apellative. He carried his strong sense of right and wrong into the practice of law, it was often said, acting charitably on behalf of poor clients with justice on their side and refusing clients he believed in the wrong. He was a physical as well as a moral athlete, capable of unbelievable feats of strength. In the secession winter of 1860–61, Lincoln joked about receiving a broadside of South Carolina's ordinance of secession for Christmas, then turned grave, according to the recollection of a political friend, and said, "I am in the Garden of Gethsemane now, and my cup of bitterness is overflowing." The visitor thought that it was only during this crisis, while he anxiously waited to assume the reins of power, that "that look of settled despondency," which many thought habitual to him, was fixed upon his face. Of course, the impressions, the stories, the conversations recorded in the reminiscences of Lincoln the President were legion.[9]

The quality of truth in the reminiscences ranged from the instantly credible through the reasonable or plausible to the implausible and downright unbelievable. Joshua Speed, Lincoln's dearest friend, described their first meeting at his store in Springfield in 1837, and the anecdote had such a ring of truth that it became a commonplace of Lincoln biography. The aspiring lawyer walked in with saddlebags—his sole possession—on an arm, introduced himself, and asked for credit to buy a bed; Speed, liking his sad face, invited him to share his double bed above the store; Lincoln climbed the stairs, looked around, dropped his saddlebags, returned, and said briskly, "Well, Speed, I am moved!"

If anyone doubted the railsplitter legend, Richard Oglesby, who started it, told how it came about at the Illinois Republican convention in Decatur in 1860. A champion of Lincoln's nomination, Oglesby heard John Hanks say that his cousin had built a cabin and split rails for a fence near Decatur upon entering Illinois thirty years before; they went out to the place and located several of the same rails; two were taken up and carried into the convention with great fanfare, after which Lincoln made a little speech saying that though he could not swear he had split these exact rails, he had split many like them. Oglesby's account, published in *Century*, carried authority on its face; some others—for instance Joe Cannon's, in which Lincoln was passed feet-first to the platform on the hands of the densely packed convention—were pleasantly embroidered.

General James B. Fry, a professional soldier who was appointed Provost Marshal General in 1863, had many opportunities to observe the President in action. Fry held that Lincoln controlled Stanton rather than the reverse, as many thought, and told several anecdotes that found their way into the history books. He saw the President read General Meade's congratulatory order after the Battle of Gettysburg. "When he came to the sentence about 'driving the invaders from our soil,' an expression of disappointment settled upon his face, his hands dropped upon his knees, and in tones of anguish he exclaimed, *'Drive the invaders from our soil! My God! Is that all?'* " Another eyewitness quoted the President as saying, "They will be ready to fight a magnificent battle when there is no enemy to fight." And Robert Lincoln, on an informant's report in 1872, said it was the only time he ever saw his father cry. Lincoln was angered both by Meade's failure to pursue the retreating rebels and by the implication that the Confederacy was not "our soil."[10]

Many plausible reminiscences, although widely believed, are suspect in the absence of corroboration or in the face of evidence to the contrary. An especially important instance concerned the Freeport Doctrine, which Stephen A. Douglas propounded in answer to Lincoln's question in one of the famous debates of 1858. Attempting to trap Douglas between his adherence to "popular sovereignty" in the territories, on the one hand, and the Dred Scott decision of the Supreme Court, on the other, the challenger asked if the people of a territory could lawfully exclude slavery prior to the formation of a state constitution. Douglas answered that they could by "local police regulations," thereby pleasing his Illinois constituents but alienating his friends in the South. Horace

White, who reported the debates for the *Chicago Tribune* and who wrote the chapter on this subject for John L. Scripps's campaign biography in 1860, credited Lincoln with a brilliant stroke. His counselors, however, had advised him against it, arguing that the Senator would stand by the principle and finesse any problem posed by Dred Scott. "But," Lincoln had replied, "if he does that, he can never be President." To his friends' rejoinder that it was the Senate seat he was seeking, Lincoln said, "No, gentlemen, *I am killing larger game.* The battle of 1860 is worth a hundred of this."

Every biography repeated the story and made it a primary exhibit of Lincoln's political sagacity. Yet it stood without corroboration until 1895, when Joseph Medill, the *Tribune*'s long-lived editor, offered his recollection of the affair. Before the debate, he and Elihu Washburne, Norman Judd, and several others were closeted with the candidate in a Freeport hotel trying to dissuade him from asking the question, which, they thought, would simply play into Douglas's hand. But to this Lincoln replied, "If he answers as you say he will, and as I now think he will, that slavery can be excluded from a Territory, he will beat me for the Senate, but he will never be elected President." And, Medill concluded, had Lincoln taken their advice, *he* would never have been elected President. Present-day scholars remain skeptical of this story in part or in its entirety.* [11]

Many reminiscences touched upon Lincoln and his cabinet. Curiously, no cabinet officer other than Welles wrote of the inner history of the administration, although Salmon P. Chase's diary proved a valuable source when finally published. Chase was the President's cross-to-bear in the cabinet, and the subject of a stream of anecdote, much of it plausible but unverifiable. Over the strenuous objections of Seward and Weed, Lincoln appointed Chase Secretary of the Treasury. The *New York Herald* reporter who covered Lincoln from Springfield to Washington in the early months of 1861 asked him at the Inaugural Ball if he had any message for the *Herald*'s editor, James Gordon Bennett, an old

*A curious footnote, especially important for its symbolism, concerns the report that Douglas, defeated in the presidential election, held Lincoln's hat during his inauguration. Seated on the platform and observing Lincoln's befuddlement over what to do with his hat as he rose to take the oath of office, his old rival graciously offered to take it. This was stated some days after the event in a Cincinnati newspaper; Holland and Arnold alluded to it in their books published in 1866; Henry Watterson vouched for it as an eyewitness in his writings. It was also included in *Diary of a Public Man*, in 1879; this anonymous work purported to be an inside history of Lincoln's assumption of power, but its authenticity has long been in dispute and is now generally discredited. As to the story of Douglas and the hat, some scholars have found it too good to be true.

enemy of the Seward-Weed duo, and Lincoln replied, "Yes, you may tell him that Thurlow Weed has found out that Seward was not nominated at Chicago." Pondering this cryptic remark on his way to the telegraph office, the reporter suddenly realized that the scepter had passed from the East to the West and Lincoln was President in fact as well as in name.[12]

One of the secret and abortive peace missions in 1864 involved a journalist and a chaplain, James R. Gilmore and James F. Jacquess respectively, passing through enemy lines to talk unofficially to Jefferson Davis. According to Gilmore, Lincoln purposely included Chase in the meeting to hear the terms of settlement to be given to Davis. Among these was a provision for compensation of slave property, to which Chase objected, as expected. Why was Chase included? Not for reasons of counsel or generosity, Gilmore said, but on the shrewd political calculation that had Davis accepted, and the Radical Republicans howled, Lincoln would have been able to name Chase a participant in the transaction. Upon his return Gilmore, with the President's approval, wrote an anonymous article about the mission for the *Atlantic Monthly*. Emphasizing that the article was unauthorized, Gilmore was all but silent on Lincoln's role, which was divulged years later.

Lucius E. Chittenden, a Vermonter, served in the Treasury throughout the war. His *Recollections of President Lincoln*, in 1891, found many readers. In it Lincoln appeared as a man of God and a man of infinite mercy. But Chittenden also told an interesting story about Chase's resignation in 1864 after he had maneuvered to take Lincoln's place as the Republican candidate for President. Chase chose a trivial issue on which to submit his resignation, and he expected, as before, that it would be declined. (When his son Robert had inquired why an earlier resignation from the disloyal Secretary was not accepted, the President snapped, according to another memorialist, "He is much safer in my Cabinet than he would be outside.") Called in to discuss plans for naming a successor, Chittenden wrote, he pleaded with Lincoln to keep Chase. But the President said Chase had fallen into two bad habits: first, of believing he was indispensable, and second, of thinking that he should be President. He was a good financier destroyed by his own vanity and ambition. The resignation, Lincoln continued, was intended either to annoy me or draw a pat on the shoulder. "I will not do it. I will take him at his word." Then he paused and reflected that Chase would make a good Chief Justice if the opportunity arose to make an appointment. When it did, later that year, Lincoln was able to neutralize Chase politically and gratify him at the same time.[13]

Many reminiscences shaded from the implausible to the unbelievable to the spurious. How did Lincoln come to study law? While sitting for his portrait in 1860, according to the artist A. J. Conant twenty-eight years later, Lincoln told him that as a storekeeper he had bought a barrel of junk for fifty cents, and when he later dug through it found Blackstone's *Commentaries* at the bottom. He read it, and the rest was history. Early biographers usually said Lincoln bought the *Commentaries* at auction in Springfield (Lamon said he borrowed it), and such is the verdict of scholarship. LaSalle Corbell Pickett, wife of Confederate General George Pickett, whom Lincoln had known briefly in Illinois, wrote in her "intimate personal recollections" that during his celebrated tour of smouldering Richmond the President searched out the Pickett family home. Finding it at last, he was greeted at the door by Mrs. Pickett carrying a baby. Announcing himself as plain Abraham Lincoln, "George's old friend," he was told that the General was still in the fight, at which he took the baby and kissed it, then remarked, "Tell your father, the rascal, that I forgive him for the sake of your mother's sweet smile and your bright eyes." No less an authority than Douglas Southall Freeman dismissed this yarn as fantasy, though another Civil War historian, Allan Nevins, found it hard to believe Mrs. Pickett had invented it out of whole cloth.

Many of the reminiscences were patently self-serving as well as incredible. In his contribution to the Rice volume and in his autobiography, Union General Benjamin F. Butler recorded a conversation he had with the President in February 1865 on the question of what to do with the freedmen. He put quotation marks around a fearful harangue in which the President spoke of Negro soldiers becoming a class of guerrillas and criminals in the South and asked Butler to draw up a plan to colonize them outside the country. Butler did so, proposing that the Negroes be deported to Panama and employed to dig a canal across the isthmus. "There is meat in that, General Butler," the President said, and he directed him to talk to Seward about the plan, which came to nought. Nothing but Butler's testimony supports this. By 1864 Lincoln had, in Hay's word, "sloughed" ideas of colonization. That he would contemplate deporting Negro soldiers while at the same time calling for their enfranchisement, or that he would entrust a project of this magnitude to a proven scoundrel like Ben Butler, beggared belief.[14]

One truly sinister reminiscence came from a recusant Catholic priest, Charles Chiniquy. Lincoln had successfully defended Father Chiniquy in a suit for slander in 1850. The trial at Urbana had been full of fireworks, and Chiniquy warned Lincoln after the verdict that it would prove a sentence of death on him. Years later in his autobiography, *Fifty Years in the Church of Rome*, Chiniquy claimed that the assassination, like the war itself, was a Roman plot. Boasting an intimacy with Lincoln in both body and soul, Chiniquy told of three visits at the White House. He filled Lincoln's mouth with anti-Catholic slander. The Pope and Jeff Davis were allied in this war against liberty; and Lincoln must fall victim to a Jesuit assassin. *The Assassination of President Lincoln*, extracted from Chiniquy's book, circulated underground for many years. This "Lincoln Lie," as it was branded in Catholic quarters, enjoyed a second life early in the twentieth century coincident with the revival of the Ku Klux Klan.[15]

Reminiscences came from politicians and soldiers, ministers, hospital workers, lawyers, actors and other people of the theater, secretaries (not only Nicolay and Hay, the principals, but William O. Stoddard), telegraph operators (David H. Bates's *Lincoln in the Telegraph Office* is a minor classic), and, above all, from journalists. Don Piatt, an Ohio journalist and poet turned soldier during the war, claimed to have started the tide of Lincoln reminiscence with an article published in the newspapers in March 1885. His impressions deviated from the popular ideal of Lincoln, which he described as "an elongated oddity of tears and gross fun." Lincoln was coarse, tough, strong, and hard. Compared to him, Stanton was a pussycat. The notion that he was emotionally burdened by the war—pacing the floor at nights—was false. "I enjoy my rations, and sleep the sleep of the innocent," Piatt quoted him saying. And because he could say that, he was a great leader.

Henry Villard, a young German immigrant, met Lincoln during the Senate canvass of 1858. Once when both were waiting on a railroad platform to flag a train, a storm came up and they took shelter in an empty freight-car. Villard, who wrote his *Memoirs* some forty years after the event, remembered Lincoln confiding to him as if he were a bosom friend. Now embarked on the campaign, Lincoln said he no longer doubted his qualifications for the Senate. "I am convinced that I am good enough for it," Villard quoted him, "but, in spite of it all, I am saying to myself every day, 'it is too big a thing for you; you will never get it.' Mary insists, however, that I am going to be Senator and President of the United States, too." At this he roared with laughter, then exclaimed, "Just think of such a sucker as me as President." Villard was equally incredulous.

The journalist Jane Swisshelm was one of a number of women who left vivid reminiscences of Lincoln. A dedicated abolitionist, she went to Washington in 1863 prejudiced against the President, and agreed to attend a White House reception only reluctantly and provided she was not presented. But his majestic presence overwhelmed her. "It was as if I had suddenly passed a turn in the road and came into full view of the Matterhorn." Simplicity, honesty, selflessness were evident at a glance. Swisshelm took his hand, and became a steadfast admirer.

Charles A. Dana broke his long association with Greeley's *Tribune* in 1862 to become Assistant Secretary of War. His *Recollections*, in 1899, contained a masterly sketch of Lincoln, which owed something to Ida Tarbell, who ghostwrote the book. He thought Lincoln "a supreme politician" whose skill arose from his extraordinary knowledge of human nature. As an instance, Dana gave the admission of Nevada to the Union. Seeing that he needed two more votes in the Senate to secure passage of the Thirteenth Amendment, Lincoln dispatched Dana to Congress to massage enough egos to win the necessary votes. Dana also cited Greeley's Niagara mission. Lincoln approved of it knowing the pesky editor would return empty-handed. Afterwards he remarked slyly to Dana, "I guess I am about even with him now." Dana also thought Lincoln a brilliant military strategist, "the greatest general we had."[16]

Noah Brooks was the journalist in most intimate relation with the President. Although they had met in Springfield, the friendship between them formed in 1862 when Brooks came to the capital as the correspondent for a California newspaper. An amiable thirty-two-year-old widower, Brooks became a frequent visitor at the White House, a friend of the family, and a trusted counselor. He was slated to succeed Nicolay in 1865 as the President's private secretary, but when Booth altered that plan, Brooks constituted himself one of Lincoln's chief remembrancers. He began with an article, "Personal Recollections of President Lincoln," in *Harper's* magazine for July 1865. Here he emphasized Lincoln's artless simplicity of manners. Thus he disliked being called "Mr. President" or even "Mr. Lincoln," preferring to be addressed only by his surname, as he addressed everyone else, and insisted on referring to the White House as "this place." Most of the esteemed elements of Lincoln's character, including the religious, found their place in this essay.

Brooks's best book, *Washington in Lincoln's Time*, appeared in 1895. In it the author glimpsed sides of Lincoln that tended to be neglected, for instance his reading of literature. He had an amazing memory, saying he "couldn't help remembering." When Brooks's cousin John Holmes Goodenow, from his native state of Maine, called at the White House, Lincoln greeted him warmly and launched into recitation of a long poem, then concluded triumphantly, "There! That poem was quoted by your grandfather Holmes in a speech which he made in the United States Senate in . . . ," and he named the precise date some thirty years before. Limited as was his acquaintance with literature, he apparently remembered everything he read, including speeches in the *Congressional*

Globe. Yet, curiously, he was seldom given to nostalgia. His sensibility toward physical nature was exquisitely Emersonian. On one of their woodland excursions, Brooks pointed to a tree wrapped by a luxuriant vine. "Yes, that is very beautiful," said his companion, "but that vine is like certain habits of men; it decorates the ruin that it makes."[17]

Some reminiscences engendered controversy. Every lawyer who remembered Lincoln felt compelled to say something about his defense of William "Duff" Armstrong in 1858. It was his most famous trial. Armstrong stood accused of the drunken murder of Preston Metzker in the vicinity of a camp meeting on an August night the previous year. Lincoln, as a friend of Armstrong's parents, volunteered to defend him, though he rarely took capital cases. The chief witness for the prosecution said the light of the moon at 10:00 p.m. enabled him to see the defendant strike the victim with "a slung shot" that killed him. Lincoln, in rebuttal, introduced in evidence an almanac showing there was very little moonlight at the hour in question; and this, it was said, secured Armstrong's acquittal. Newspapers publicized the story after Lincoln's nomination in 1860.

Before long Democratic newspapers were claiming that the almanac Lincoln had introduced was for the wrong year. In 1872 Lamon's *Life* flatly asserted he had won the trial by trickery. An English law book, *Ram on Facts*, subsequently cited Lincoln's action in the case as an instance of the commission of fraud. As a young attorney, Abram Bergen, who later became a judge in Kansas, had witnessed the trial in the Beardstown courtroom; he testified in 1898 that the correct almanac had been introduced, and moreover that the jury foreman later swore in an affidavit he had personally verified the year of the almanac. Unfortunately, the almanac supposed to have been used had disappeared. But Boy Scouts later, and repeatedly, verified the dimness of the light at the same spot in the same lunar period. The novelist Edward Eggleston, in 1887, offered a fictional treatment of the Armstrong case in *The Graysons: A Story of Illinois*. He moved the trial back twenty years and freely altered the facts while, of course, keeping Lincoln the hero. Toward the end of his life Armstrong told his own story. Lincoln had been right: there was no moon. "But it was light enough for everybody to see the fight. The fight took place in front of one of the bars, and each bar had two or three candles on it. I had no slungshot. . . . It was only a fist fight, and if I killed 'Pres' Metzker I killed him with my naked fist." He regretted the conviction of his co-defendant, sentenced to eight years, in a separate trial. The course of reminiscence seemed to obfuscate rather than clarify this story.[18]

A controversy of another kind was set off in 1891 when Colonel Alexander K. McClure, editor of the *Philadelphia Times*, asserted upon the death of Hannibal Hamlin, the Vice-President in Lincoln's first term, that he had been dropped from the Republican ticket in 1864 in favor of Andrew Johnson at the President's explicit request. Hamlin, a respected elder of the party, long believed Lincoln had been faithful to him, and Johnson had been the candidate of Lincoln's enemies. But here was McClure, a Pennsylvania Republican leader, saying the President sent for him the day before the Baltimore convention to urge him, with other leaders, to nominate Johnson because he was southern, thereby disarming the Union's enemies abroad and signifying the success of the war against the rebellion. Nicolay and Hay had said in their *History* that Lincoln was rigidly non-committal in this matter, while entirely friendly to Hamlin.

In an affair of this kind the line between neutrality and deceit was narrow, and the implication of McClure's story was that Honest Abe had been less than honest. Nicolay disliked McClure, thinking he traded too much on a limited connection with Lincoln. He promptly wrote a public letter branding McClure's claim fraudulent. The Hamlin family praised this "manly and noble act." McClure filed a contemptuous rebuttal. Republican newspapers chose sides as the party washed its dirty linen in public. "The whole thing is growing very ridiculous," Hay wrote. "Every dead-beat politician in the country is coming forward to protest that he was the depository of Lincoln's inmost secrets and the engineer of his campaign." McClure, who Hay dubbed "Smart Aleck," restated his position and published the testimony he had gathered in an appendix to his new book, *Lincoln, and Men of War Times*, in 1892. Nicolay's "review" of the controversy appeared as a supplement in the 1899 biography of Hamlin written by his grandson. Interestingly, Hamlin's scholarly biographer seventy years later concluded that McClure was right. Lincoln had indeed abandoned Hamlin for Johnson and sought to appear neutral lest he alienate New England Republicans.[19]

Noah Brooks, writing for the *New York Times* in 1898, summed up the war generation's magnificent remembrance and reminiscence of Lincoln.

> It is questionable if material relating to the human existence of any person has ever been so thoroughly explored, sifted, and analyzed as the material relating to the humble birth and obscure youth and manhood of Abraham Lincoln has been. What rummaging! What minute scrutiny! What indefatigable questioning of every person who had the slightest acquaintance with Lincoln, his friends, and his neighbors!

In addition to the flood of reminiscence, the last ten years had witnessed the publication of *Herndon's Lincoln,* with other biographies, and Nicolay and Hay's great history, together with their edition of Lincoln's writings. Well might Brooks wonder if it was not at an end. "There can be no new 'Lincoln stories'. . . . The stories are all told." Oh, on occasion, a new reminiscence might drop into the sea of print or a new masterly estimate be offered of Lincoln's character, like Carl Schurz's recent essay. "But for the most part," said Brooks, "the mental figure of Lincoln, as it will appear to future generations of men, has already begun to take permanent shape." However that may be, Brooks's "final estimate" failed to reckon with three things. First, that the new generation would be as absorbed with Lincoln as the generation that elected him President, fought the war, and mourned at his funeral. Second, that scholarly inquiry had not even begun at the century's close. And third, that "the mental figure," although it had begun to take permanent shape, would be subtly reformed and reshaped as history advanced.[20]

Aspects of Character

Nothing perhaps better defined Lincoln than his humorous disposition. When the cabinet assembled at the hour of noon, September 22, 1862, the President, at the head of the table, looked around and asked, "Gentlemen, have you ever read anything by Artemus Ward?" And he proceeded quite deliberately to read from a small book a ludicrous story called "A High Handed Outrage in Utica." At its conclusion he laughed heartily. According to Chase, everyone but Stanton managed a smile. Laying the book aside, Lincoln said, "Gentlemen, why don't you laugh? With the fearful strain that is upon me night and day, if I did not laugh I should die, and you need this medicine as much as I do." Then, gravely, he laid before the cabinet the Proclamation of Emancipation, which it approved for preliminary issue. The incongruity of the comic with the solemn on this historic occasion epitomized the action of humor in Lincoln's character and politics.[21]

The stream of droll humor coexisted with Lincoln's life, changing as his life changed. In youth, as modern students have said, he told stories for the fun of it, at the bar he employed wit and anecdote as tools of the trade, in politics as weapons of satire, and in the presidency as a kind of therapy. What were the sources of this unending stream? The first was his father, acknowledged by the son when he began, "As my old father used to say . . . ." He picked up stories and honed his wit in the company of circuit-riding lawyers and in the yearly rounds of stump speaking on the prairies. He read joke books, especially *Joe Miller's Complete Jest Book*, published in 1845; and he read the humorous writings of Seba Smith (Major Jack Downing), George W. Harris (Sut Luvingood), and others. As President he was an enthusiastic reader of Petroleum V. Nasby and Orpheus C. Kerr, in addition to Ward. The broad and earthy native strain in Lincoln's humor led Constance Rourke, in the twentieth century, to consider it the high climax of an invincible national habit of comic storytelling, one which combined western ebullience with the laconic Yankee lineage. In this respect, certainly, Lincoln's humor proclaimed his Americanness. It was more than that, however. *Aesop's Fables*, which he read as a child, never left him.* He dealt in fables as well as fun, especially in his mature years, employing humor to enforce useful truths. In this respect, his humor, based upon a keen knowledge of human nature, was universal. "When he spoke," Charles Sumner observed, "the recent West seemed to vie with the ancient East in apologue and fable."[22]

Lincoln's humor became less clownish and more restrained as he matured. His friends, looking back, dated the change from the 1850s. As President he was well aware of his reputation as a funny man and sought to check it. Not everyone appreciated his storytelling habit; some associated it with incompetency. Books appeared: *Old Abe's Joker*, in 1863, which was hostile, countered by *Old Abe's Jokes* the following year. Anecdotes about Lincoln—not to be confused with his own anecdotes—often dealt with his fun-making. His own favorite among the stories told about him, it was said, reported the conversation of two Quaker women overheard on a train speculating on the outcome of the war:

> "I think that Jefferson will succeed," one said.
> "Why does thee think so?" asked the other.
> "Because Jefferson is a praying man."
> "And so is Abraham a praying man."
> "Yes, but the Lord will think Abraham is only joking."[23]

*Joshua Speed, in his reminiscence of his friend, said that when he was advised by a distinguished Virginian in March 1861 to surrender to the Confederacy both the disputed forts, Pickens and Sumter, along with other federal property, the President asked the adviser if he recalled the fable of "The Lion and the Woodman's Daughter." The lion fell in love with the daughter and applied to the woodman to marry her. "Your teeth are too long," the lion was told. He went to a dentist and had them extracted. He returned, but the woodman again rejected his proposal, saying, "Your claws are too sharp." The lion went away and had them removed. Again he returned, and the woodman broke open his head. "May it not be so with me," Lincoln finished, "if I give up all that is asked?" Joshua F. Speed, *Reminiscences of Abraham Lincoln* (Louisville, 1896), 31-32; also see the report in *Conversations with Lincoln*, Charles W. Segal, ed. (New York, 1961), 89.

With his death, Lincoln abruptly ceased to be a subject of humor; indeed it became an unwritten rule among humorists and in public discourse generally that jokes about him were forbidden. Mark Twain, whom William Dean Howells crowned "The Lincoln of our literature," never wrote or uttered more than a casual word, comic or serious, about his western contemporary in the storytelling line. He was not allowed to treat Lincoln humorously, and if he treated him seriously no one would believe him.[24]

While the apotheosis was chiefly responsible for the reverence that hedged Lincoln, many people also recognized that his humor lay close to the bone; it was the corollary of his melancholy, and therefore went to the fundamental mystery of the man. Judge David Davis said his friend's humor was intended to "whistle off sadness." The writer Francis Grierson, who grew up in Illinois during the war years, called Lincoln "one of the great melancholics," belonging to the fortunate class who laugh because they can (rather than those who languish without laughter) and for whom humor is the safety valve of genius. Lincoln spoke of humor as his lifeline. But why was he so sad as to need a lifeline? The agonies of the war provided only a partial answer, for the sadness was of long duration. So men pondered this mystery of his personality. William H. Herndon thought the key lay in Lincoln's dark ancestry. He emphasized, too, the heartbreak of lost love, while others pointed to the early death of his mother and his sister as well. Grierson thought it was endemic to the prairie pioneers. Physiological explanations, such as chronic constipation, were also offered.[25]

The uses of Lincoln's humor were many and various. It was never, at least in the later years, employed frivolously, and rarely without purpose. Nor was it always on display. Greeley knew Lincoln for sixteen years and never heard him tell a story. And it was a side of the President that Frederick Douglass never observed. Other than as a relief valve, Lincoln used humor to cut to the heart of the matter and illustrate a truth in commonplace terms. A New Yorker, on an errand for the Governor in 1863, was driven one night by a drunken major to see the President at his refuge on the outskirts of the city, the Soldiers' Home. He got out of bed and, in a tired and disheveled state, transacted the official business with his callers; as they were about to leave, the major slapped Lincoln on the knee and asked him to tell one of his famous stories, at which he rose and with becoming dignity said:

> I believe I have the popular reputation of being a story-teller, but I do not deserve the name in a general sense; for it is not the story itself, but its purpose, or effect, that interests me. I often avoid a long and useless dis-

> cussion by others or a laborious explanation on my own part by a short story that illustrates my point of view. So, too, the sharpness of a refusal or the edge of a rebuke may be blunted by an appropriate story, so as to save wounded feelings and yet serve the purpose. No, I am not simply a story-teller, but story-telling as an emollient saves me much friction and distress.

This states the strategy of Lincoln's humor exactly. He used stories as arguments, or as rejoinders to or diversions from arguments. And in a sense, as Frederick Seward said, his stories decided questions of state. But, more than argument, stories were a mode of conversation with Lincoln. Or perhaps a mode of learning: Josiah Holland thought that experiences lodged in Lincoln's mind as stories of common people in the common vicissitudes of life. He invested invincible habit with democratic purpose. So he told Chauncey Depew: "They say I tell a great many stories; I reckon I do, but I have found in the course of a long experience that common people"—and he repeated it—"common people, take them as they run, are more easily influenced and informed through the medium of broad illustration than in any other way."[26]

"That reminds me of a story" might well have been Lincoln's by-line. Among the stories treasured in memory are the following:

> Apropos of politicians' adherence to fixed ideas: A lad plowing a field on the prairie asked his father in what direction he should strike for the furrows. He was told to head for the oxen at the far end of the field. The father left, and as the lad started, the oxen started too; he kept steering toward them, all around the field, until he had gone in a circle instead of a straight line.
>
> An old-line Whig, torn about the war until the tide shifted to the Union in 1863, came to the President looking for an office. Lincoln was reminded of the time he went to his first dance in Springfield. He bought a new suit of clothes and a new hat. He had such a good time at the dance, he was among the last to leave. When he went to retrieve his hat, he was handed an old, worn one. "This is not mine; I had a new hat," he said. The reply came back: "Mr. Lincoln, the new ones were all gone two hours ago."
>
> After the fall of Vicksburg, a delegation visited the President protesting General Grant's paroling the defeated soldiers of General Pemberton's army. Lincoln was reminded of the story of the bullying man and his bulldog. One day the village boys wrapped a cartridge of gunpowder in a piece of meat attached to a long fuse and placed it in the road. The dog came, ate the meat, and exploded. The master looked over the carnage and lamented, "Bill was a good dog; but, as a dog, I reckon his usefulness is over."
>
> At the time Lincoln replaced his first Secretary of War, Simon Cameron, a Senate delegation called and recommended a total overhaul of the cabi-

> net. The President replied with the story of a farmer bedeviled by skunks in the hen house. One night he went out with a shotgun; seven skunks came along, and he shot one. Upon returning to the house, his wife asked why he hadn't shot all of them. And he replied, "The one skunk made such a stink that I let the other six go."
>
> When the Confederate commissioners at Hampton Roads complained that emancipation would bring ruin, that the Negroes would not work, and everybody would starve, Lincoln was reminded of an Illinois farmer who hit upon the plan of feeding his hogs by simply turning them loose in a field of potatoes. A neighbor, upon being told of the plan, said it was fine now in the summer, but wondered how the hogs would manage when the ground froze. The farmer scratched his head and said, "Well, it will be 'root, hog, or die!' "[27]

One or more of those stories might have been classified as coarse and vulgar in Victorian America. Presumably Lincoln would not have told them in a lady's drawing room. Nor would he have publicly epigrammatized his patronage woes as "too many pigs for the tits." The tradition that he told "smutty" stories—the Democrats charged this against him in 1864—was supported by Villard and Piatt, among others, and is now generally accepted; but since they would not bear printing and clashed with saintly stereotypes, they are not part of the literature of reminiscence. Of course, many stories were attributed to Lincoln that he never told, and to separate the genuine from the spurious has proved impossible.

Lincoln himself played down his accomplishments as a humorist. Thus he insisted to Brooks that he had no invention and was "only a retail dealer." Blessed with an excellent memory, he remembered the jokes he heard and read. If, as some said, he kept a file or notebook of stories, no such record survived. Despite his disclaimer, Lincoln possessed the faculty of humorous invention, though he could never have been a great performer on the order of Artemus Ward and Mark Twin. He was spirited in telling his stories and got personal enjoyment from them. "When he told a particularly good story, and the time came to laugh," a congressman recalled, "he would sometimes throw his left foot across his right knee, and clenching his foot with both hands and bending forward, his whole frame seemed to be convulsed with the effort to give expression to his sensations." Another referred to his habit of drawing his knees, with arms around them, up to his face. The pleasure was more his than the auditors.[28]

Lincoln was a wit as well as a storyteller. He liked figures of speech. A drink of water was "Adam's ale." Douglas's argument for popular sovereignty, Lincoln quipped in the famous debate at Quincy, "was as thin as the homeopathic soup that was made by boiling the shadow of a pigeon that had starved to death." Occasionally, he indulged in puns. Lamon told the story of his own appearance in court with a split in the rear seam of his pants; in the ensuing hilarity, his fellows at the bar circulated a petition to raise money for a new pair, to which Lincoln responded, "I can contribute nothing to the end in view." Some of Lincoln's epigrams became stereotyped in public discourse. From the reelection campaign came the admonition "that it is not best to swap horses when crossing a stream." From one of Lincoln's dreams, according to Hay, came the aphorism "The Lord prefers the common people. That is the reason he makes so many of them." (In Hay's original the expression is "common-looking people.") He was credited with the saying "It is true that you can fool all the people some of the time; you can even fool some of the people all of the time; but you can't fool all of the people all the time." However, there is no credible evidence Lincoln ever said it.[29]

Lincoln's wit was alert, swift, ingenious, and cutting. At Hampton Roads one of the rebel commissioners referred the President to the historic correspondence between Charles I and Parliament as a constitutional precedent for his dealings with the South. He replied, "Upon questions of history I must refer you to Mr. Seward. . . . But my only distinct recollection of the matter is that Charles lost his head." When one of a group of bankers in the capital to negotiate a loan to the government said to him, "You know, Mr. President, where the treasure is, there will be the heart also," he returned, "I should not wonder if another text would not fit the case better, 'Where the carcass is, there will be the eagles gathered together.' " David R. Locke, the creator of Nasby, met Lincoln in 1858. Speaking of a recently deceased politician of inordinate vanity, he remarked to Locke, "If General——had known how big a funeral he would have had, he would have died years ago." To an Englishman (often named as Lord Lyons, the British minister) who expressed surprise that American gentlemen blacked their own boots, Lincoln, no less astonished that English gentlemen did not, asked, "Well! Whose boots do they black?"

His most celebrated retort, first reported in 1864, was made to a temperance delegation that requested General Grant be relieved of command because he drank too much whiskey. Lincoln wondered if any of them knew Grant's brand of whiskey, "because, if I can find out, I will send every general in the field a barrel of it!" Unfortunately, although Lincoln may have said it, modern research has identified the riposte as a switch on Joe Miller's story of George III remarking on General Wolfe and his habits. On the other hand, no analogue has been found for the laconic response Lincoln made when asked how he felt after losing the senatorial contest to Douglas: "Well, I feel just like the boy who stubbed his toe—too damn hurt to laugh and too damn proud to cry." In 1952 another Illinoisan, Adlai Stevenson, had occasion to recall this pleasantry.[30]

WHAT HAS BEEN called "the gentle legend" of Lincoln was deeply rooted in the war and its memories. "There is scarcely a hamlet in the loyal states," said one of the martyr's biographers in 1865, "that does not contain some witness of his clemency and lenity." Stories of his pardons and reprieves, of his acts of kindness to soldiers and mothers, lovers and widows, were well known at the time of his death; they increased manyfold afterwards. From the Amnesty Proclamation of 1863 to the last grand chord of the Second Inaugural Address—"with malice toward none; with charity for all"—he sought to make the seat of power a seat of mercy. It was the brutal irony of his assassination that its perpetrator, in the words of the poets, "killed him in his kindness" and "murdered Mercy."

> I knew the man. I see him, as he stands
> With gifts of mercy in his outstretched hands;
> A kindly light within his gentle eyes,
> Sad as the toil in which his heart grew wise;
> His lips half parted with a constant smile
> That kindled truth, but failed the deepest guile;
> His head bent forward, and his willing ear
> Divinely patient right and wrong to hear.[31]

In the story of *The Sleeping Sentinel*, Lincoln witnessed the creation of his own legend. Like most legends, this one was founded on truth. William Scott, a young Vermont volunteer, fell asleep on sentry duty at a camp outside the capital in 1861; he was arrested, tried, convicted, and condemned to death, but at the last instant was saved by the President's pardon, only later to die in battle for the Union. Francis De Haas Janvier, a government clerk, wrote a narrative poem based on Scott's story. It was first read publicly by James E. Murdoch, the celebrated actor and elocutionist, to a select circle at the White House headed by the President and First Lady. Later that same day, in January 1863, the performance was repeated before a distinguished audience in the Senate chamber, the Lincolns again being present. Murdoch prefaced the poem with Portia's speech from *The Merchant of Venice* in which "the quality of mercy" is called "twice blessed."

> It blesseth him that gives and him that takes.
> Tis mightiest in the mightiest: it becomes
> The throned monarch better than his crown.

The poem, after describing Scott's life and predicament, shifts to the President.

> The war of thirty millions filled his burdened heart with grief;
> Embattled hosts, on land and sea, acknowledged him their chief;
> And yet, amid the din of war, he heard the plaintive cry
> Of that poor soldier, as he lay in prison doomed to die!

Scott is marched before the firing squad, when suddenly a coach rolls up in a cloud of dust, and the President emerges to save the stricken soul as the shouts of a thousand soldier voices rend the air. Scott lives to demonstrate his bravery but, alas, is mortally wounded in an infantry charge.

> While yet his voice grew tremulous, and death bedimmed his eye—
> He called his comrades to attest, he had not feared to die!
> And, in his expiring breath, a prayer to heaven was sent—
> That God, with His unfailing grace, would bless our President!

Murdoch went on to recite *The Sleeping Sentinel* before large audiences throughout the North, and it was read by thousands who never saw the patriotic performance. The story remained popular after the war, though its truth was brought into question. That Scott was sentenced, pardoned, and later died in battle was a matter of record. Lincoln's part was not. In his *Personal Recollections*, Lucius Chittenden vouched for the story's historical authenticity. A fellow Vermonter, Chittenden said he had himself pressed Scott's plea on the President and had later informed him of the soldier's dying words, in which he thanked his chief for giving him "the chance to fall like a soldier in battle, and not like a coward by the hands of my comrades." Lincoln was deeply touched.[32]

In this case, as in so many others, Lincoln showed the generous and sympathetic character that had been a badge all his life. His knowledge of human nature told him that American farm boys stuffed into a soldier's uniform and armed with a musket were scarcely fit subjects for the received code of military justice. He took it upon himself as Commander in Chief to review every court-martial death sentence for desertion and similar crimes. The Reverend Newman Hall of London, who would raise the first monument to Lincoln abroad, told of a Union officer who in the first week of his command forwarded to the President warrants for the execution of twenty-four deserters. Lincoln refused to sign any of them. To the officer's plea that an example must be made, he replied, "General, there are already too many weeping widows in the United States. For God's sake, don't ask me to add to the number, for I won't do it." Desertion was a terrible problem. Visitors to the White House sometimes heard him talk gloomily about it. "To fill up the Army is like undertaking to shovel fleas. You take up a shovelful"—suiting the word with a comic gesture—"but before you can dump them anywhere they are gone." Limits on the President's mercy seemed not to limit his compassion. A Union general, John Eaton, recalled an interview in the President's office. After a discharge of muskets was heard, Lincoln rose, walked to the window, and looked across to the Virginia shore. "This is the day when they shoot deserters," he said mournfully. Eaton felt that "for a man so relentless with himself in the performance of his own duty, Lincoln's charity toward others was little short of phenomenal."[33]

The President's acts of clemency extended beyond court-martial victims to all the casualties of war. Many of the anecdotes turned on a duel between Lincoln and Stanton, the latter upholding harsh rules of justice and considerations of military exigency against the President. General Grant, in his *Memoirs*, said they were "the very opposites of each other," and judged that Stanton's firmness, however unpleasant, was necessary to keep Lincoln from being imposed upon. A Union major, Nathan Goff, who had been a prisoner-of-war at Libby and Salisbury but was exchanged in August 1864, undertook at the President's request to write an account of his experience. Lincoln was shocked; showing the report to Stanton, he said, "I told you so. . . . Mr. Stanton, this is terrible. . . . Mr. Stanton, don't you think the boys had better come home?" The Secretary lectured in reply that in any large-scale exchange the Union would be returning twenty-five thousand strong men and receiving that many walking skeletons. Lincoln was firm: "Mr. Stanton, the boys are coming home." The Secretary stormed out of the office. Goff was pleased when the government adopted a more relaxed policy toward prisoner exchange.

Lincoln's tenderness was shown in his visits to the wounded and dying in hospitals. The history of the Eighteenth New Hampshire Volunteers contained the vivid account of the President, at City Point in 1865, stopping at the cot of a captain whose left leg had been amputated. He placed his hand on the captain's forehead, bent, and kissed his cheek. They talked briefly; Lincoln asked the surgeon to lift the covering so he might see the wound. Aghast, he exclaimed, "Oh, this awful, awful war!" Then, tears streaming down his cheeks, he turned to the captain,

"Poor boy! Poor boy! You must live! You must!" And he moved on to the next cot. Confederate soldiers, too, received Lincoln's clemency. Lieutenant Waller R. Bullock was wounded in battle and became a prisoner-of-war on Johnson's Island in the Ohio River. Fearing for his life, his younger brother, a lad of fifteen, went to Washington and, thanks to the President's "open door" policy, got an interview in which he begged for parole of his wounded brother. Told that Walter would not take the mandatory oath of loyalty, Lincoln said he could do nothing. The lad persisted. Lincoln fell into deep meditation, elbows on his knees, head in hands, gazing into the fire. Then, suddenly, he sprang to his feet, straightened to his full six-foot-four, brought a clenched fist down on the desk, and, staring directly at young Bullock, declared, "I'll do it; I'll do it." He scribbled on a card that the rebel lieutenant should be paroled to his parents' custody in Baltimore. And he was.

The Unionist governor of Virginia, Francis Pierpont, often told the story he heard from a mother on the streets of Washington in 1862. Her son Willie had left his unit in the Shenandoah Valley for a nocturnal visit to his sweetheart; alerted by the approach of Union soldiers, Willie dressed in her father's civilian clothes in order to make his escape, but he was caught and sentenced to death as a spy. The President had been adamant with the mother. When Pierpont returned with her, offering personal assurances Willie was not a spy, Lincoln pardoned him.[34]

To Herndon, Lamon, Piatt, and some others, the emphasis on Lincoln's acts of clemency smacked of sentimentality, and they were annoyed by it. (Lamon later wilted, conceding in his own reminiscences that Lincoln was "by nature singularly merciful.") It projected the image of a man guided by his heart rather than his head, one who placed mercy above justice, and wallowed in Christian love and benevolence. This was, in fact, one-sided. Lincoln's inflexibility, his respect for law, his commitment to reason and clarity were not easily reconciled with the gentle legend. "There was no flabby philanthropy about Abraham Lincoln," Dana declared. "He was all solid, hard, keen." His charity was not inexhaustible, as both history and reminiscence testified; and on a more balanced view the character revealed by his actions was more complex than at first appeared.[35]

Lincoln could be stern and tough when he believed the cause required it. After all, 267 men were executed by Union military authorities during the war, 141 of them for desertion. When Lincoln paused at the sound of firing squads across the Potomac, he knew that he was ultimately responsible for the death of these soldiers. It was a terrible burden however he bore it. In 1880 newspapers carried the story of a father who had gone to Washington to plead for his son, sentenced to death for desertion, but the President refused to see him, saying through a secretary that he had studied the papers carefully and concluded that the boy, who had deserted three times, was incorrigible. Mary Livermore of the Sanitary Commission witnessed two applications to Lincoln for pardon, one for a Virginian captured as a spy and one for a Union officer who killed his captain in a brawl, both rejected. One crime—a civil crime—toward which Lincoln felt no mercy was slave-trading. In Rice's *Reminiscences*, John B. Alley, a Massachusetts congressman, told of going to the President with a petition of pardon from a convicted but repentant slave-trader. Acknowledging that he was naturally inclined to pardon even the worst crime the mind of man could conceive, Lincoln continued, "but any man who, for paltry gain and stimulated only by avarice, can rob Africa of her children to sell into interminable bondage, I never will pardon, and he may stay and rot in jail before he will ever get relief from me."[36]

Lincoln was daily faced with appeals to his mercy, and the struggle weighed heavily upon him. There is no better evidence of it than the recollection of Cordelia Harvey of Wisconsin, an agent of the Sanitary Commission, whose good works earned her the title "The Angel of Wisconsin." She went to Lincoln on behalf of sick and wounded western soldiers rotting, so she believed, in hospitals on the lower Mississippi River. Bring the men north, and with the tonic of northern air, you will get ten healthy men in a year where you get only one now, she told the President. He was skeptical. Bring them north and they will all desert, he said. The dialogue between them continued for several days. Lincoln grew fretful, yet continued to listen to this pertinacious woman. In a moment of despair, he threatened to discharge all of the soldiers and be done with the problem. Patiently, she reminded him of their loyalty and promised that if he would grant her petition he would be glad as long as he lived. The President bowed his head and with indescribable sadness said, "I shall never be glad any more." Finally, Mrs. Harvey prevailed; she got her Wisconsin hospital. Reflecting on this ordeal, it seemed to her she had conversed with half a dozen different men. "Lincoln blended in his character the most yielding flexibility with the most unflinching firmness, childlike simplicity and statesmanlike wisdom and masterly strength, but over and around all was thrown the mantle of unquestioned integrity."[37]

One class of war victims with reason to remember Lincoln as both unmerciful and unjust was the thousands of civilians who suffered military arrest, trial, and imprisonment. They claimed denial of the rights of habeas corpus and the liberties guaranteed by the First, Fifth, and Sixth Amendments to the Constitution. Many of the victims were antiwar Democrats. The most notorious was Clement Vallandingham of Ohio. In 1863 a military court sentenced this former congressman to confinement for the duration of the war. Lincoln amended the sentence to banishment behind Confederate lines, which added the spice of humor to the unconstitutional proceedings. It was with regard to Vallandingham and his ilk that Lincoln poignantly posed his dilemma: "Must I shoot a simple-minded soldier boy who deserts, while I must not touch a hair of a wily agitator who induces him to desert?" To Lincoln, clearly, the preservation of the Union committed to freedom justified the emergency exercise of whatever powers proved necessary. The testimony of political prisoners against this view appeared in an extraordinary book in 1869, *American Bastille: A History of the Illegal Arrests and Imprisonment of American Citizens During the Late Civil War.* In some seventy case histories, Lincoln appeared as a tyrant, as malignant as Louis XI and Lucretia Borgia. Not only did he allow summary military arrest and imprisonment, but on the word of several of these victims he turned a deaf ear to all their appeals for justice. "We might as well undertake to justify the assassination by Booth," said one, "as to defend the acts of . . . tyranny of the Lincoln administration."[38]

Perhaps the most striking case of Lincoln's denial of clemency was that of Captain John Y. Beall, a thirty-year-old Virginian of a prominent family and a graduate of the University of Virginia, hanged from the gallows in New York harbor on February 24, 1865. Active in fomenting sabotage and resistance across Lake Erie from Canada, Beall had been captured in civilian garb near Buffalo. He was tried and convicted of being a guerrilla and a spy. "It is murder!" he cried. General John A. Dix, at the head of military affairs in New York, said that humanity itself demanded Beall's execution. A petition for commutation was carried to Washington. Many congressmen signed it. Some of Lincoln's most trusted advisers recommended clemency. But to the end he refused to override General Dix.

A record of the trial and proceedings leading up to the execution was published after Lincoln's death. And that would have been the end of it but for a rumor set afoot that John Wilkes Booth assassinated Lincoln out of revenge for the execution of his bosom friend Beall. In 1876 John W. Forney, a well-known Republican newspaper editor, felt compelled to stamp "fabrication" on a report that he had accompanied Booth to the White House to plead for Beall's life and, moreover, that Lincoln had granted the plea but reneged after Seward objected.

By 1900 Beall was being recognized as a Confederate martyr. In the pages of the *Confederate Veteran* magazine and in the *Christian Observer*, a religious journal published in Louisville, men who had known Beall and vouched for Booth's friendship backed up the theory of the assassin's revenge for Lincoln's perfidy. To the editor of the *Observer* this theory, which fixed on Booth's deranged affections, undercut the northern theory that Booth acted for the South. Such claims led Isaac Markens, an enthusiastic Lincolnian, to examine with care this "weird and lurid story." He found no evidence that Booth's deed was inspired by a broken promise, nor even that he and Beall were acquainted; and, of course, it was common knowledge that Booth's designs antedated the Beall case. Markens could not establish the genesis of the story but suspected the infamous Copperhead editor "Brick" Pomeroy. The author thought, nonetheless, that the denial of clemency to Beall placed Lincoln "in a light that refutes most forcibly the popular impression of his pliancy, lacking in backbone and easily swayed by appeals for mercy." Here, strikingly, he turned a deaf ear to powerful supplications from all quarters.* [39]

AS LINCOLN WAS A man of humor and a man of mercy, so was he in a sense a man of letters. Emerson prophesied the fame that lay before him in this regard at the time of his death.

> He is the author of a multitude of good sayings, so disguised as pleasantries that it is certain they had no reputation at first but as jests; and only later, by the very acceptance and adoption they find in the mouths of millions, turn out to be the wisdom of the hour. I am sure if this man had ruled in a period of less facility of printing, he would have become mythological in a very few years, like Aesop . . . by his fables and proverbs. But the weight and penetration of many passages in his letters, messages and speeches, hidden now by the very closeness of their application to the moment, are destined hereafter to wide fame. What pregnant definitions; what unerring common sense; what foresight; and, on great occasion, what lofty, and more than national, what humane tone! [40]

The Gettysburg Address was already being recognized as a masterpiece of English prose. The first small collection of Lincoln's utterances and writings selected for their literary merit was published in Boston in the year of his apotheosis. By 1900 the fame Emerson had foreseen had arrived. Selections from Lincoln's pen were routinely included in hand-

*A contrasting case, that of the Confederate spy Belle Boyd, was better known in the South. She had been twice captured and twice imprisoned in Washington. A legend grew up that, after the second arrest in 1863, she was sentenced to death but that Lincoln gallantly intervened to save her. No evidence supports this. But in 1891 a southern writer, J. V. Ryals, commended Lincoln for an act of southern chivalry and said he hoped "to travel to his grave some day, and . . . leave a flower in commemoration of that noble deed."

some sets bearing such titles as *The World's Great Classics* and *Library of the World's Best Literature.* In 1891 John Earle, professor of Anglo-Saxon literature at Oxford University chose one of Lincoln's letters (to James C. Conkling, August 26, 1863) for his *English Prose,* a standard text. That a poor, unschooled, prairie-grown son of pioneers should attain such eminence, unmatched by any English statesman of the nineteenth century, was wonderful. Oddly enough, the first published scholarly monograph on Lincoln was not by a historian but by a literary scholar interested in the evolution of his style. It inaugurated the series of University of Illinois Studies in 1900. By then Lincoln's writings were being gathered in readers for use in the schools. From similar books—preceptors, as they were then called—the boy Lincoln had studied correct models of English composition. How astonished he would have been to find himself among them![41]

By the age of twenty-one, according to early biographers, Lincoln had read a spare one-foot shelf of literature: the King James Bible, *Aesop's Fables*, Bunyan's *Pilgrim's Progress*, Defoe's *Robinson Crusoe*, Franklin's *Autobiography*, Weems's *Life of Washington*, with several others. Two of these books particularly, the Bible and Aesop, influenced Lincoln's style. The rhythmic cadences of the former resound in the Second Inaugural Address, for instance, and his fable-telling had a source in Aesop. As Lincoln said in his sketchy autobiography, what he had in the way of education he had "picked up." He never became a learned man, but of his eagerness for books and learning there could be no doubt. The tradition of his walking thirty or forty miles to borrow a book, of reading crouched before the fire, as in Eastman Johnson's painting of 1868, of improvising a slate from a blackened shovel—these stories made up the exemplary legend passed through generations of children. Reminiscences gathered from surviving friends of his childhood testified to his ardor. "Abe was all'ers much given to larnin," said an aged Hoosier, once Lincoln's playmate.

All his life he "picked up" odds and ends of learning. When a German immigrant came to reside in Springfield, it was said, Lincoln organized a class to study the language. In view of what he became, it was difficult to fault his education. Holland wrote that "Abraham's poverty of books was the wealth of his life," suggesting the benefits of narrow but concentrated study of select texts. Robert Ingersoll, in his oration on Lincoln, spoke of his good fortune in graduating from "the University of Nature." Formal university education would have spoiled him. "Why, if Shakespeare had gone to Oxford they would have made him a quibbling lawyer or a hypocritical preacher," said Ingersoll, and so it would have been with Lincoln. As with other self-educated men, there were

Boyhood of Lincoln. Painting by Eastman Johnson, 1868. (University of Michigan Museum of Art, Bequest of Henry C. Lewis 1895.90)

large blank spaces in Lincoln's learning, and he struggled with spelling all his life. One day, not long before he died, he confessed to David Davis he had just learned how to spell "maintenance."[12]

Lincoln's love of poetry came with maturity. He tried his own hand at verse, but his modest efforts were forgotten; the poetry for which he was remembered belonged to his prose compositions. Among the English poets, none equaled Shakespeare in his eye. According to Herndon, he read Shakespeare constantly. Shakespeare was his great love on the stage as well. Joseph Jefferson, the famous actor, recounted in his *Autobiography* Lincoln's campaign to get the city fathers of Springfield to license a theater. As President, on Hay's testimony, he read Shakespeare more than all other authors together, and he rarely missed a performance of one of the plays. Obviously, his tragic sense of the Civil War owed something to his familiarity with Shakespeare's historical tragedies. Well might he recite the lament of Richard II:

> Let's talk of graves, of worms, and epitaphs;
> Make dust our paper, and with rainy eyes
> Write sorrow on the bosom of the earth.
>
> . . .
>
> For God's sake, let us sit upon the ground
> And tell sad stories of the death of kings.[43]

In the case of Robert Burns, the Scots' great bard, it was a question not only of literary influence but also of resemblance of mind, character, and national spirit. Joseph H. Barrett made the point:

> What Robert Burns has, proverbially, been to the people of his native land, and, to a certain extent, of all lands, as a poet, Abraham Lincoln early became to us as a statesman and patriot, by his intimate relations alike with the humbler and the higher walks of life. . . . The experiences of the "toiling millions," whether of gladness or of sorrow, had been his experiences. He had an identity with them, such as common trials and common emotions produced. He had become in person, no less than in principle, a genuine representative man in the cause of free labor.

The spirit of Burns was in Lincoln's humor, in his compassion, above all in the democracy of "a man's a man for a' that." Lincoln toasted Burns at a centennial dinner in the poet's honor in Springfield in 1859. "He could quote Burns by the hour," recalled one of his friends at the bar. Andrew Carnegie, who hailed from Dunfermline, credited Lincoln's love of Burns to a Scottish schoolmaster and said he later lectured on the bard to Illinois audiences, but for all Carnegie's searching, no trace of the lecture had been found. (Nor of the Scottish schoolmaster!) Carnegie liked to imagine Burns and Lincoln, " 'their patents of nobility direct from the hand of Almighty God,' " together in life as in heaven. The pairing of Burns and Lincoln was not uncommon. Through Lincoln in America, as through Burns in Scotland, said a popular literary critic in 1898, the life of the people "found a voice, vibrating, pathetic, and beautiful beyond most voices of the time."[44]

Lincoln's taste in poetry was quite eclectic. It included the funereal, as shown by his liking for William Knox's melancholy verses. It also included Ossian (James Macpherson), whose wild strains Lincoln recited to Arnold during a White House visit; and Lord Byron, whose *Don Juan*, in a handsome octavo edition, and well thumbed, was observed by Caroline Dall in Lincoln's old law office, where Herndon told her it was "the office copy"; and such American romantics as Fitz-Greene Halleck, Oliver Wendell Holmes, and Henry Wadsworth Longfellow. Holmes's melancholy "The Last Leaf" was a favorite, as was Longfellow's prophetic "The Building of the Ship." According to Noah Brooks, Lincoln encountered both poems in the columns of newspapers, where he did most of his reading, and neither knew nor much cared what the names of the authors were. He was moved by the pathos of one and the patriotism of the other. When the journalist recited the whole of Longfellow's poem, in which the Union is imaged as a sturdy ship, he said the President's eyes filled with tears at the final lines:

> Our hearts, our hopes, our prayers, our tears,
> Our faith triumphant o'er our fears
> Are all with thee,—are all with thee! * [45]

Lincoln's writings include speeches and addresses, letters, proclamations, and sundry state papers. Nicolay and Hay edited them in two fat volumes in 1894. Such a body of material did not lend itself to easy stylistic definition. Conceding Lincoln's excellence as a writer, of what did that excellence consist? It was clear, forceful, and purposeful, largely barren of figure and metaphor, some said. Thus Frederick Douglass attributed the power of Lincoln's words to the power of accurate statement, without refined logic or rhetorical embellishment. "He had a happy faculty of stating a proposition . . . so that it needed no argument. It was a rough kind of reasoning, but it went right to the point." Lincoln's assertion of the wrong of slavery from the proposition of equality in the Declaration of Independence was a case in point. The language of his most famous state paper, the Emancipation Proclamation, was strictly utilitarian. He did not choose to add an eloquent preamble, perhaps because of the moral anomalies and limitations of the act, with the result that, in the words of a recent historian, "it had all the moral grandeur of a bill of lading." The only patch of rhetoric, at the close, was provided by Chase.[46]

On the other hand, Lincoln was a man of poetic sensibility with a passion for self-expression, whose writings had the character of literary art. This was true of the celebrated opening of the "House Divided" speech, and the Cooper Union Address, for all the sheer power of its exposition of the founders' intentions about slavery, was a triumph of art as well as of thought. Even so, it could not be placed in the same category with the Second Inaugural and the Gettysburg Address, which were

---

*Certain words, once read, lingered in Lincoln's memory only later to be called into service, quite unconsciously, in his speeches and writings. An instance spotted by the critic Montgomery Schuyler concerned the President's special message to Congress in 1861 and Longfellow's *Hyperion*. The peroration of the address reads, "And having thus chosen our course, without guile, and with pure purpose, let us renew our trust in God, and go forward without fear, and with manly hearts," which compares to the poet's "Go forth to meet the Shadowy Future without fear, and with a manly heart"—a transcription so literal, Schuyler thought, as to be beyond the reproach of plagiarism. "Lincoln's English," *Forum*, v. 41 (1909), 125.

national hymns, or the touching letter of condolence to Mrs. Bixby, thought by some the most beautiful ever written, or the Farewell at Springfield, with its pathetical metered cadence. Lincoln did not think of himself as an orator, and there was little appreciation of him in this light. After his early years on the stump, he never indulged in florid or extravagant speech and, with rare exceptions, in homespun humor. "He spoke," said a memorialist of the 1850s, "in an almost conversational tone, but with such earnestness and such deep feeling upon the questions of the day that he struck the hearts of all his hearers."[47]

The Address Delivered at the Dedication of the Cemetery at Gettysburg was the consummation of Lincoln's literary artistry. Here was argument by statement compressed in lofty yet powerful words. What pregnant definitions indeed! Starting with the idea of a nation "brought forth" in 1776, "conceived in Liberty, and dedicated to the proposition that all men are created equal," Lincoln employed the myth of birth, death, and renewal to comprehend the terrible sacrifice on this battlefield and to illuminate the nation's destiny. The musical cadence of the prose was easily deflected into poetic measure: "But in a larger sense / We cannot dedicate / We cannot consecrate / We cannot hallow this ground." No living poet could have equaled that.[*48]

On this distinguished occasion, the President followed the orator of the day, Edward Everett. He had read and admired Everett's oration in advance of delivery and had no thought of competing with it. He had been invited in his official capacity; he was expected to be brief, even perfunctory. Lincoln spoke in a clear, full voice, naturally high-pitched, without animation or gesture, observers said, though memories divided on whether he read from a manuscript or spoke from memory. According to the Associated Press report, he was interrupted by applause six times. Yet he had hardly begun before he finished, so that many in the audience did not take in what had been said. A tradition arose that the address was greeted by stunned silence, that it was a failure and Lincoln and everybody else knew it. Ward Lamon, an eyewitness, put his authority behind this tradition in his *Recollections*. This was wrong, though

---

*In 1899 a minor poet took a couple of lines from Joshua Speed's report of an alleged last interview with Lincoln in 1865, as published by Herndon, and with very little change of wording produced the rhymed verse:

> Die when I may, let it be said
> By friends who knew me best:
> He plucked a thistle from its bed
> And set a flower in its stead
> Wher'er a flower could rest.

from the eminence the address later attained the immediate response necessarily seemed tepid. The day following, Everett hastened to congratulate the President, saying, "I should be glad, if I could flatter myself that I came as near to the central idea of the occasion, in two hours, as you did in two minutes." The fact that Lincoln himself prepared so many manuscript copies for admirers suggested his own pride in the address. These copies also made it virtually impossible to determine an authoritative original text.[49]

Another and related tradition concerns the actual writing of the address. Isaac Arnold, in his early book, *Lincoln and the Overthrow of Slavery*, stated that the President wrote it hurriedly on the train going to Gettysburg. This was embellished by Ben: Perley Poore and by John P. Usher, Lincoln's Secretary of the Interior, both of whom said that Lincoln wrote the address on a piece of cardboard balanced on his knee as the train rolled toward Gettysburg.[50]

In 1906 Mary R. Shipman Andrews, hearing these things from her school-age son, decided to weave a story around them. Whether they were true or false did not concern her. The result was "The Perfect Tribute," published originally in *Scribner's* and subsequently as a brochure, which sold more copies than any other title on or about Lincoln and for several decades was commonly taught in the schools. Although Andrews added two or three details about the composition, delivery, and reception of the address, for instance that the audience tittered on first hearing Lincoln's shrill voice, her story lies in the sequel. Lincoln returns to Washington disappointed. While on a solitary stroll, he encounters a boy who is frantically searching for a lawyer to draft a will for his dying brother, a rebel captain, in a prison hospital. Lincoln, without revealing his identity, accompanies the boy. The dying captain insists upon talking about President Lincoln, who had made a great speech at Gettysburg. In fact, he had been told, the speech went so deeply to the hearts of the audience that it held its breath in reverent silence at the close. The captain read the speech, then remarked that while it might be wrong from the southern point of view, it would be recited by schoolboys, both northern and southern, in years to come. Suddenly, struck by a spasm, he gripped Lincoln's hand, then died. Mrs. Andrews, born and raised in the South, lived her adult life in the North. Her story was not only a tribute to Lincoln but part of the current of reconciliation that ran through American fiction at the turn of the century.[51]

The Gettysburg Cemetery began as a Pennsylvania memorial but became a national park with a forest of monuments in 1895. It soon included Lincoln's bust with his immortal words written in bronze. In commemorative observances from year to year, Lincoln could not be forgotten. "The world will little note, nor long remember what we say here," he declared. But he was wrong. Indeed, the fame of the address eclipsed the fame of the battle. Which was as it should be, Charles Sumner thought: "The battle itself was less important than the speech. Ideas are more than battles." Lincoln at Gettysburg was a tough act to follow. Bayard Taylor, in his "Gettysburg Ode" of 1869, asked,

> What voice may fitly break
> The silence, doubly hallowed, left by him?

Taylor proceeded with a rhymed paraphrase of the "Nation's litany" before venturing his own voice. Richard Watson Gilder, on the twenty-fifth anniversary of the battle, invoked the shade of Lincoln—"dear, majestic ghost"—to bless the sublime accord of the Blue and the Gray reunited on this sacred ground. The voice of the war's greatest captain resounded across the years at Gettysburg.[52]

The First Culmination

John George Nicolay and John Milton Hay were, in the words of a modern scholar, "the only licensed transmitters of the Lincoln era." Their monumental work *Abraham Lincoln: A History*, published in ten volumes in 1890, bore the imprimatur of Robert Todd Lincoln and claimed, as no other work could, to be based upon the presidential papers. The authors were contemptuous of reminiscence and made much of grounding everything in documentary records. They were not trained historians—far from it—and they wrote in the patriotic conviction that the perpetuation of Lincoln's fame depended upon placing it before the rising generation, and generations yet unborn, in the true light of the men, ideas, and events that shaped his life. They called their biography a history, for they agreed with Emerson, "There is properly no history, only biography," and they approved his application of the aphorism to Lincoln when he said, "He is the true history of the American people in his time."[53]

Nicolay and Hay brought impressive qualifications to their task. Nicolay, the elder of the two, was an immigrant child from Bavaria, whose family settled in Pike County, Illinois. His only college was the office of a newspaper, the *Free Press*, published in the county seat, Pittsfield, of which he became both proprietor and editor at the age of twenty-two. In Pittsfield he met young Hay, who came from Indiana to prepare in the local academy for entrance to Brown University. After graduation from Brown, Hay studied law in the office of his uncle, Milton Hay, in

Nicolay and Hay with Lincoln. Photo by Alexander Gardner, 1863. (The Lincoln Museum, Fort Wayne, a part of Lincoln National Corporation)

Springfield. Here he resumed his budding friendship with Nicolay, now the clerk in the office of the Illinois Secretary of State and a political lieutenant of the town's leading citizen, Lincoln. When he was nominated for the presidency, Lincoln named Nicolay his private secretary, and continued him in this position upon his election. At Nicolay's suggestion, Hay was invited to go with the presidential party to Washington as assistant secretary. He jumped at the opportunity to leave the dust of the prairie and the dust of the law behind him forever. In their personalities, Nicolay and Hay were quite different, one grave and methodical, the other witty, carefree, and high-spirited. Together they served the President for over four years, performing a variety of duties that belied their humble titles, with scarcely a ripple between them. Their working friendship laid the basis for a remarkable literary collaboration.[54]

At the close of Lincoln's first term, both secretaries took up diplomatic posts in Paris. Nicolay returned after four years, while Hay, enjoying the cosmopolitan life, served in Vienna and Madrid as well, returning to the United States finally in 1872. They conceived the idea of a biography early in the Lincoln administration. Hay kept a diary with this in view; Nicolay saved interesting papers and made memoranda of conversations. No other men had a closer day-to-day relationship with the President than "the boys," as he called them. Between themselves they called him jestingly "the Ancient" and "the Tycoon," but they never presumed to familiarity, nor did Lincoln invite it. They idolized him from the first, and he grew constantly in their esteem until, after his death, their worship was unbounded and unabashed. "He is our ideal hero," said Nicolay. To Hay, quite simply, he was "the greatest character since Christ." Obviously, the time had come now that both were resettled in the United States to embark on the Lincoln work.

The disaster of Lamon's biography underscored the need and the opportunity. Milton Hay wrote to Nicolay, "God-fearing men make up the reading public. They want a model for all the good little boys to follow, and Billy Herndon's model won't do." One thing more was required: that Robert Lincoln make available his father's papers. He did so with the provision that he be allowed to review the manuscript prior to publication. The papers were entrusted to Nicolay in Washington, where in his new office as marshal of the Supreme Court he was ideally situated for literary work. Hay, meanwhile, married the daughter of a Cleveland capitalist, which put him at his leisure; in 1878, being appointed Assistant Secretary of State, he relocated to Washington. And so between Hay's new mansion on Lafayette Square and Nicolay's humble home on Capitol Hill the *History* was written.[55]

Nicolay and Hay never divulged the division of labor between them; and so successfully did they submerge their identities in the serviceable but pedestrian prose of the *History* that readers found it impossible to apportion authorship. It was understood, however, that Hay was responsible for most of the first volume, taking Lincoln to 1856. Here the portrait was thin because of the author's aversion to non-documentary sources. The mass of reminiscences were false, though strikingly, Hay noted, they all ran in the direction of truth, thereby enhancing Lincoln's reputation. The authors conceded their own partiality, not only with regard to the hero but toward those who opposed him, from Stephen A. Douglas through Jefferson Davis and Robert E. Lee, both of whom had deserved to hang. Politically, Nicolay and Hay were devout Republicans. As Lincoln had preserved the Union, so was the Republican party of Grant and Hayes and Garfield its custodian. Writing at a time when historical research was still a gentlemanly endeavor, and the ideal of objectivity had yet to be established, Nicolay and Hay were not exemplary historians. Yet they searched out and used historical records in an impressive way. Some of these were in the Lincoln Papers, or in the *Official Records of the War* rolling off the presses in endless volumes, or got from executive departments, often with difficulty, or from the private papers of Lincoln's contemporaries. Part of the plan from the first was to compile Lincoln's writings with a view to editing a complete edition. So the *Complete Writings* progressed with the *History*, although it was published several years later.[56]

As the manuscript piled up, the authors looked about for a publisher. In 1882 Richard Watson Gilder, the new editor of *Century* magazine, successor to *Scribner's Monthly*, expressed an interest in serial publication of part of the work followed by subscription publication of the whole in six or more volumes. The overture was turned aside, but Gilder persisted. *Century*, a slick-paper magazine, prided itself on fine illustrations and took a special interest in Lincoln and the Civil War. It published reminiscences, articles on portraiture, and fiction touching on Lincoln, for instance Eggleston's *The Graysons*. In 1884 the magazine launched its "War Series," devoted to battles and commanders, and over two years watched the circulation double to 250,000. Nicolay and Hay's *Lincoln* was a logical sequel. The senior author, while vacationing in New Hampshire in 1885, invited the editor to read the manuscript, which had reached half a million words. Gilder was enthralled. The president of the Century Company, Roswell Smith, decided the book was "the most important literary venture of the time," and easily outbid the leading competitor, the prestigious house of Harper. Nicolay and Hay were offered the unprecedented sum of $50,000 for serial rights, plus royalties on book sales. It was a magnificent coup, Gilder boasted: an "inside history" of the war and the only "authorized" life of the greatest man the country had produced, "at least since Washington"—the usual caveat—though in human interest he surpassed Washington. The book will hold the national audience for the *Century* for the next several years, said Gilder. "In addition to this, the work will have a great moral and political effect in that it will help unite the North and South as never before, around the story and experiences of the great President."[57]

The editor's desire to make the *History* a vehicle of reconciliation was not shared by the authors. This became a source of contention as the work went into publication. Gilder undertook the abridgement. Somewhat less than one-half of the *History* appeared in the magazine. It began in November 1886 and continued through February 1890, by which time *Century* was losing readers. The work was too ponderous for a popular magazine. Gilder wished to lighten it with more attention to

Lincoln's personal and domestic side, less to the conduct of the war, where it overlapped the War Series. But he was most distressed by what he called the "partisan" tone, and he begged the authors to show a little of Lincoln's magnanimity toward the South. His friend Augustus Saint-Gaudens, the sculptor, whom he had asked to read the articles prior to publication, agreed with him. But Nicolay retorted, "We deny it is partisanship to use the multiplication table, revere the Decalogue, and obey the Constitution of the United States." According to Nicolay's daughter, to whom he dictated large parts of the work as his eyesight failed, some objectionable adjectives were scratched in the interest of more balanced presentation.

Of course, the work was submitted to Robert Lincoln for his scrutiny. His only concern, it seems, was with the details of his father's paternity and early life, the source of so much trouble with Herndon. Sending him the first several hundred pages of manuscript, Hay remarked, "I need not tell you that every line has been written in the spirit of reverence and regard." Still, if anything did not suit, Lincoln should strike it. "I will adopt your view in all cases whether I agree with it or not." Only one correction has been positively identified: a story from Lamon's biography of Lincoln sewing up the eyes of hogs in order to drive them aboard a flatboat. But there were others, matters of taste for the most part, which Hay anticipated and corrected himself. If the work was written in a spirit of filiopiety, this was because the authors felt it rather than because the son demanded it. Robert Lincoln thought his father exceedingly fortunate in his biographers.[58]

Eight of the ten volumes of the *History* were devoted to the Civil War. To the knowledge of Lincoln's life and career before his election to the presidency, Nicolay and Hay added nothing of consequence. They gave Lincoln a respectable lineage, taking care to prove his legitimacy and to show his mother's moral and intellectual superiority over the crude environment. They associated Lincoln with the great westward march of conquest and settlement now drawing to a close. In 1893, three years after the *History* was published, a young scholar, Frederick Jackson Turner, derived a theory of American history from this movement, one which postulated a distinctive western character; but Nicolay and Hay, though westerners themselves with a keen appreciation of pioneer habits and values, declined to characterize Lincoln in these terms.

In an earlier literary work, *Pike County Ballads*, Hay had created memorable folk characters in "Jim Bludso" and "Little Breeches"; but they represented a crude, barbarous culture, and he had no wish to portray Lincoln in this light. He did suggest a pioneering explanation for Lincoln's melancholy, saying that "this constitutional sadness was endemic among the early settlers of the West." However, the explanation was offered defensively, as an alternative to Herndon's, which turned on lost love. Hay dismissed Ann Rutledge in two sentences and dated the "splendid maturity" of Lincoln's mind and character to the time of his marriage. He credited his hero with a driving political ambition. But why, Hay asked, had he chosen to pursue it in a party, the Whig party, that was a perennial minority in Illinois? Because it was the party of the better sort of people, and young Lincoln meant to associate with the better sort rather than the Jacksonian masses.[59]

From his closeness to Lincoln during the secession winter, Nicolay probably wrote this part of the book. The President-elect was portrayed as sharply opposed to President Buchanan's vacillation. While exercising "prudent reserve," Lincoln nevertheless made his position crystal clear. The power of the national government to coerce a state was settled by Andrew Jackson. When told that Buchanan intended to surrender the forts in South Carolina, Lincoln responded in Nicolay's presence, as Jackson had responded to John C. Calhoun, "If that is true, they ought to hang him!" It was a good line, yet Nicolay did not use it in the *History*, preferring to say on his own responsibility that Buchanan "consented to national suicide." A parallel between Lincoln and Davis was scathing toward the latter. If the fates had changed their parents' paths of emigration, Lincoln would still never have been a slaveholding aristocrat, nor Davis a champion of humanity. The Illinoisan's view of the crisis was best summed up, Nicolay thought, in a letter found in the son's papers. "We [the Republicans] have just carried an election on principles fairly stated to the people. Now we are told in advance, the government shall be broken up unless we surrender to those we have beaten. . . . In this way they are either attempting to play upon us, or they are in dead earnest. Either way, if we surrender, it is the end of us, and of the government."[60]

Although privy to the formation of Lincoln's cabinet, Nicolay offered only a cursory account of it. The inclusion of all the President's principal Republican rivals had been a cause of wonderment. It made manifest his determination to have a cabinet at once distinguished and representative; and it made manifest his confidence in his ability to control it. "In weaker hands such a Cabinet would have been a hot-bed of strife; under him it became a tower of strength." Apprehensions for the cabinet centered on the well-known hostility between Seward and Chase. At the time of the inauguration, Seward, fearing Chase's ascendancy, submitted his resignation. Lincoln refused it, saying to his private secretary, Nicolay, as he handed him the reply, "I can't afford to let Seward take the first trick." In this instance Nicolay threw caution to the winds and introduced testimony only he could vouch for. Seward immediately withdrew the resignation. Lincoln had asked him to critique the draft of the Inaugural Address. Seward read it with admiration, suggesting only one substantive change, but proposed a new and more sonorous peroration. From Frederick W. Seward, his father's biographer, Nicolay and Hay received copies of these documents. Lincoln adopted Seward's peroration, yet made it his own by substituting his concise and condensed diction for the New Yorker's vague and diffuse expression and, in the last sentence, turning the germ of a poetic thought into poetry itself.*

Literary mastery was accompanied by political mastery, as the authors proceeded to show. The Secretary of State had led Confederate authorities to believe that Fort Sumter would be evacuated. The President with a majority of the cabinet determined to reinforce it. An expedition was mounted. This painted Seward into a corner, and he acted boldly, if foolishly, to extricate himself. On April 1 he handed Lincoln a paper bearing the title "Some Thoughts for the President's Consideration," in which he took issue with the Sumter policy, proposed to provoke war abroad in order to preserve peace at home, criticized the lack of executive leadership, and nominated himself as the virtual premier of the gov-

---

* Roy P. Basler, *Abraham Lincoln: His Speeches and Addresses* (Cleveland, 1946), p. 48, gives Seward's paragraph (left column) as altered by Lincoln (right):

I close	I am loth to close.
We are not, we must not be, aliens or enemies, but fellow-countrymen and brethren.	We are not enemies, but friends. We must not be enemies.
Although passion has strained our bonds of affection too hardly, they must not, I am sure they will not, be broken.	Though passion may have strained, it must not break our bonds of affection.
The mystic chords which, proceeding from so many battlefields and so many patriot graves, pass through all the hearts and all the hearths in this broad continent of ours, will yet again harmonize in their ancient music when breathed upon by the guardian angel of the nation.	The mystic chords of memory, stretching from every battlefield and patriot grave, to every living heart and hearth-stone, all over this broad land, will yet swell the chorus of the Union when again touched, as surely they will be, by the better angels of our nature.

ernment. The President replied at once to these extraordinary propositions, rejecting them, and insisting that whatever was done, "*I* must do it."

These papers from the Lincoln collection were a revelation. In 1873 Robert Lincoln had proposed giving them to Welles to use in his controversy with Adams, but Nicolay dissuaded him with a plea not to steal his and Hay's thunder. In addition, the authors discussed the by now celebrated diplomatic dispatch to Adams, corrected by Lincoln, as further evidence of the President's mastery. Nicolay and Hay felt no personal animosity toward Seward. Quite the contrary. They owed their diplomatic appointments to him; and they emphasized in the *History* his diligent, efficient, and selfless service to the President after a rocky start. Their feelings toward Chase were different. He was a man of gall, unable to put aside his presidential ambitions; his attitude toward Lincoln wavered between "active hostility and benevolent contempt." The surest way into his confidence was to say something derogatory about Lincoln, the authors said. The President was well aware of his secret and sullen opposition, which came to a head in 1864, but tolerated it because of Chase's stature and ability.[61]

In the whirl of confusion, discord, and impatience of the administration's early months, Nicolay and Hay viewed Lincoln's steadiness and common sense as "a rock of safety to the nation." His Sumter policy was successful. In coming to the defense of the beleaguered fort, he discharged his high constitutional duty, and if the Confederacy resisted, as he expected, the people would take care of the rest. "He was looking through Sumter to the loyal states; beyond the insulted flag to the avenging nation." In their detailed treatment of Lincoln as Commander in Chief of the Union armies, they credited him with a military genius not far short of his political genius.

Hay, in particular, admired the Comte de Paris's lengthy *History of the Civil War*, despite its "outrageous prejudice" against Lincoln, who was blamed for frightful disasters, and its partiality for General McClellan. After assuming command of the Army of the Potomac in July 1861, this gallant "Little Napoleon" showed tremendous organizational energy and cast a spell over civilians and soldiers alike. He created a magnificent army. Unfortunately, said Nicolay and Hay, drawing upon memoirs and letters published over a quarter-century, he suffered "a sudden and fatal degeneration of mind" during the winter months in Washington. Of course, this was a literary rather than a clinical diagnosis, but the authors knew not how else to account for McClellan's subsequent behavior, which included delusions of glory, panic-stricken fear, and paranoia. They recorded an incident in November, noted in Hay's diary, after McClellan had succeeded doddering old Winfield Scott as General in Chief. Lincoln made an evening call at the General's house on some business of importance. Finding him out, the President waited an hour for him to return; when he did, McClellan marched upstairs and thirty minutes later sent word by a servant he had gone to bed. The President chose to ignore the snub. Hay noted some weeks later, however, "He stopped going to McClellan's and sent for the General to come to him."*

Disappointed with the inertia of his generals, Lincoln became his own general-in-chief. He pored over books on military science, planned campaigns, and issued war orders. Yet he deferred to McClellan's plan for the Peninsular Campaign. It failed, not because of any fault of the plan, but because the commander vastly overestimated the forces of the enemy and lacked the will to fight. McClellan's exculpatory defense, in a *Century* article and in the posthumous *McClellan's Own Story*, was labeled a "fantasy" by Nicolay and Hay. (This was an edited volume; McClellan wrote much more derisively of the President in letters to his wife, which the authors did not see.) He blamed Lincoln for the failure or, more exactly, Stanton and the Radicals around the President who poisoned his mind toward his General in Chief. The same charge was made by the Comte de Paris, and by George T. Curtis and other Democrats whose partiality for McClellan led them to portray Lincoln as a bigoted Republican. The President demoted McClellan after the disaster on the peninsula between the York and James rivers, but soon gave him an opportunity to redeem himself, partially fulfilled at Antietam. Then the old dilatory infirmity returned, and Lincoln abandoned him for good.[62]

The ten volumes of the *History* defy quick and easy summary. Developing the theme of emancipation, the authors observed a logical progression terminating in the Thirteenth Amendment. The issue was never one between the Union and emancipation, they insisted, but how these twin objectives could be made to serve each other. And so there was no inconsistency between the President's revocation of General John C. Frémont's decree of emancipation for Missouri in 1861 and his own national proclamation a year later. The former action saved Kentucky for the Union—a priority at the time. The policy of gradual compen-

---

*Another story first recorded in the *History* (IV, 469n) became part of the folklore. F. A. Mitchel, the son of General O. M. Mitchel, told in a communication to the authors how his father, in conference with Lincoln, was stood up by McClellan. Mitchel expressed amazement at such insolence, but the President responded calmly, "Never mind; I will hold McClellan's horse if he will only bring us success." A later Supreme Commander, Dwight D. Eisenhower, liked to tell this story after he became Commander in Chief, thinking it showed Lincoln's single-minded dedication to the task at hand (*Lincoln Lore*, no. 307 [1954]).

sated emancipation and colonization laid before Congress in December 1861 deliberately appealed to border-state sentiment and opened a *via media* that must be tried, the prudent President thought, before choosing the more extreme route. It met with a discouraging response; this and the collapse of the Peninsula Campaign caused Lincoln to propose the Emancipation Proclamation sooner that he had expected. It was surrounded by difficulties and perplexities. Lincoln left no record of how he came to this momentous decision—and it was eminently *his* decision—nor could Nicolay and Hay explain it. The proclamation was grounded in military necessity. As slavery was the backbone of the rebellion, it was necessary to free the slaves in order to save the Union, which was to say *ourselves*. As Lincoln put it, "In giving freedom to the slave we assure freedom to the free." Emancipation was thus a profoundly national act.[63]

Among the Lincoln Papers was a letter to a Kentucky newspaper editor who had called on the President in 1864 to discuss the enlistment of Negro troops in that state. Here in recapitulating the little speech he had made during the interview, the authors thought, Lincoln best summed up his philosophy on the subject.

> I am naturally anti-slavery. If slavery is not wrong, nothing is wrong. I can not remember when I did not so think, and feel. And yet I have never understood that the Presidency conferred upon me an unrestricted right to act officially upon this judgment and feeling. . . . And I aver that, to this day, I have done no official act in mere deference to my abstract judgment and feeling on slavery. I did understand however, that my oath to preserve the constitution to the best of my ability, imposed upon me the duty of preserving, by every indispensable means, that government—that nation—of which that constitution was the organic law. Was it possible to lose the nation, and yet preserve the constitution? . . . I felt that measures, otherwise unconstitutional, might become lawful, by becoming indispensable to the preservation of the constitution, through the preservation of the nation. Right or wrong, I assumed this ground, and now avow it. I could not feel that, to the best of my ability, I had even tried to preserve the constitution, if, to save slavery, or any minor matter, I should permit the wreck of government, country, and Constitution all together.

He went on to say that failure of gradual emancipation proposals drove him to the bolder measure.[64]

One of the revelations of the *History* was the terrible gloom that enveloped Lincoln over the uncertain outcome of the election of 1864. He had not been alarmed by Chase's plot to wrest the Republican nomination from him. Hay introduced a statement he had recorded in his diary:

"I have determined to shut my eyes, so far as possible, to everything of that sort." This threat was dead by March, and Lincoln was renominated at Baltimore without opposition. But the Democrats nominated a man on horseback, McClellan, still a military hero, and capitalized on northern defeatism in the campaign for his election. The President's alleged rebuff of peace overtures at Niagara was used against him. By the end of August, such was the pessimism of his political advisers, Lincoln prepared himself for defeat. He wrote a memorandum saying that in this eventuality it would be his duty to cooperate fully with the President-elect during the interim between the election and the inauguration. At his request members of the cabinet signed the document on the verso not knowing what it said. Lincoln then placed it under seal in his desk. After his reelection, he opened the envelope and explained the memorandum to the cabinet. Hay was familiar with this episode. He had recorded it in his diary and made the copy in the Lincoln Papers.[65]

Fifty or more people read the *Century* abridgment to every one who read the ten-volume *History*. Five thousand copies were sold by subscription and perhaps two thousand in addition. Nicolay, in 1902, made his own abridgment in one volume. It sold thirty-five thousand copies. Ponderous though the *History* was, the judgment of Hay's biographer stands: no other American historical work reached so many readers in so short a time. The authors had reason to be gratified by the critical reception as well. Hay was surprised that the reviewers raised so few objections. "Laws-a-mercy!" he laughed to his friend Henry Adams. "If I had the criticizing of that book, what a skinning I could give it."[66]

Carl Schurz, in the *Atlantic Monthly*, turned his review into a reflective essay on Lincoln, later published separately and much admired. Literary critics thought that the history submerged the biography, while historians rated the work better as biography than as history. One of the latter, Frederic Bancroft, said that the authors were so overwhelmed by their materials they were unable to develop a coherent narrative, and their blind devotion to Lincoln robbed the work of scholarly value. Nicolay and Hay were especially pleased by Howells's rave review in *Harper's*. He regretted the neglect of Lincoln's private life, but thought that the outcome if not the progress of the work *was* biographical. At Gilder's behest, Hay wrote an article on Lincoln's life in the White House, which Howells appreciated; and Helen Nicolay, after her father's death, used the notes he had assembled for her book *Personal Traits of Abraham Lincoln*, published by *Century* in 1912.

Whatever its failings, the *History* enshrined the memory of the Civil War President and transmitted his legend to a new generation with the stamp of authority. Yet its fame soon faded. Curiously, at the same time

Abraham Lincoln: A History was published, Henry Adams brought out his nine-volume *History of the United States During the Administrations of Jefferson and Madison*, and although the work lacked any magazine readership and sold precious few copies, it became a classic of American historical literature, still studied and esteemed one hundred years later, while Nicolay and Hay's *History* gathers dust in the recesses of a thousand libraries.[67]

WILLIAM H. HERNDON, ruminating at his farm, Chinkapin Hill, on the Sangamon River, read the early chapters of Nicolay and Hay's *History* and predicted, "It will result in delineating the real Lincoln about as well as does a wax figure in the museum." They suppressed many important facts, including "the finest story in Lincoln's life," his love of Ann Rutledge. Perhaps they dared not write about such truths. "Nicolay and Hay handle things with silken gloves and 'a camel hair pencil'; they do not write with an iron pen." Herndon, sixty-eight years old in 1886, had again embarked on the biography he had envisioned twenty years earlier. It would, of course, be written with "an iron pen."[68]

Disappointed by the failure of Lamon's *Life* and assailed for his views on Lincoln's religion, Herndon had retired to the obscurity of his farm north of Springfield, where he made a precarious living and cared for his young wife and children. He corresponded sporadically with such Illinois luminaries as Isaac Arnold and Henry C. Whitney. Occasionally, his name surfaced in newspapers. It was reported that he had attempted to commit suicide, which he denied, and some years later that he had become a common drunkard and pauper. To this report, in 1882, Herndon replied in a printed broadside. The reason for these libels, he said, was that he had dared to tell the truth about Lincoln's infidelity bordering on atheism.

In the same year, Herndon returned to the platform, delivering lectures at Petersburg and Pekin. According to an advertisement, he discussed such topics as the Rutledge romance and the courtship and marriage to Mary Todd. That poor woman had at last returned to Springfield to die. In 1875 her behavior had become so alarming to her only surviving son, Robert, that he filed a petition to adjudge her insane under Illinois law. She was committed to a private sanatorium and released, following a second trial, six months later. The proceedings created a sensation in the press. Many people could not forget the contribution Billy Herndon had made to Mrs. Lincoln's pain and suffering. She died on July 16, 1882.[69]

A last opportunity to write the long-meditated biography was more than Herndon had any right to expect; miraculously, it materialized in

William H. Herndon. (Illinois State Historical Library)

the person of Jesse W. Weik of Greencastle, Indiana. A graduate of the local university (now DePauw), he had become an avid student of Lincoln and the Civil War, which led to a desultory correspondence with Herndon. In 1882 this intensely serious young man took up residence in Springfield as federal pension agent. Here he extended his inquiries about Lincoln and got better acquainted with Herndon. After about three years Weik resigned his position for reasons of health and returned to Greencastle with intentions of writing. He sought Herndon's assistance on an article; they agreed to collaborate, but with the volcanic eruption of Herndon's mind on Lincoln, the article was scrapped and the idea of a biography rose in its place.

During five months Herndon wrote thirty-five long letters full of opinion and reminiscence. He told how his partner used to take Euclid's *Elements* on circuit and study its problems by candlelight before bed. He told how Lincoln loved the game of "fives"—a version of handball—

and was a good chess player. He speculated on the causes of Lincoln's melancholy. He described Mrs. Lincoln as "*the female wildcat of the age*," and said the poor husband's home was "hell on earth." "Jesus, what a home Lincoln's was! What a wife!" After a time Herndon remarked, "My letters to you are half the Biography—ready to hand." In January 1886, he declared a halt. "I feel that I am pumped dry." The letters continued, but at a slackened pace.

In November he sent to Weik the original documents, withheld from Lamon and Black, of the Lincoln Record. Weik's task was to synthesize this mass of data and turn it into a biography. That was no easy task, for Herndon mixed the trivial with the momentous, voiced outrageous opinions, and from one page to the next flatly contradicted himself on Lincoln's character. In August 1887, he traveled to Greencastle, and in the second-floor room where Weik had made his office, above his father's store, the two men labored day after day for five hot weeks on the biography. When published two years later, the book was entitled *Herndon's Life of Lincoln*, and in the apportionment of authorship Weik was accorded second billing. The book was a genuine collaboration, yet the title did not lie. This was Herndon's Lincoln finally brought to fruition by the listening and compositional tales of Jesse Weik.[70]

The authors were unable to decide how to treat the matter of Lincoln's paternity, so Herndon laid it before several of the Springfield luminaries upon his return home. He had grudgingly accepted the documentary evidence, discovered since Lamon's biography, of the marriage of Nancy Hanks and Thomas Lincoln on June 12, 1806. But he still credited the Kentucky gossip that Lincoln was the bastard son of Abraham Inlow (or Enlow); in addition, on the word of Dennis Hanks, he believed that Thomas Lincoln was impotent as a result of mumps, though whether this occurred before or after Nancy's pregnancy he could not be sure. There was the related question of Nancy's illegitimacy, which Lincoln had raised with Herndon on a buggy ride in 1850. Herndon wavered between telling the whole truth—his credo—and suppressing it, as he accused Nicolay and Hay and others of doing. The counsel he sought was divided. After weighing it he came down on the side of Richard Oglesby and suppression. "I say bow down to the inevitable," he wrote cynically to Weik. "Success is what we want. . . . Do what is necessary to gain that end, short of lying, or fraud. Please Lincoln's friends, the publishers and all mankind, past, present, and the future." Weik, meanwhile, went to Chicago and got the same advice from Leonard Swett. Herndon, as if conscience-stricken, continued to waver, toying with ways to fudge the issue. In the end, the biography perpetuated the legend of

Nancy's illegitimacy by the reported statement of her son to Herndon, and it planted doubts of the son's legitimacy by repeating the Kentucky gossip, even if it was consigned to a footnote.[71]

Herndon's Lincoln is an exceedingly personal biography—much of it is written in the first person singular—and any estimate of it must reckon with the reliability of the author's impressions and memories. The biography is essentially a sequence of reminiscences and anecdotes. About one-half of these are derived from the author's own knowledge and observation, mixed with a good deal of intuition and more than a little magnification of his own importance in Lincoln's career. It is often said that Herndon did not lie. Yet he sometimes erred and sometimes contradicted himself, as on the question of whether Lincoln discovered the poem "Mortality" before or after Ann Rutledge's death. The reader of Herndon's Preface may be astonished by the bald assertion that he, like Weik, visited Lincoln's Kentucky birthplace. He did not. Most of the other reminiscences were collected by Herndon. The question of their reliability poses more difficult problems, earlier touched upon. In terms of its contents, the biography may be divided into three parts. The first runs to 1844, when the partnership of Lincoln and Herndon was formed. Here the book follows the outlines of Lamon's *Life*. The reminiscences are mainly those gathered by Herndon after Lincoln's death. The second part, from 1844 to 1861, is the heart of the biography. For this period, when Herndon saw more of Lincoln than anyone but his wife, he offers a vivid and close-up portrait. The third part, on the presidential years, is no more than an afterword. Its only purpose is the mandatory one of conveying Lincoln to his death.

Herndon's Lincoln was particularly significant for what came to be called "the folklore Lincoln." At least as early as Caroline Dall, Americans had looked to Herndon for an authentic western portrait of the Martyr Chief. He attempted to respond but was not sufficiently attuned to folk traditions to succeed. He wrote of pioneering hardships, crudities, frolics and superstitions. To the last of these, along with early Baptist train ing, he traced Lincoln's fatalism and belief in the significance of dreams and visions. He called his adolescent lampoon "The Chronicles of Reuben," which Lamon had introduced to the public, "a memorable chapter of backwoods lore," for Herndon heard the verses recited from memory thirty years later in southern Indiana. One of his informants, Abner Y. Ellis, had given him a report of Lincoln's first stump speech in 1832. (This, too, was in Lamon's book.) "My politics are short and sweet, like an old woman's dance," declared the tall, gawky youth, who went on to repeat the litany of Henry Clay's platform. The speech, being only a remembrance, is still considered apocryphal by scholars. Honest Abe's stories had a western tang, of course. Herndon said that his fame as a storyteller preceded his fame as a lawyer and politician. These were all indicia of a "folklore Lincoln," but they scarcely made a complete portrait.[72]

Herndon's Lincoln is a man scarred by loss. The most damaging loss, emotionally and psychologically, was Ann Rutledge. Since Herndon's first disclosure of this blighted love, it had entered the stream of both biography and folklore. Now the author added several new touches to the story. He claimed to have known Ann Rutledge himself, though he was only sixteen at the time of her death. The overwhelming significance of the romance, as told by him, was that Lincoln never got over her death but brooded about it the rest of his life. Herndon quoted his partner once saying to him, after speaking of Ann's death and her grave, "My heart lies buried there." He introduced the testimony of Isaac Cogsdale [Cogdal], a New Salem pioneer, who came to see the President-elect before his farewell to Springfield. Asked a direct question about Ann Rutledge, Lincoln replied yes, he had loved her and thought of her still. (Herndon did not notice them, but stories occasionally appeared in Illinois newspapers of Lincoln the President acknowledging his love of Ann. In one of these, a Kentucky mother and daughter called at the White House to plead for the pardon of a rebel son and brother; the daughter went to the piano and played a popular ballad, *Gentle Annie*, at which the President drew a large red handkerchief from his pocket to wipe away the tears.) Herndon devoted a chapter to Lincoln's affair with Mary Owens. He had tracked her down, Mrs. Vineyard, in Weston, Missouri. The affair was more amusing than sad, but the author used it to support his view that Lincoln was clumsy and uncomfortable with women.

With his romantic imagination, Herndon painted a dramatic picture of "the fatal first of January," 1841—the announced date of his wedding to Mary Todd. The Edwards mansion, where she resided in her sister's family, was all in readiness, the wedding supper was set, the guests gathered, and "the bride, bedecked in veil and silken gown," awaited the bridegroom; but he had disappeared. Herndon got this story from Elizabeth Edwards, Mary's sister, forty years after the event, and added his own embellishments. Again, as upon Ann's death, Lincoln suffered a fit of insanity. Some nineteen months later, Lincoln and Mary Todd married. Herndon was puzzled to explain why. Not for love, certainly. He married her "to save his honor"—because he had promised to marry her—and in doing so he sacrificed domestic felicity. But why, after being jilted, did she marry him? For revenge, said Herndon. Sadly, she exacted it every day of his life.[73]

Herndon described Lincoln as a successful lawyer who was deficient in technical learning and zeal. Aside from his own observation, he drew upon the reminiscences of Henry C. Whitney, of the Chicago bar, and others. Herndon's Lincoln truly excelled as a politician. "His ambition was a little engine that knew no rest. The vicissitudes of a political campaign brought into play all his tact and management and developed to the fullest extent his latent industry." The junior partner, as previously observed, preened himself on being Lincoln's political mentor and guide. In the instance of the "House Divided" speech, he took credit for one of the most famous utterances in American history. Lincoln, he recalled, read to him a draft of the speech paragraph by paragraph. They paused upon the line "a house divided against itself cannot stand," but decided to keep it because it was true and needed saying. Later he tried out the speech with a group of friends in the State House. Several objected to the assertion of this doctrine. But Herndon said, "Lincoln, deliver that speech as read and it will make you President." Historians have been skeptical about this prophetic statement, but two eyewitness accounts substantiate Herndon.[74]

The biography is distinguished by inimitable vignettes which etched and colored the public image of the man.

> He exercised no government of any kind over his household. His children did much as they pleased. Many of their antics he approved, and he restrained them in nothing. He never reproved them or gave them a fatherly frown. He was in the habit, when at home on Sunday, of bringing his two boys, Willie and Thomas—or "Tad"—down to the office to remain while his wife attended church. He seldom accompanied her there. The boys were absolutely unrestrained in their amusement. If they pulled down all the books from the shelves, bent the point of all the pens, overturned inkstands, scattered law-papers over the floor, or threw the pencils into the spittoon, it never disturbed the serenity of their father's good-nature.*
>
> No man had greater power of application than he. Once fixing his mind on any subject, nothing could interfere with or disturb him. Frequently I

*The original of this in Herndon's letter to Weik was as follows: "He, Lincoln, used to come down to our office on Sunday when Mrs. Lincoln had gone to church, *to show her new bonnet*, leaving Lincoln to care and attend to the children. Lincoln would turn Willie and Tad loose in our office, and they soon gutted the room, gutted the shelves of books, rifled the drawers, and riddled boxes, battered the points of my gold pens against the stairs, turned over the inkstands on the papers, scattered letters over the office, and danced over them and the like. I have felt many a time that I wanted to wring the necks of these brats and pitch them out of the windows, but out of respect for Lincoln and knowing that he was abstracted, I shut my mouth, bit my lips, and left for parts unknown." Quoted in *The Hidden Lincoln, from the Papers of William H. Herndon*, Emanuel Hertz, ed. (New York, 1938), 128–29.

would go out on the circuit with him. We, usually, at the little country inns occupied the same bed. In most cases the beds were too short for him, and his feet would hang over the floorboard, thus exposing a limited expanse of shin bone. Placing a candle on a chair at the head of the bed, he would read and study for hours. I have known him to study in this position till two o'clock in the morning. Meanwhile, I and others who chanced to occupy the same room would be safely and soundly asleep. On the circuit in this way he studied Euclid until he could with ease demonstrate all the propositions in the six books.

In the role of a story-teller I am prone to regard Mr. Lincoln as without an equal. I have seen him surrounded by a crowd numbering as many as two and in some cases three hundred persons, all deeply interested in the outcome of a story which, when he had finished it, speedily found repetition in every grocery and lounging place within reach. His power of mimicry . . . and his manner of recital, were in many respects unique, if not remarkable. His contenance and all his features seemed to take part in the performance. As he neared the pith or point of the joke or story every vestige of seriousness disappeared from his face. His little gray eyes sparkled; a smile seemed to gather up, curtain like, the corners of his mouth; his frame quivered with suppressed excitement; and when the point—or "nub" of the story, as he called it—came, no one's laugh was heartier than his.

A brief description of Mr. Lincoln's appearance on the stump and of his manner of speaking may not be without interest. When standing erect he was six feet four inches high. He was lean in flesh and ungainly in figure. Aside from the sad, pained look due to habitual melancholy, his face had no characteristic or fixed expression. He was thin through the chest, and hence slightly stoop-shouldered. When he arose to address courts, juries, or crowds of people, his body inclined forward to a slight degree. At first he was very awkward, and it seemed a real labor to adjust himself to his surroundings. . . . When he began speaking, his voice was shrill, piping, and unpleasant. His manner, his attitude, his dark, yellow face, wrinkled and dry, his oddity of pose, his different movements—everything seemed to be against him, but only for a short time. After having arisen, he generally placed his hands behind him, the back of his left hand in the palm of his right, the thumb and fingers of his right hand clasped around the left arm at the wrist. For a few moments he displayed the combination of awkwardness, sensitiveness, and diffidence. As he proceeded he became somewhat animated, and to keep in harmony with his growing warmth his hands relaxed their grasp and fell to his side. Presently he clasped them in front of him, interlocking his fingers, one thumb meanwhile chasing another. His speech now requiring more emphatic utterance, his fingers unlocked and his hands fell apart. His left arm was thrown behind, the back of his hand resting against his body, his right hand seeking his side. By this time he had gained sufficient composure, and his real speech began.[75]

Unfortunately for the reception of *Herndon's Lincoln*, it was published by a Chicago house, Belford, Clarke & Company, on the brink of bankruptcy. The authors earned nothing from it. The disappointment to Herndon seemed a reprise of the Lamon biography. A Chicago newspaper greeted the book with an effusion of gall. "It is one of the most infamous books ever written. . . . It vilely distorts the image of an ideal statesman, patriot and martyr. It clothes him in vulgarity and grossness. . . . It brings out all that should be hidden. . . . In all this book is so bad it could hardly have been worse." Herndon heard from a friend in Paris, who was unable to find a copy in the bookstores, that—so he was told—the United States Ambassador to Great Britain, Robert Todd Lincoln, had bought up the entire English edition and burned it. Herndon was only too willing to believe the worst of "Bob." "He is a Todd and not a Lincoln," he wrote to Weik. Happily, the reviews in the principal magazines applauded the work. The critic in the *Nation*, who would be lukewarm to Nicolay and Hay, said the book should take its place as the most authoritative and trustworthy life of the subject to 1861. It was candid, vivid, and honest. Acknowledging its source in reminiscence, he called the book, not a biography, but "personal recollections." The *Atlantic Monthly* generally agreed, and commended the book for bringing into relief the western aspect of Lincoln's character.[76]

The book would have languished but for Horace White, Herndon's old Illinois friend, now associated with the *New York Evening Post*. He found an eastern publisher, Scribner's, to rescue the book and bring out a new edition. And when Charles Scribner changed his mind, after hearing that Robert Lincoln objected to the book, White went to Appleton's. The publisher acquired the plates wrote new contracts, and issued a new edition in 1892. By then Herndon was dead. White's service to the book, of which he held a high opinion, went further. He recommended editorial changes (removal of doubts cast upon Lincoln's legitimacy, omission of the raucous "Chronicles of Reuben," for instance) and he contributed two chapters on Lincoln in Illinois politics in the 1850s. The new edition sold five thousand copies in a little more than three years. *Herndon's Lincoln* was on its way to becoming the first classic of the literature.[77]

For all its merits, it is a truncated biography. "It is a statue without a head," as David Donald has said. The growth of Lincoln's mind and character was important, after all, because of what he became as President. Without carrying the story to its fulfillment, without tying the cords of the amazing growth into the life of the President, the book failed to achieve the historical significance it might have attained. The principal reason for his failure is that Herndon's mind and archive did not reach the presidency. Another reason, rarely noticed, is that neither author had a good sense of history or of Lincoln's place in it. And so fascination with his character took the place of fascination with his shaping power in history.[78]

IN THE HOUSE of Lincoln were many mansions. There were biographies of all sizes, shapes, and descriptions. Horace Greeley remarked with pardonable hyperbole, "There have been ten thousand attempts at the life of Abraham Lincoln, whereof that of Wilkes Booth was perhaps the most atrocious; yet it stands by no means alone." From the Illinois group came Isaac Arnold's biography and Whitney's *Life on the Circuit with Lincoln*. The former had an aura of authority but was cold and dry. The latter was a book of rambling reminiscences, Herndonian in spirit, often unreliable, and virtually unreadable.

In 1893 John T. Morse, Jr., a Boston gentleman-historian, waded in with a two-volume life in the American Statesmen series under his general editorship. While the work made no significant contribution to Lincoln biography, it contained striking insights. Morse thought Lincoln such a weird combination of the uncouth and the noble, of western grotesqueries and spiritual grandeur—of what he was and what he became—that the biographer must necessarily navigate between the extremes of unsparing realism and idealism. He found Herndon's portrait distasteful, yet could not blink its truth. Lincoln's preeminent characteristic as a statesman was his "close and sympathetic touch with the people," said Morse. For a long time, he could not rise above them; but when he did, and his people expanded to encompass the nation, he remained one with them. "The very quality which made Lincoln, as a young man, not much superior to his coarse surroundings was precisely the same quality which, ripening and expanding rapidly and grandly with maturing years and a greater circle of humanity, made him what he was in later life." This was the key, the only key, said Morse, to solve the enigma of "the most singular life, taken end to end, which has ever been witnessed among men."

Carl Schurz, the German-American statesman who had been Lincoln's political associate, also emphasized his solitary and enigmatic individuality in a long biographical essay. As sympathy was the strongest cord of his nature, he attracted the common people to him, and the bond between them ensured that even in a great crisis like the Civil War his leadership would be democratic. Washington's popularity rested on veneration and awe; Andrew Jackson's was a thing of party; but the people were bound to Lincoln by sentimental attachment. "It was . . . an affair of the heart, independent of mere reasoning."[79]

The landmark of juvenile biography remained William M. Thayer's *The Pioneer Boy and How He Became President*. This Massachusetts minister and schoolmaster became a prolific author of books for young readers, which combined ministrations of the Christian Gospel with the gospel of success. He revised, expanded, and reissued the Lincoln work in 1882, and again it enjoyed wide popularity. Making a fictionalized story of Lincoln's early life, Thayer constructed dramatic scenes and invented dialogue. He added entirely new episodes to Lincoln's biography, for instance Thomas Lincoln's sale of his Kentucky farm for ten barrels of whiskey loaded on a flatboat and lost in the river on the way to Indiana. The story, unproven then as now, entered into the Lincoln legend. Thayer wrote with the didactic purpose of showing his youthful readers that Lincoln's success was the result, not of advantages or of luck, but of solid virtues like honesty, industry, and temperance; and if they would practice these virtues they, too, might become President.[80]

This moral didacticism was the hallmark of Lincoln juvenilia. Within two years of his death, the Sunday-School Union in New York published Z. A. Mudge's *The Forest Boy*, specifically for use in Sabbath schools. For all his virtues, Lincoln was reproved in this book for one grave fault: he did not publicly avow his Christian faith. Horatio Alger, in *Abraham Lincoln, the Backwoods Boy* (1883), portrayed him as a real-life Ragged Dick whose virtues were rewarded. No other American life, said the author, may be studied with more profit by American youth. Here were the legendary stories of young Lincoln's scrupulous honesty, such as the long walk to return six and one-half cents overcharged a woman customer in Offut's store. "If I were a capitalist," Alger wrote, "I would be willing to lend money to such a young man without security." Lincoln's success at the bar was ascribed to his sense of justice. "He wanted justice to triumph, however it affected his own interests." His benevolence was not limited to fellow human beings but extended to animals. Here was the story of the rescue of the hog from the mire. "Emulate that tenderness of heart . . . and, like him," the author preached with the subtlety of a sledgehammer, "you will win the respect and attachment of the best men and women!" Honest Abe's kindness to animals was a common theme in the writing for juveniles. "Lincoln's nature was that of a champion for the right. He was a born knight, and, strangely enough, his first battles in life were in defense of turtles and terrapins."

Noah Brooks, in his biography for young people, made him a real Polly-Andrew: "He never played cards, nor gambled, nor smoked, nor used profane language, nor addicted himself to any of the rude vices of the time." In *St. Nicholas*, the premier children's magazine, Brooks wrote warmly of the affection between the President and his younger son, Tad. Tad, it seems, was a little tyrant in the White House. He threw a tantrum when he realized that a turkey affectionately named Jack was about to be sacrificed for Christmas dinner. Running to his father, he pleaded for Jack's life; the President, as soft toward turkeys as toward sleeping sentinels, issued a reprieve. Some of the verse and fiction in *St Nicholas* was written for beginning readers. As early as 1870, Bayard Taylor published a child's picture book, *The Ballad of Abraham Lincoln*. Put aside the fairy tales of old, the poet said; I tell of one whose name you know:

> His arm was strong, his heart was bold,
> His deeds were wise and true;
> He did not live in days of old,
> But here at home with you.

In jerky rhyme, he went on to sketch Lincoln's life from birth to death.[81]

Curiously, in view of Lincoln's mythopoeic qualities, he was rarely the subject of historical fiction and drama before 1900. "Why, if the old Greeks had had this man," Walt Whitman mused, "what trilogies of plays—what epics would have been made out of him!" But late nineteenth-century America was not Attica, and Lincoln and the war were too near for literary epic. He did appear as himself in such novels as Eggleston's *The Graysons*, Joseph Kirkland's *The McVeys*, and Winston Churchill's *The Crisis*. The last of these is a Civil War romance. The young Yankee hero, Stephen Brice, is sent on an errand to Lincoln in 1858 and finds him in Freeport pondering "The Question" to be put to Douglas. Brice thinks him a buffoon. Later, as a Union Officer he goes to Lincoln to plea for the life of a close relative of his fiancée, Virginia Carvel, who is a rebel. The pardon is issued, and their marriage is in a sense sealed by the Christ-like compassion of President Lincoln.

The best piece of fiction came from the pen of Joel Chandler Harris, well known for his Uncle Remus stories. In "The Kidnapping of President Lincoln," an odd pair of Georgians are sent on a mission to kidnap Lincoln. Posing as escorts for a female spy, Elise Clopton, to be returned to the South, they manage to meet their prey. Before long they are swapping stories with him and having a hilarious time. Lincoln is described as a man "with Melancholy at one elbow and Mirth at the other." He tells the Georgians, "I know no North and no South. . . . I would, if I could, take the South in my arms and soothe all her troubles." Elise is quartered in the White House. She, who had hated Lincoln, has fallen in love with him. In their banter, the Georgians' kidnap plot is revealed to the President. They, of course, have abandoned it, but he prevails upon them to go through with a mock kidnapping in order to play a joke on Stanton. This is done to perfection. Lincoln enjoys a moment of high glee at Stanton's expense. The Georgians return home convinced the war wouldn't last a day if the people of the South knew Lincoln as they knew him.[82]

Many Americans learned about Lincoln not from books but from the lecture platform. Several of the Illinoisans, like Arnold and Swett and Senator Shelby Cullom, were prominent lecturers. Schuyler Colfax of Indiana, formerly Speaker of the House and Vice-President, was celebrated for his lecture on Lincoln. In 1865 he had delivered a eulogy of the Martyr President at South Bend. Nine years later he was asked to repeat it at the dedication of the monument in Springfield. A gratifying popular reception there led to other invitations; he was soon besieged and receiving up to $2,500 per engagement. A blend of reminiscence and anecdote and Unionist sentiment, the address encapsulated the essential Lincoln for a listening audience.

Robert Ingersoll, the famous agnostic and orator, having grown up an Illinois Democrat, became a Lincoln convert in 1862, and ever after considered him the greatest statesman in all history. While he sometimes wrote about Lincoln as a freethinker, his lectures skirted this controversial topic and focused on the historical figure. Lincoln's achievement marked the triumph of character over training. The orator liked to contrast him with Everett at Gettysburg, the Bostonian all drawing-room smoothness, Lincoln all native power. "Abraham Lincoln was not a type: he stands alone—no ancestors, no followers, no successors." True character, Ingersoll maintained, is only finally disclosed by the exercise of power. And here Lincoln proved himself, for he used power for noble ends and never abused it except on the side of mercy. With the Emancipation Proclamation, he made the cause of the American Republic sacred. His memory, said Ingersoll, was "the most precious treasure of the Great Republic."

"Marse Henry" Watterson was another famous lecturer. His grand theme was reconciliation and reunion under Lincoln's mantle. His lecture before Chicago's Lincoln Union in 1895 was repeated countless times, with variations, as Watterson toured the country. Withdrawing from editorship of the *Louisville Courier-Journal*, he hoped to write a biography of his hero that, among other things, would vindicate the honor of his ancestry, as against Herndon, Lamon, and company, and show that he was a southerner as well as a northerner, contrary to Nicolay and Hay.[83]

Walt Whitman's lecture, "The Death of President Lincoln," proved a favorite with literary audiences. First delivered in New York City on the fourteenth anniversary of Lincoln's death in 1879, it was repeated nine times in the next eleven years, the last being in Philadelphia two years before the poet's death. In Lincoln he perceived a man consonant with his own ideal: earthy, fraternal, democratic. "After my dear, dear mother," he mused, "I guess Lincoln gets almost nearer to me than anybody else." In the reminiscence contributed to Allen T. Rice's volume, Whitman recalled his notebook observation of the President in Washington, clothed in "plain black, somewhat rusty and dusty . . . as the commonest man," astride an "easy-going gray horse," always wearing a latent sadness on his dark, furrowed face. "We have got so that we change bows, and very cordial ones." They did not meet. Whitman later heard that Lincoln had read *Leaves of Grass*, having borrowed Herndon's copy, and he liked to tell the story, probably imaginary, of the President observing the sauntering poet from his office window and remarking, "Well, *he* looks like a man." Whitman, too, toyed with writing a book about Lincoln. The lecture in 1879 was got up as a benefit for the poet by Gilder and his literary friends. Again, as in the poems, he dwelt on the death of Lincoln, wishing to give that "most dramatic reminiscence" of our history an adequate memorial, yet feeling unequal to the tragic splendor of the subject.

The venerable poet was no orator either in manner or appearance. Clad in a suit of gray and a shirt with a broad open collar, with flowing white hair and beard and a round red face, he put on spectacles and read from a manuscript in a low but clear voice. He described the assassination in some detail, then groped for its meaning. "Strange (is it not?) that battles, martyrs, agonies, blood, even assassination, should so condense—perhaps only really, lastingly condense—a Nationality." The ultimate uses of Lincoln, as of other great men, lay not in his deeds but in the fabric and color of ideals passed on to future generations. And here Lincoln was precious not only to nationality but to democracy and humanity. Whitman usually closed with a reading of "O Captain! My Captain!" as he did on the occasion of the most celebrated of these public lectures, in New York on April 14, 1881. In the audience were James Russell Lowell, John Burroughs, John Hay, Augustus Saint-Gaudens, Andrew Carnegie, and other notables. One was the Cuban poet and liberator José Martí, who said that the luminaries listened to the lecture "in religious silence, for its sudden grace notes, vibrant tones, hymnlike progress, and Olympian familiarity seemed at times the whispering of the stars."[84]

What made Whitman's lecture memorable was not the content, which on the printed page was slight and unremarkable, but the rhapsodic sympathy between "the good gray poet" and Abraham Lincoln. As Edmund Stedman wrote, "Something of Lincoln himself seemed to pass into this man who had loved and studied him." Whitman communicated the dramatic import of Lincoln's death for the nation's future. Stedman's eyes filled with tears, and he wished that Whitman, like the troubadours of old, might go from town to town rhapsodizing about Lincoln. A later poet, James Oppenheim, was similarly moved by the spiritual kinship between Whitman and Lincoln.

> And the one brother leaned and whispered:
> "I put my strength in a book,
> And in that book my love . . .
> This, with my love, I give to America . . ."
> And the other brother leaned and murmured:
> "I put my strength in a life,
> And in that life my love,
> This, with my love, I give to America."

He laid a sprig of lilac on both their coffins, and begged America to keep green the strange, deep memory that made brothers of the poet and the statesman.[85]

4

To the Afterwar Generation

As THE CIVIL WAR faded into the past and new generations rose to maturity, Abraham Lincoln became—more than an affectionate memory—a sacred possession of the nation. "The most striking fact of our time, of a psychological kind," Horace White observed, "is the growth of Lincoln's fame since the earth closed over his remains." It was a busy, busy world, after all, and Americans saluted the past, as Whitman remarked, "like a corpse going out the door." Yet Lincoln was not only saluted but sanctified. No American's words were more often quoted as gospel than his. "Everybody seems to think that a quotation from him is a knock-down argument," as White said. To the new generation Lincoln loomed as the Titanic expression of national genius in pursuit of a national ideal. In his poem "The Master," Edwin Arlington Robinson wittily, subtly, caught the alteration of the image.

> Shrewd, hallowed, harassed, and among
> The mysteries that are untold,
> The face we see was never young
> Nor could it wholly have been old.
> For he, to whom we had applied
> Our shopman's test of age and worth,
> Was elemental when he died,
> As he was ancient at his birth:
> The saddest among kings of earth,
> Bowed with a galling crown, this man
> Met rancor with a cryptic mirth,
> Laconic—and Olympian.

Several years later southern-born John Gould Fletcher hit upon the same theme in a poem that, in the words of Amy Lowell, "raised Lincoln to the veiled awe of a national legend." He, too, spoke of "a darkness in this man" and of "roots stretched down into the earth / Towards old things." He, too, reflected that the man "whom we mocked . . . descended like a god to his rest." Fletcher concluded with a eulogy over Lincoln's grave which is also a tribute to the healing power of the myth.

> Strew over him flowers:
> Blue forget-me-nots from the north and the bright pink arbutus
> From the east, and from the west rich orange blossom,
> But from the heart of the land take the passion-flower;
> Rayed, violet, dim,
> With the nails that pierced, the cross that he bore and the circlet,
> And beside it there lay also one lonely snow-white magnolia,
> Bitter for remembrance of the healing which has passed.[1]

The Lincoln Centennial of 1909 epitomized these tendencies. No saint or hero was ever more liberally commemorated. Of course, in praising Lincoln, the nation was also praising itself, with the institutions and ideals personified by him.

> There runs a simple argument
> That, with the power to give a great man birth,
> The insight and exaltation
> To judge him at his splendid worth
> Best proves the vigor of a continent,
> The blood that pulses in a nation.

Increasingly recognized as a democratic touchstone by liberal statesmen everywhere, Lincoln became an object of study for his political thought and practice. In 1895 his *Complete Works,* as edited by Nicolay and Hay, was made the basis of a graduate seminar in constitutional history at the University of Pennsylvania. Lincoln's speeches and writings, said the professor, were "authoritative and classic," essential for understanding the Civil War era, and "an equipment for American citizenship." Thus did he enter the academy.[2]

Curiously, no state was named for Lincoln (it had been proposed for Wyoming and one of the Dakotas), though twenty-two counties and thirty-five cities and towns were named for him, and his birth was officially celebrated in eight states at the time of the Centennial. The Lincoln Association of Jersey City, founded on February 12, 1867, was the first such body to celebrate his birthday yearly. Some years later, Republican clubs in major cities began the tradition of festive dinners on the natal day of the party's patron saint. The most widely publicized were the banquets of the Republican Club of New York City. The partisanship of these affairs embarrassed some otherwise loyal Republicans. At Delmonico's in 1891, the venerable Hannibal Hamlin deplored any suggestion that Lincoln was less than a hero of all the people and called upon Congress to make his birthday a legal holiday.

As early as 1875 petitioners had memorialized Congress to this end, but to no avail. The southern representation was opposed, and the presence on the calendar of Washington's birthday, already widely observed, in the same month as Lincoln's, gave rise to other objections. What was the point, it was asked, of adding to the leisure of bank clerks? By 1896 five states (Illinois, Minnesota, Washington, New York, and New Jersey) had legislated the new holiday.[3]

Appropriate civic events followed as a matter of course. By 1900 Congregationalists across the land annually observed Lincoln Memorial Sabbath—the Sunday nearest his birthday—in order to identify its missionary work, especially Negro schools in the South, with the Great Emancipator. At Carnegie Hall the following year was held the first big benefit for Lincoln Memorial University, established at the Cumberland Gap in Tennessee.* J. P. Morgan and Andrew Carnegie himself were in the audience; Mark Twain presided, and Henry Watterson was the principal speaker. The famous author, in proximity to the Lincoln cult for the first time, joked about how he and Watterson were both reconstructed rebels. Nothing, said one newspaper editor, better exhibited popular feeling about Lincoln than the annual flood of writing and oratory on his birthday. Nor was it likely the future course of historical scholarship would alter the popular judgment. "Historians will never quarrel on him as they do on Cromwell and Napoleon." Never would there be a Clarendon or a Taine writing down Lincoln. Meanwhile, periodic efforts to make his anniversary a legal holiday nationally failed in Congress. The consensus on Lincoln's glorification remained incomplete.[4]

---

*According to General Oliver O. Howard, the father of the institution, Lincoln one day in 1863 spoke to him of the wonderful loyalty of the people of eastern Tennessee, then said with great seriousness, "General, if you come out of this horror and misery alive, and I pray to God that you may, I want you to do something for these mountain people. . . . If I live I will do all I can to aid, and between us perhaps we can do the justice they deserve." Howard subsequently commanded in the area and, having already given his name to Howard University, presided over the founding of this new Appalachian college in 1897.

Lincolniana: The Collectors and Ida Tarbell

By the turn of the century the multifaceted activity of collecting and preserving letters, books, relics, photographs, and other memorabilia of Lincoln had grown from an avocation into an industry. A unique derivative, "Lincolniana," had been coined to describe it, and before long it entered the dictionaries. It all began with the Railsplitter's rails. After the Decatur convention in 1860, John Hanks sold the two rails exhibited there for five dollars; seeing a good thing, he dug up some more, which he sold for a dollar each. In 1865 Hanks, Oglesby, and others authenticated the log cabin on the site as the one Lincoln had built in 1830. It was dismantled, reassembled and exhibited at the Great Sanitary Fair in Chicago in May, then during the summer set up on Boston Common, open to the curious with a ten-cent ticket, after which it went to the Barnum Museum in New York, then disappeared from view. Even the maul allegedly used by the Railsplitter survived to become a collector's item.[5]

Osborn H. Oldroyd, the first dedicated collector of Lincolniana, was smitten with the Illinois Republican as a boy hawking newspapers in Mt. Vernon, Ohio, before the war. He enlisted in the Union army at eighteen; stationed at Memphis when he learned of the President's death, Captain Oldroyd vowed to devote his life to preserving the martyr's memory. In 1883 he took up residence in the homestead at Springfield

Osborne H. Oldroyd's Museum in the Lincoln Home, Springfield, Illinois, after 1883. (Illinois State Historical Library)

and turned it into a museum of Lincolniana. His collection included clothing, furniture, books, portraits, caricatures, autograph letters, sheet music, medals, and so forth. Oldroyd induced Robert Todd Lincoln to give the house to the state. This was done, and the Captain was appointed custodian. Robert Lincoln fretted over the custodian's vulgarities and gaucheries, for instance placing a picture of John Wilkes Booth on the mantel. Oldroyd found little support for his plan to build a memorial hall in Springfield; and when John P. Altgeld, the first Democratic governor of Illinois since the war, removed him from office in 1893, the Captain decamped with his collection of some two thousand items to Washington.

The District of Columbia Memorial Association had leased the house where Lincoln died, opposite Ford's Theater, and Oldroyd was invited to exhibit his collection there. Eventually, the government acquired the house; meanwhile, Oldroyd, its unpaid caretaker, found a job in the War Department to support himself. He continued to add to the collection, which he called a "sacred vocation." Especially prized objects were the old family Bible, the rocking chair occupied by the President when he was shot, and, oddly, a feather of the Wisconsin war eagle, "Old Abe," dropped in the midst of battle. In 1926 Congress, having already acquired the house where it was exhibited, purchased the collection for $50,000.[6]

From time to time there were special exhibits of Lincolniana. In 1886 two Springfield businessmen, J. M. Keyes and S. B. Munson, with the assistance of William H. Herndon, got up a "Lincoln Memorial Collection" for exhibit in Chicago during the winter of 1886–87. Patronage was disappointing, and although the collection was afterward shown in Milwaukee, plans to take it to other cities were canceled. The great Chicago World's Fair of 1893 had a similar exhibit. It included Lincoln's law books, his law office table, the hickory chair in which he was sitting when he received news of his nomination for President, his fee book, autograph letters, and the like. Sold at auction in Philadelphia the following year, the collection netted $2,566.25.[7]

The first exhibit devoted to a particular antiquarian interest was in the field of numismatics. Here the pioneer collector was Andrew C. Zabriski of New York. As early as 1873 he had published a catalogue describing 189 medals. And he was responsible for organizing an exhibit of Lincoln medals at the meeting of the American Numismatic and Archaeological Society in 1893,to which he also contributed a paper, "Medalic Memorials of Abraham Lincoln," published in the society's *Proceedings*. He was now able to describe about 350 medals, divided into three groups: 1860, 1864, and posthumous. Curiously, all the medals honoring Lincoln as Emancipator—and there were a great many—appeared after his death.[8]

In 1899 New York City's Grolier Club mounted a remarkable exhibit of engraved and other portraits of Lincoln. As this was the first show of its kind, it attracted considerable interest, and Charles Henry Hart's *Catalogue* was the pioneer study of Lincoln portraiture. Most of the 159 items possessed little artistic merit, Hart acknowledged, yet they were valuable human and historical documents. A certain sameness in the engraved images suggested their derivation from a few photographs. Fourteen were shown to permit comparison. The exhibit again focused attention on Lincoln's physiognomy. To the notion that he was ugly, Hart said, "No man with the soulful expression that flowed from his eyes could be ugly, and that evanescent, subtle quality which was supreme in him, is most difficult to catch and fix." Truman H. Bartlett, another early student, as well as collector, of Lincoln portraits, attributed the excellence of his appearance first to the face, especially the eyes, and second to the long-limbed grace of movement. Frederick Hill Meserve, the foremost collector of Civil War photographs, also became the foremost collector of Lincoln photographs. In 1911 he gathered one hundred of these in a volume which, being limited to 102 copies, instantly became a collector's item itself. Two supplements followed. Meserve established the numbered chronological series of Lincoln photographs long in use.[9]

Collectors of books, pamphlets, and broadsides were immensely aided by the publication of bibliographies. The first of these, compiled by William V. Spencer and published in Boston in 1865, listed some three hundred sermons, eulogies, articles, and letters occasioned by the President's death, and for its title apparently invented the word "Lincolniana." Charles H. Hart, already encountered as a student of portraiture, was a brilliant and indefatigable Americanist whose interest in Lincoln began in his college years during the war. In December 1865, he read in a Philadelphia newspaper the account of Herndon's first lecture and struck up a correspondence with him. Before long he proposed to compile a bibliography about Lincoln to be published as an appendix to Herndon's biography. Herndon encouraged the youth. As the years passed and no biography materialized, Hart decided to publish *Bibliographia Lincolniana* independently. It appeared, actually, as Part I of a comprehensive bibliography as of 1870, Part II being the titles added by Andrew Boyd, a maker of directories in Albany and a Lincoln collector. This *Memorial Lincoln Bibliography* became the manual of collectors for a quarter-century or longer. It was finally superseded by the *Lincoln Bib-*

liography compiled by Daniel Fish and published as Volume XI of the expanded edition of the *Complete Works* in 1906.[10]

Judge Fish, of Minneapolis, was one of the so-called Big Five among Lincoln collectors at the turn of the century. A veteran of the Union army and Adjutant General of the Grand Army of the Republic, Fish exerted unusual influence through his numbered bibliography, for it provided the standard—a more rigorous standard than had heretofore prevailed—against which other collectors measured themselves. The most scholarly of the Big Five, William H. Lambert of Philadelphia, an insurance company executive, was also a Union veteran. At the time of his death in 1912, his collection, rich in both books and manuscripts, was esteemed the best in the country. Charles W. McLellan had been a twenty-four-year-old bank clerk in Springfield, Illinois, in 1860, when he asked the advice of the town's leading citizen, just nominated for President, about taking a better job in the Deep South; regrettably, he disregarded the advice, took the job, and within a year found himself a conscript in the Confederate army. After the war McLellan made his fortune in New York, then retired to his estate on Lake Champlain and devoted himself to his growing Lincoln collection. At his death in 1918 it consisted of 1,921 bound volumes, 1,348 pamphlets, 1,100 engravings, 479 pieces of sheet music, and some manuscripts, in all about 6,000 items, many of them related to Lincoln only by association with the Civil War. Rounding out this inner circle were two younger collectors, Judd Stewart of Plainfield, New Jersey, and Joseph B. Oakleaf of Moline, Illinois.

A friendly competition prevailed among the Big Five, who "possessed a 'corner' in the field," in which there were hundreds of players. Indeed, they had reckoned without John E. Burton of Lake Geneva, Wisconsin, whose collection rivaled their own. Several of the Big Five formed the nucleus of the Lincoln Fellowship organized in New York City in 1908. A rare exhibit gave the public a peek at the great collections; the display of Major Lambert's at the Historical Society of Pennsylvania in 1909 was especially notable. The Lambert and Burton collections were dispersed at auction upon decease of the owners. Fortunately, the others were preserved intact and ended up in research libraries. McLellan's was sold to John D. Rockefeller, Jr., who gave it to Brown University; the Lincoln Library and Museum, at Fort Wayne, Indiana, acquired the Fish collection; Stewart's was purchased by the Huntington Library; and Oakleaf's, the last, went to Indiana University in 1942.[11]

Autograph manuscripts were the jewels of historical collectors. An earlier generation had rummaged the nation's attics for Revolutionary autographs. The sale of Thomas Addis Emmet's magnificent collection to New York City's Lenox Library for $50,000 in 1896 was a landmark. The Civil War opened a new field, and Lincolniana was its glory. Prices rapidly advanced. Historically significant autograph letters that sold for $30 in 1890 were bringing up to $1,000 by 1909. Books and other imprints also rose in value. One needed a guide to keep up. The first *Priced Lincoln Bibliography* appeared in 1906; better yet, prices for manuscripts could be followed monthly in Walter R. Benjamin's magazine, *The Collector*. Autograph copies of the Gettysburg Address led the big-ticket items. One of the five belonged to John Hay. The magnifico J. P. Morgan, an enthusiast for Lincoln autographs, reportedly offered $50,000 for it. Had the offer been accepted—and it was not—it would have been the highest price ever paid for a literary manuscript. Lincoln's own copy of the Emancipation Proclamation had been consumed in the great Chicago fire. In addition to the autograph copies, Lincoln and Seward had signed fifty printed copies. At the Lambert sale one of these went for $1,900, while a rarer autograph of the Thirteenth Amendment brought $3,250, an early Lincoln notebook commanded $2,250, and a lock of hair $330.[12]

As prices increased, so did forgeries. The most notorious forgers of Lincolniana were Joseph Cosey and Charles Weinburg. Cosey, perhaps the most notorious autograph forger of modern times, specialized in legal briefs; and Weinburg made a trade of Lincoln endorsements—a signature under a line or two on a letter addressed to him—supposedly just discovered in a trunk in the attic. Embarrassingly, some forgeries found their way into collections of Lincoln's writings. A most improbable letter to an old Illinois Democrat, Colonel E. D. Taylor, in 1864, which paid tribute to him as the father of "greenback" currency, first appeared in newspapers, then was copied into the National Edition of Lincoln's *Works* in 1906. Henry W. Cleveland, the biographer of Alexander H. Stephens, forged a Lincoln letter to the Georgian dated January 19, 1860. Considered an important discovery when it appeared in a volume of uncollected Lincoln letters in 1917, it was exposed as fraudulent eleven years later. One letter turned up repeatedly, that addressed to Governor Hahn in 1864, remembered for the eloquent line in advocacy of Negro suffrage, "to keep the jewel of liberty within the family of freedom." The problem in this instance was not fraud but the difficulty of distinguishing lithographed facsimiles Hahn had made for his friends from the original, which had disappeared. "So," in the words of a dealer, "the Hahn letter is forever rising to the surface, like a hungry trout to the fly."[13]

THE MOST POPULAR writer on Lincoln in 1900 was Ida M. Tarbell. Her literary career, just then taking off, earned for her a leading place in the history of magazine publishing in the United States and, as one of the first Muckrakers, a place in the history of American reform as well. She stumbled into the Lincoln field in 1895; often thereafter she wandered from it, yet never for long, and to the end Lincoln was the thread that held her career together. Some years before her death in 1944, one of her fellow laborers in the vineyard saluted Miss Tarbell as "the great pioneer in the field of Lincolniana." The homage was not misplaced, for Tarbell fashioned a more vivid and lifelike Lincoln from documents, portraits, relics, and memorabilia previously overlooked. In her later years, when she had enough money, she became a collector herself. She was an unrelenting researcher. Just when Noah Brooks declared "the stories were all told," Tarbell uncovered whole new layers of stories. This was valuable, but placed at the service of her literary talent, it also had unusual popular appeal.[14]

Tarbell was born in 1857 in the Oil Region of Pennsylvania. The first great event in her life, though she was too young to remember it, was Edwin Drake's striking oil in 1859, and the second, which she remembered well, was her father trudging home on a spring day, head bowed, then speaking quietly to her mother, who suddenly buried her face in her apron and ran to the bedroom sobbing as if her heart would break. The house was shut up, crepe was put on the doors, and seven-year-old Ida learned that Abraham Lincoln was dead. "From that time the name spelt tragedy and mystery," she wrote in her autobiography.[15]

After graduating from Allegheny College, in nearby Meadville, Tarbell was caught up in the Chautauqua movement and helped to edit its magazine for eight years, after which she went to Paris to study and to write. She was fascinated with Madame Roland—with the theme of women and revolution—and wrote a biography of her. She fell in love with France. The course of her life was changed in 1892, however, when S. S. McClure walked into it. Brisk, eager, and enthusiastic, he was recruiting writers for McClure's Syndicate and talked grandly of a new kind of monthly magazine. *McClure's* was launched in 1893. The following year the publisher called Tarbell to New York to join its staff with a first assignment to write a biography of Napoleon. Seeking to compete with *Century* and *Harper's*, *McClure's* emphasized a fresh and vivid approach, turning its subjects into "human documents." The centerpiece of the Napoleon project was a choice collection of pictures and portraits owned by a prominent Washingtonian, Gardiner Green Hubbard, father-in-law of Alexander Graham Bell—and of the telephone, as Tarbell noted—and her task was to furnish a well-researched biographical narrative to accompany them. "Napoleon" was a big success, boosting the magazine's circulation to one hundred thousand. Everyone at

Ida M. Tarbell (Allegheny College)

McClure's, including the author, agreed "it was the pictures that had done it"; still, the text had skillfully framed and interpreted the pictures, and that was no small accomplishment.[16]

McClure now proposed that Tarbell do something similar on Lincoln. In his opinion, Lincoln was "the most vital factor in our life since the Civil War." No American magazine worth its salt could fail to treat his character and influence. McClure refused to believe that the wells of memory had been exhausted. A resourceful researcher would turn up fresh reminiscences and new truths about Lincoln, which when stitched together and richly illustrated would present a compelling portrait of the man who rose to the presidency. The idea, then, was not a biography but a kind of documentary of which Ida Tarbell would be the creative editor. She felt some trepidation toward the project. What, after all, did she know about American history? But an annual salary of $5,000 overcame all doubts. Since she lived in Washington, she decided to approach John G. Nicolay for aid and counsel. He flatly rebuffed her. "You are invading my field," he barked. "You write a popular Life of Lincoln and you do just so much to decrease the value of my property." Anyway, there was nothing more to be known and said about Lincoln. He and Hay had said it all. Nicolay's tragedy, she later reflected, lay in not having found a fresh field, whereas his friend Hay had pursued a rich and varied public career.[17]

In February 1895, Tarbell headed for Lincoln country in Kentucky. The fruits of this initial foray were disappointing. She hired a researcher (curiously, the aforementioned Henry W. Cleveland) and went on to Chicago. There a socially prominent acquaintance had promised her an interview with Robert Lincoln, the president of the Pullman Company. Over teacups in a resplendent drawing room she searched the face of this plump man, perfectly groomed, admirably poised, for any resemblance to his father, but concluded with so many others that he was, indeed, "all Todd." He did not see how he could help her; he had almost nothing from his father's pre-presidential years. Then, under his hostess's prodding to give Miss Tarbell something, he mentioned a daguerreotype he believed to be the earliest portrait of his father. The picture, as it turned out, was worth more than cabinets of papers. The earliest known portrait, Alexander Hesler's photograph showing him in all the fright of his tousled hair, as the subject had remarked, dated from 1857. Robert Lincoln's photograph dated from eight or ten years earlier—precisely when it was taken, where, and by whom was still a mystery—but it showed a handsome, dignified, sensitive, and youthful Lincoln, reminiscent of the young Emerson. The picture, Tarbell sensed when it finally came to hand, shattered the tradition of Lincoln the

The Young Lincoln. Daguerreotype by N. H. Shepherd, 1846, first reproduced in *McClure's*, December 1895. (The Lincoln Museum, Fort Wayne, a part of Lincoln National Corporation)

rude frontiersman. "It was another Lincoln, and one that took me by storm."[18]

On her travels Tarbell met and interviewed a number of people who had known Lincoln. A woman tall and gracious, with a warm and generous personality, she was not one to invite umbrage; yet other authors, in addition to Nicolay, considered her an upstart. Henry C. Whitney called her a "bluestocking" and complained that she had "*sponged* more or less Lincoln afflatus" during three days at his home. Meanwhile, *McClure's* published an appeal for materials about Lincoln. The response to the "Lincoln Bureau" kept Tarbell busy following up leads and checking facts. She employed investigators in Illinois and Indiana, J. McCan Davis and Anna O'Flynn respectively, to search newspapers, courthouse records, and other sources. As the work progressed, it changed its character. The fragmentary materials required an author to shape and interpret them; moreover, Tarbell realized, there was no stopping place short of a complete biography.[19]

The Lincoln series began in the December 1895 number of *McClure's* and continued through March of the following year. (Not until the last number was Tarbell credited as author rather than editor of the work.) The early portrait made a stunning frontispiece to the first installment and "set the key for the series," in Tarbell's words. Seldom, if ever, had a portrait been so effectively used as a historical document. It framed the biography and informed every line. It astonished many readers, some of whom wrote to the magazine of their reaction. Murat Halstead, the prominent journalist who had known the subject, said the portrait showed "a new Lincoln" more attractive than anything the public has possessed. "The head is magnificent, the eyes deep and generous, the mouth sensitive, the whole expression something delicate, tender, pathetic, poetic." John T. Morse, having shown the portrait to several persons without identifying it, said some supposed it was of a poet, others of a philosopher—a resemblance to Emerson was noticed—and none named it correctly. Woodrow Wilson remarked on an expression of "dreaminess" in a face familiar for its sadness. As circulation of *McClure's* soared, the publisher used the portrait in an advertising brochure in which he boasted that the magazine, capitalized at $7,000, had attained a circulation of three hundred thousand, thanks to the Lincoln series. Richard Watson Gilder of *Century* had laughed, upon hearing of *McClure's* venture into a field claimed for itself, "They got a girl to write a Life of Lincoln." But the joke was on him. The Lincoln series, as Ray Stannard Baker said, turned into "a legendary magazine success."[20]

The first series, taking Lincoln to 1858, was followed by a second, to his death, completed in September 1899. All the while Tarbell was adding new materials and revising the work. Her two-volume *Life of Abraham Lincoln* appeared the following year. It was remarkable for its illustrations and also for its appendix of some two hundred pages devoted to newly discovered or uncollected Lincoln documents: speeches, writings in newspapers, autograph letters (including the earliest surviving letter), and hundreds of telegrams obtained with "infinite trouble" from War Department files. Even Robert Lincoln was impressed. "I consider it an indispensable adjunct to the work of Nicolay and Hay," he wrote to Tarbell.[21]

Aside from the documents, both written and pictorial, what was the biography's contribution to the Lincoln image? First, it cast a much brighter light upon Lincoln's parentage, childhood, and youth. Tarbell relied upon a newly published genealogical study of the Hanks family to establish Nancy's legitimacy, gave Thomas Lincoln a respectable lineage, and denied that their child was seriously disadvantaged. Nor was the family's environment as backward and impoverished as generally depicted. "There was nothing ignoble or mean in this Indiana pioneer life," Tarbell wrote. "It was rude, but only with the rudeness which the ambitious are willing to endure in order to push on to better conditions. . . . These people did not accept their hardships apathetically. They did not regard them as permanent. . . . For this reason they endured hopefully all that was hard."

Second, although she learned much from Herndon, and capitulated to the charms of the Rutledge romance, the tendency of her work ran toward discrediting him. She was untroubled by doubts about Lincoln's legitimacy; she believed that he became a Christian in his grief over Willie's death in 1862. Where she directly took aim at Herndon, however, was on the subject of the "broken engagement" to Mary Todd. From Herndon's "sensational" account one must infer that Lincoln was a coward or insane. It was all untrue. No marriage license had been issued. No one had ever heard of the "wedding party" on the first of January before Herndon invented it. Tarbell suggested that his own marginal position in the Springfield aristocracy had something to do with the libel. This angered Whitney. "The idea of an obscure Bohemian insisting that Billy Herndon was a *pariah* in Springfield society is an inexcusable outrage besides being a d——d lie." When the "bluestocking" had asked him about the story, he confessed to having no direct knowledge but insisted that if Herndon said it, it must be true.

Third, Tarbell's resourcefulness and perseverance as a researcher showed how much more could be learned about Lincoln. Some of her informants, like Austin Gollaher, the only living person from Lincoln's Kentucky boyhood, although known to others, was put to better use by her.

Tarbell made important discoveries in newspapers, not only those in Springfield, but in Massachusetts where she found reports of the forgotten speeches made in the 1848 presidential campaign. She got some of Carl Schurz's reminiscences in advance of his autobiography; and she extorted from Edward Everett Hale his memorandum of a conversation with Charles Sumner about an interview with Lincoln touching on emancipation.*[22]

Tarbell's weakness, as with other biographers who specialized in reminiscences, lay in her credulity. She was too willing to believe good stories, like A. J. Conant's about Lincoln's finding Blackstone's *Commentaries* in a barrel of junk. She was too accepting of Caroline Hitchcock's explanation of Nancy Hanks's parentage—this would dog her for decades. She also put her authority behind Whitney's reconstruction of Lincoln's so-called Lost Speech of 1856. This speech in Bloomington at the birth of the Republican party in Illinois had a legendary reputation. Lincoln spoke "like a giant inspired." So powerful was his oratory that the reporters, as if hypnotized, laid down their pencils, and when it was over no one could recall what had been said. They simply reported the fact of the speech to their newspapers. Whitney had heard the speech and coolly taken notes from which he now, forty years later, reconstructed it. He sent his handiwork to Joseph Medill of the *Chicago Tribune*, who had been one of the enthralled reporters, and Medill vouched for its closeness to the original. Tarbell, who had interviewed Medill in Chicago, was carried away by the discovery. She wrote a newspaper article about it in June 1896, and "Lincoln's Lost Speech" appeared in *McClure's* some months later. In its contents the speech resembled others Lincoln made during these years. It may be identified by the peroration as Whitney gave it: "We will say to the Southern disunionists, we won't go out of the Union, and you shan't!!" Many knowledgeable persons greeted Whitney's version of the Lost Speech with skepticism. Nicolay thought it devoid of Lincoln's style and pronounced it a forgery. Robert Lincoln concurred. A generation later the Lincoln scholar Paul M. Angle, while regretting the embarrassment to Ida Tarbell, exposed it as the "fabrication" of a scamp.[23]

Tarbell's next big book was the Muckraking classic *The History of the Standard Oil Company*, in 1904. It lay close to her roots in the Oil

---

*The unpublished memorandum was dated April 26, 1862. Sumner said that in a wide-ranging discussion, he and the President found themselves in agreement on most subjects, even on emancipation. "Well, Mr. Sumner," the President said, "the only difference between you and me on the subject is a difference of a month or six weeks in time." If so, the senator replied, he would not say another word about it unless the President brought it up. Hale included this in his *Memories of a Hundred Years* (New York, 1902).

Region, where her father had battled the monopolists and lost in the 1870s. Jesse Weik asked her if she had dropped Lincoln for good. "Of course, I have not dropped Lincoln," she replied. "I intend to keep hold of him as long as I live." And she did, though with frequent excursions into the public life of her times.

Other than as an ideal of democracy and brotherhood, Lincoln had little to do with her reform activities. But the more she knew Lincoln, the better she liked him. "He is companionable as no public mind that I've ever known. . . . You feel at home with him, he never high hats you and he never bores you, which is more than I can say of any public man living or dead with whom I have tried to get well acquainted." So, as she later wrote in her autobiography, "I have kept him always on my work bench."

As the Centennial approached, she created a folk character, Billy Brown, and set him talking about Old Abe in a number of stories. The first of these, "He Knew Lincoln," has Billy reminiscing with fellow-townsmen around the stove of the drug store Lincoln used to patronize. He tells how one day in court the judge found Lincoln in contempt for disorderly conduct and levied a fine on him, then, after demanding to hear the joke that caused the ruckus, laughed heartily and remitted the fine. This was an improvement on the same anecdote as told in the biography, where the judge was David Davis and it was the clerk who was fined for repeating one of Lincoln's jokes. Tarbell later sold the stage rights to "He Knew Lincoln" for almost $10,000 to Charles A. "Chic" Sale for his Broadway revue *Gay Paree.* It became a part of the comedian's repertoire. Al Jolson made a radio classic of Billy Brown in the 1930s.

Continuing her career as a popularizer, Tarbell wrote *The Boy Scout's Life of Lincoln* in 1921 and *In the Footsteps of Lincoln* three years later. Both were much admired and much read, though by different audiences. The latter book plunged her into controversy over Lincoln's ancestry. Such was the esteem for her, however, that nothing upset her charmed place among Lincolnians, for whom she was a wise counselor and friend. Her middle name was not Minerva for nothing. She was the goddess of Lincolniana.[24]

The Political Lincoln

The Republican party's domination of the national government guaranteed that Lincoln's name would be kept before the American people. The party boasted, of course, that Lincoln was its founder and leader. "His eulogy is its encomium," said George Boutwell, "and therefore when we set forth the character and services of Mr. Lincoln we set forth as well the claims of the Republican party to the gratitude and confidence of the country, and the favorable opinion of mankind." From this it followed that the Democrats, being of the party that opposed Lincoln, had no claim on the country's gratitude. Until the end of the century the Republicans owned Lincoln's birthday. They gazed upon his visage and listened to speeches extolling him at their annual banquets. Occasionally the "bloody shirt" of the war would be waved in his name; on the whole, however, such was the sanctity of the name, it was not abused in this way. A virtual auxiliary of the party was the Grand Army of the Republic, with membership of some four hundred thousand Union veterans.[25]

Lincoln's political legacy was twofold: first, the supremacy of the national government, and second, its responsibility to advance the freedom and equality of all citizens. In his conception the Constitution became the instrument for realizing the promises of the Declaration of Independence. With respect to the wartime issues of union and emancipation, that was well enough understood; toward the issues of an increasingly complex industrial society after the war, it was not understood at all. The Republican rulers lopped off responsibility to the freedmen and ceased to challenge white supremacy in the South. Instead of turning Lincoln's legacy toward lifting the oppressions of a ruthless industrial order from the backs of the common people, the party submitted to the power and greed of bankers, manufacturers, and railroadmen. It was a different party from the one Lincoln had led. Republican appeals to his name and authority became hollow. Various political factions or parties attempted to lay hold of the Lincoln symbol for their own purposes, always without much success. Somehow the vision contained in the symbol could not be released for significant re-creations and reenactments on the political stage.

Several examples of "tying up with Lincoln" may be noticed. Carl Schurz, among others, sought to link the martyr's name and influence to civil service reform. He spoke of a personal conversation when the President was besieged by officeseekers. "Do you observe this?" Lincoln exploded. "The rebellion is hard enough to overcome, but there you see something which in the course of time, will become a greater danger to the Republic than the rebellion itself." Unfortunately, it was impossible to discover anything in Lincoln's writings or in his practice to warrant invoking him as a prophet of civil service reform. Whatever the abuses of the patronage power, including his own, no President had employed it more effectively.[26]

Republicans found plenty of evidence to associate Lincoln with the protective tariff. He had been a Henry Clay Whig, elected President on a protectionist platform, and had supported higher and higher duties during the war. So it is surprising that Republicans should have felt any need to put words in Lincoln's mouth on this subject. The *Republican Campaign Text-book* of 1904 quoted him as follows: "I am not posted on the tariff, but I know that if I give my wife twenty dollars to buy a cloak and she brings [home] one made in free-trade England, we have the cloak but England has the twenty dollars; while if she buys a cloak made in the protected United States, we have the cloak and the twenty dollars." The quotation, or some variant, was repeated so often, according to the economist F. W. Taussig, "that it has come to be associated with Lincoln almost as much as the cherry tree with Washington." Taussig, a free-trader, set out to discover the origin of the "Lincoln tariff myth," and after seven years thought he had found it in a speech by Robert Ingersoll, only later to amend this to an "accidental collocation" of the speaker's words.* Of course, Lincoln's support for high duties in wartime offered no true precedent for peacetime protectionism; moreover, nothing in the image readily sanctioned the policy.

Democrats, Mugwumps, and assorted reformers conscripted Lincoln in the campaign against America's imperialist venture in the Philippines. In Congress and the press, Representative Abraham Lincoln's great speech of January 12, 1848, against the Mexican War was cited in arguments against the Spanish-American War. Leading anti-imperialists, like Moorfield Storey, invoked Lincoln the Emancipator on behalf of the freedom of the Filipinos. Did not he assert, "No man is good enough to govern another without his consent?" And were not the same arguments used by the McKinley administration to justify domination of the Philippines that were earlier employed to justify slavery? While it could be argued, and was, that the cases were not analogous, or even that Lincoln's subjugation of the South offered a precedent for domination of the Philippines, nevertheless, this was as good an instance of the instructiveness of Lincoln's teaching as politics afforded.[27]

In 1901 Democrats in number celebrated Lincoln's birthday for the

*It was probably not "accidental." In his famous lecture of 1894, Ingersoll said that Lincoln's maiden political speech in 1832 (of which there is only a reminiscence) was in favor of the protective tariff, for he knew that a nation that manufactured for itself had more wealth and brains than one that did not. The lecturer continued: "If we purchase a ton of steel rails from England for twenty dollars, then we have the rails and England the money. But if we buy a ton of steel rails from an American for twenty-five dollars, then America has both the rails and the money." Judging from the current economic depression, said Ingersoll, Lincoln "stood on solid rock and was absolutely right." (Robert G. Ingersoll, *Works* [New York, 1915], III, 127–28.) Clearly the intention was to ascribe the argument, even the words, to Lincoln.

first time under the auspices of the newly formed Jefferson-Jackson-Lincoln League, which incorporated the first Republican President into the distinguished lineage of the Democratic party. The McKinley Republicans had forfeited all rights to him, said Ohio Congressman John J. Lentz. The league believed that just as the nation could not endure half slave and half free, so it could not endure the capitalist exploitation of the many by the few; that labor is prior to capital, as Lincoln had said, and that the government belongs to the people and must be made responsive to them. The development reflected the social and political ferment among farmers, workers, and reformers at the end of the century. In Illinois, John P. Altgeld, the state's first post-war Democratic governor, represented in the eyes of his followers a return to Lincoln's humane ideals. With Jane Addams instructing immigrant children at Hull House in Chicago to take Lincoln as their guide, with the young poet Vachel Lindsay singing Lincoln's name in Springfield, and with Altgeld in the governor's mansion, Illinois seemed to be experiencing a rebirth of the Lincoln spirit.

Nationally, these stirrings revolved around the figure of William Jennings Bryan. "The Boy Orator of the Platte," in fact, had lived the first twenty-seven years of life in Illinois. In his first campaign for the presidency in 1896—and in 1900 as well—he quoted Lincoln's words on subjects ranging from bimetallism to the finality of Supreme Court decisions, but most of all on the rights of labor. His maiden speech in the East, at Madison Square Garden, recurred to the philosophy of free labor asserted by Lincoln in his first annual message to Congress within the context of the war against a system founded on slave labor. He there extolled "those who toil up from poverty" and warned, "Let them beware of surrendering a political power which they already possess, and which power, if surrendered, will surely be used to close the doors of advancement against such as they, and to fix new disabilities and burdens upon them till all of liberty shall be lost." Bryan voiced Lincoln's warning in a political climate rife with fears of class war. His Republican foes labeled it "demagogic and strife-breeding." He had distorted Lincoln's message, they said. The lesson of his life was the lesson of the insignificance of class distinctions in America. Bryan ought to be ashamed, wrote an angry editorialist in the *New York Times*. "Every word of that noble man ought to be a rebuke."

It would be a mistake, however, to leave the impression that Lincoln was a political talisman for Bryan. That service was performed by Thomas Jefferson, in whom Bryan, with many Democrats, found a political code readily deciphered for application to modern problems. Lincoln, from the unique circumstances of his life and leadership, left nothing like

"HOW ABOUT ME?"

Cartoon by Power O'Malley, *c.* 1910. (Library of Congress)

Drawing by Dan Beard in his *Moonblight*, 1903.

SPEECH ON FREE LABOR, DELIVERED SEPTEMBER, 1859.

WHAT ABE LINCOLN DID.—"I hold, if the Almighty had ever made a set of men that should do all the eating and none of the work, he would have made them with mouths only, and no hands; and if he had ever made another class that he had intended should do all the work and none of the eating, he would have made them without mouths and with all hands."

that, and so while Bryan might enroll him in the Jefferson tradition, he found only marginal use for his name and authority.[28]

Recognizing Lincoln's friendly views toward labor, both urban and agrarian radicals were anxious to make him an oracle of their cause. Thus it was that "Lincoln's Prophesy" not long before his death came into circulation during the 1896 campaign. The texts varied, but all ascribed to Lincoln an apocalyptic vision of capitalist tyranny.

> As a result of the war, corporations have been enthroned, and an era of corruption in high places will follow, and the money power of the country will endeavor to prolong its reign by working on the prejudices of the people until all wealth is aggregated in a few hands, and the republic is destroyed. I feel at this moment more anxiety for the safety of my country than ever before even in the midst of war. God grant that my suspicions may prove groundless.

Nicolay immediately repudiated the prophesy, first spotted in 1888, as "a bold, unblushing forgery." He was correct, though this did not end its use. It was quoted by respected journalists, like Henry Demarest Lloyd, by clergymen, and by congressmen, and it even found its way into the *Lincoln Encyclopedia* published in 1950.

The socialist William J. Ghent, after deftly exposing an expanded text of the "money power" prophesy in 1905, concluded in a second article that Lincoln had no conception of the industrial change that transformed America after the war and that he was "fundamentally a Jeffersonian, though his Jeffersonianism was qualified by a good deal of what today would be called socialism." It was easy to understand Lincoln's appeal to social radicals, said Ghent, for he held very advanced views of the rights of labor. As early as 1847 he had written, "To secure to each labourer the whole product of his labour, or as nearly as possible, is a most worthy object of any good government," which was remarkable for a prairie lawyer of that time. Speaking in New England in 1860, he praised the right to strike, as then being exercised by the shoemakers of Lynn. His clear assertion of the labor theory of value in the 1861 message—"Labor is prior to, and . . . superior to capital"—and his answers to the addresses of workingmen abroad and at home gave a color of Marxism to his thinking. He was, surely, the best friend labor ever had in the White House. Nevertheless, he was no prophet. Imprisoned in the democratic-capitalist ideology of nineteenth-century America, he believed the free laborer toiled up from poverty to become a capitalist in his own right. Individual opportunity, not class struggle, was his message.[29]

The perils and pitfalls of wielding the Lincoln symbol as a scepter and guide in contemporary politics suggested that his better service to posterity lay in the model of presidential leadership he provided. The model was implicit in Nicolay and Hay's sprawling *History*. Hundreds of tributes, reflections, and reminiscences supported it. James Russell Lowell had written of the force of mind and character that had enabled Lincoln to transcend provincial boundaries and become, after Washington, the first truly national leader, though with an Americanism unknown to Washington. John Motley had discerned Lincoln's "power of placid deliberation in the midst of a sea of troubles, the rare sagacity which ever seemed to divine the right course amidst conflicting opinions and passions, the gift to compare the judgments of the wisest and best informed and yet retain his own," and he called it genius. Charles A. Dana had thought the key to that genius lay in Lincoln's unrivaled capacity to control men by wit, reason, and persuasion. Responding to the question "Was Lincoln an able politician?" Shelby Cullom replied:

> The shrewdest I ever knew. . . . He was always on the alert. He was the leader of the party in the State by unanimous consent, but this did not prevent him from always advocating great measures, great actions, and right actions. . . . He did not try to club men into line. . . . It was not a case of force. It was a case of persuasion. People gave him their support because they came to believe he was right, and he showed them this was so by his reason. He was the best judge of public sentiment the country ever produced. He would never act until he believed the country was ripe for action.

Of that, of course, the Emancipation Proclamation was the great example. The publication of Gideon Welles's *Diary*, first as a serial in the *Atlantic Monthly* in the centennial year, added to Lincoln's towering stature as Chief Executive. Through him, Ida Tarbell wrote, the American people were coming to realize what it meant to be an effective leader of a democracy. "Lincoln actually believed that popular government was practical. He actually listened to the people. He knew them so well that he understood what they said when he listened." He was, indeed, "the world's best guide in government by the people."[30]

A fascinating, sometimes brilliant study, Alonzo Rothschild's *Lincoln, Master of Men*, in 1906, cast new light on the political genius. The author was a forty-four-year-old German Jew, son of a Forty-Eighter, who achieved financial independence at an early age, entered Harvard at twenty-eight, settled south of Boston, and, although active in various causes, basically devoted his life to the study of Lincoln. "A spirit of mastery moved Abraham Lincoln at an early age," the book commenced. Even as a boy he had a passion to lead and excel. This counted for more, much more, than poverty. After working hours on the Indiana farm, he often walked to the village store where he was the "oracle of the grocery," entertaining the neighbors with crude discourse and humor. The youth's sense of mental superiority was reinforced by his physical prowess. Full grown at seventeen, he out-wrestled all comers and performed Herculean feats of strength, like holding an ax at arm's length, that became part of his legend. Elected captain during the Black Hawk War, he was confirmed in his self-esteem and learned many lessons in the rudiments of leadership. Rothschild interpreted the episode of the proposed duel with James Shields, dismissed by most biographers as a lark, as a revelation of Lincoln the Hotspur. The humiliation he experienced in this affair taught him to channel his "spirit of mastery" in more constructive directions.

Beginning with the challenge to Douglas, Rothschild's book became a chapter-by-chapter analysis of Lincoln's mastery over his leading rivals. He achieved dominance, even in defeat, over Douglas in 1858. The chapter concluded with the picture of Douglas holding Lincoln's hat as he delivered his Inaugural Address. Rothschild studied the President's management of Seward, Chase, and Stanton in fine detail. Chase offered the case of a man who encountered a force stronger than his at every turn, yet never recognized its existence. Lincoln finally gave him up, but only after the Secretary's use to him and the administration had ended. Although he gave the impression, for his own reasons, that Stanton controlled him, "this truculent lieutenant" never successfully opposed his will, said Rothschild. "That the President controlled so turbulent a force without sacrificing aught of its energy was perhaps his highest achievement in the field of mastership." Chapters on Lincoln's handling of two refractory generals, Frémont and McClellan, concluded the book.[31]

What lessons Lincoln's example of mastery might hold for present-day statesman was problematic. Doubtless he was an inspiration, as the poets kept insisting.

> O, rulers of this mighty land!
> O, selfish leaders, great and small!
> Let Lincoln teach you how to rule;
> He is the model for us all.

Perhaps he belonged more to the poets than to the statesmen. There could be no "Second Lincoln"—the very thought was blasphemous.

President Theodore Roosevelt was creating a model of vigorous executive leadership after decades of congressional ascendancy in the government, and he took Lincoln as his example. At the high tide of Progressivism, it was tempting to reformulate the old issue of the People versus the Slave Power in terms of the People versus the Money Power. But the reformulation, like the Lincoln appeal to labor, ran squarely up against the historical reality that the new plutocracy was outside of Lincoln's vision. The socialist writer Rose Strunsky, whose biography of Lincoln appeared in 1914, understood this. Ironically, the triumph of northern industrial capitalism in the war he led plowed under his dream of a freeholder's democracy. Although his name and ideals were constantly invoked, by all parties in the political arena, it was quite futile. The cult, the hero worship, of Lincoln was a species of nostalgia. "Except for the inspiration of his ideal of equal economic opportunity," said Strunsky, "Lincoln can no longer help us."[32]

The poet most enchanted by Lincoln's vision was Springfield's own Vachel Lindsay. His father was a Democrat out of Kentucky, opposed to Lincoln all his life; but his mother's sentiments were different, and schooling, together with Lincoln's ghostlike presence in the town, overcame the father's prejudice. Altgeld was the young poet's living hero; and he wrote his lyrical epitaph. Lindsay was a marvelously gifted artist. Part poet, part minstrel, part preacher—son of the prairie where the ghosts of buffaloes roamed—for thirty years he conjured Lincoln on behalf of his dreamy Utopia set in the town he knew. He hated much of what was Springfield—"City of my Discontent"—but loved what it might become. "I believe in the power of Springfield as incarnated in the Lincoln tradition," Lindsay wrote. He recited "On the Building of Springfield," written in the aftermath of the terrible race riots of 1908, hundreds of times. Let the city be beautiful, cultured, and wise, the poet chanted. Let it be the epiphany of Lincoln.

> We must have many Lincoln-hearted men.
> A city is not builded in a day.
> And they must do their work, and come and go,
> While countless generations pass away.

Lincoln's "imperial soul" haunted Lindsay's imagination. One of his finest lyrics, "Abraham Lincoln Walks at Midnight," was inspired by the shock of the Great War in Europe. In his hometown the "mourning figure" walks without rest, near the old courthouse, by his homestead, through the market.

A bronzed, lank man! His suit of ancient black,
A famous high top-hat and plain worn shawl
Make him the quaint figure that men love,
The prairie-lawyer, master of us all.

He cannot sleep upon his hillside now.
He is among us:—as in times before!
And we who toss and lie awake for long.
Breathe deep, and start, to see him pass the door.

. . .

It breaks his heart that kings must murder still,
That all his hours of travail here for men
Seem yet in vain. And who will bring the white peace
That he may sleep upon his hill again?[33]

The hopes of Progressivism, which shriveled with the war, had been borne, somewhat uneasily, by Presidents Theodore Roosevelt and Woodrow Wilson. Both professed to be students of Lincoln, often invoked his name, and claimed to be guided by him. In Roosevelt's case, although he rose to fame as a New York Republican, his remembrance of Lincoln was little more than a pious sentiment before he became President. An amateur historian himself, his view of the Civil War and its legacy was thoroughly orthodox. He admired James Ford Rhodes's great history of the era and approved its high estimate of the sixteenth President. He enjoyed listening to John Hay, his Secretary of State, tell tales about Lincoln, recalling how the tall, gaunt form clad in white, unable to sleep, passed along the dimly lit corridor of the White House to Hay's room, sat on his bed, and sought solace with him by reading something from the Bible or Shakespeare or Tom Hood. No wonder the twenty-sixth President saw Lincoln's ghost at every turn. "As I suppose you know, Lincoln is my hero," he wrote to George M. Trevelyan, the English historian. The night before his inauguration on March 4, 1905, Hay gave him a ring keeping a snip of Lincoln's hair. Roosevelt wore it the next day, and treasured the memento. A reporter observed a large portrait of Lincoln above the mantel behind the President's desk. Asked about it, he said, "When I am confronted with a great problem, I look up to that picture, and I do as I believe Lincoln would have done. I have always felt that if I could do as he would have done were he in my place, I would not be far from right." This was a pungent form of immortality, though how well Roosevelt read Lincoln's mind it is difficult to say.[34]

Roosevelt advocated what he called the "Jackson-Lincoln theory of the Presidency." By it he meant strong, energetic, and positive leadership, independent of Congress if need be. To some problems, like that of the Negro, Lincoln's guidance was reasonably clear; to others, like the Trusts, where it was obscure, his values and his example remained relevant. In a political tradition divided between the Hamiltonians, who were aristocrats, and Jeffersonians, who were demagogues, Lincoln "struck the right average," Roosevelt thought. In his speeches he praised Lincoln for the clarity of his vision, for his political shrewdness, for being a democrat without a trace of demagoguery, for seeking the balance between extremes. He admired his tenacity of purpose and his grace in the presence of suffering. Drawing upon *Pilgrim's Progress*, Roosevelt called Lincoln "the ideal Great Heart" of American politics. How well this emphasis on the values of tolerance, charity, and moderation comported with Roosevelt's bullying personality or, for that matter, with the "Jackson-Lincoln theory" was a question.[35]

Roosevelt sounded a bolder note in 1910, as he enunciated the creed of the New Nationalism and prepared to regain the presidency from William Howard Taft. Standing for a "square deal," he quoted Lincoln on labor's superiority to capital, observing philosophically, "If that remark was original with me, I should be even more strongly denounced as a Communist agitator that I shall be anyhow." He called Lincoln a Progressive in his time and invoked his name on behalf of direct election of senators and the President. He defended his most radical proposal, for the recall of judicial decisions, as simply the application of the principle set forth by Lincoln in opposition to the Dred Scott decision. The people, not the courts, were the true masters of the country. "It was Lincoln who appealed to the people against the judges when the judges went wrong, who advocated and secured what was practically the recall of the Dred Scott decision, and who trusted the Constitution as a living force of righteousness." Many of Roosevelt's former Republican friends dissented from the illumination. What Lincoln sought with regard to Dred Scott was a judicial reversal, not a democratic one, as they pointed out.

During the election of 1912, "What Would Lincoln Do?" was a question posed in countless speeches, articles, and editorials. Of the three major presidential candidates, Roosevelt was the most eager to turn him into a political prophet. He had a bully time entertaining reporters with nuggets from Lincoln's writings. He made a pilgrimage to Springfield, sat in the pew Lincoln had occupied in the First Presbyterian Church, and laid a wreath at his tomb. The Bull Moose Progressives of 1912 were the true heirs of the Republicans of 1860, Roosevelt argued, and their aim was simply to apply Lincoln's "tempered radicalism" to the problems and conditions of the twentieth century.[36]

Woodrow Wilson, victor in the election, came to admire Lincoln as a student, then a professor of American history and politics. Although southern-born, Wilson was free of the cant of the Lost Cause. As earlier noted, he felt no regrets over the defeat of the Confederacy, indeed rejoiced in the vindication of American nationhood in the Civil War. This did not prevent him from criticizing the work of northern historians, like Rhodes, for whom the South might as well be "a foreign country." In his own historical writing, he not only struck a more balanced view but subordinated the North-South conflict to the common experience of mastering the American West. Wilson agreed with Frederick Jackson Turner that "The Proper Perspective of American History"—the title of one of Turner's essays—was not the Atlantic coast but the West. And who better than Lincoln embodied the spirit of the West, hence the spirit of America? The poet's verdict in 1865 was now the verdict of history. "It is as if Nature had made a typical American, and then added with liberal hand the royal quality of genius to show us what the type could be." Lincoln exemplified the active, practical, full-hearted spirit of the West; but he was all growth, said Wilson, and like the figure emerging from a block of marble under Robin's chisel the whole country was finally summed up in him: "the rude western strength, tempered with shrewdness and a broad humane wit; the Eastern conservatism, regardful of law and devoted to fixed standards of duty. He even understood the South, as no other Northern man of his generation did."[37]

As the president of Princeton University, Wilson audaciously held up Lincoln as the ideal toward which liberal education should aim. Growth was part of the ideal. So was the generalist's talent of turning his hand to the changing tasks that came before him. Modern industrial society required generalists over specialists, in Wilson's opinion. Lincoln was the type. Wilson liked the story the President told when sending a man on a delicate mission. After the briefing, the man asked nervously if anything had been overlooked, which reminded Lincoln of a little girl in Springfield who was so fascinated with her alphabet blocks that she took them to bed with her. One night as she was about to doze off, she remembered her prayers, rolled out of bed, knelt, and exclaimed, "O Lord, I am too sleepy to pray, but there are the letters, spell it out for yourself!" That was the meaning of liberal education: it offered the blocks of knowledge with which one should be able to spell things out for oneself. In his educational addresses, Wilson repeatedly posed the question: "Could Abraham Lincoln have been of more or less service to this country had he attended one of our modern universities?" And confessed, "It is a question I hesitate to answer." Well he might, too, since the argument seemed to lead to the conclusion that to be like Lincoln one need not and should not go to college. Wilson only fortified the conclusion by his criticism of the clubby class spirit of the universities, including Princeton, and his call for "democratic regeneration."[38]

Not until the 1912 presidential campaign did Wilson invoke Lincoln's name and authority in contemporary politics. Even then, compared to Roosevelt or even Taft, he was quite reserved. Lincoln stood for freedom and opportunity; to that extent he offered an ideal for the reform of economics affairs in America. But Wilson never suggested he offered any solutions. He, too, made the mandatory pilgrimage to Springfield. Despite his somewhat Lincolnesque appearance, no one looked upon him as a political heir. Oddly, he had not thought of Lincoln as a strong and inspiring presidential leader. Wilson's important first book, *Congressional Government*, expressed a low opinion of the American presidency as an institution and made no exception for the sixteenth president. Later, in his own presidency, like Roosevelt in his, Wilson came to feel a kinship with Lincoln and consciously walked in the shadow he cast over the White House.[39]

The Negroes' Lincoln

Lincoln's hometown was the scene of a frightful race riot during two days of August 1908. A hoodlum mob marched on the jail to demand surrender of a Negro charged with assault on a white woman. The prisoner was clandestinely removed to Peoria. This only enraged the mob. "Lincoln freed you, we'll show you where you belong," rioters shouted as they set about terrorizing Springfield's Negroes—about 10 percent of the population—and driving them from the city. Two Negroes were lynched, a total of six persons killed, more than fifty wounded; many homes and stores were sacked and burned; almost two thousand blacks fled the city. Order was finally restored by the National Guard. The riot was especially shocking because of the town's associations with the Great Emancipator. A young Kentucky-born socialist, William E. Walling, hurried to Springfield to report the event. His article, "The Race War in the North," in the New York *Independent*, was an eye-opener. Probing beneath the surface, he said that Springfield was a city without shame. "She stood for the action of the mob. She hoped the rest of the negroes might flee." The awful truth was, said Walling, "that a large part of the white population of Lincoln's home . . . have initiated a permanent warfare with the negro race."[40]

The Springfield riot came at a time when most of the country seemed to be at war with the Negro. There had been twenty-five lynchings in the past sixty days. State after state in the South was disfranchising the Negro. Race riots had occurred in places as far apart as Atlanta and

Boston. In the Berea College case, the Supreme Court upheld a Kentucky statute that levied a ruinous fine for operating an integrated school. The infamous Brownsville Affair still embarrassed the Roosevelt administration, more particularly the President's claim to the mantle of Lincoln; in it, Roosevelt had upheld the dishonorable discharge of three companies of colored infantry for provoking riot and disorder at the Texas fort in the face of mounting evidence of their innocence. Against this background, Walling's hard-hitting article on the Springfield riot prompted interracial leaders to issue a call for a "Lincoln Conference on the Negro Question." The call went out on the one hundredth anniversary of Lincoln's birth. It spoke of the disheartening failure of the country to fulfill the promise of emancipation and invoked his name in a renewed struggle for civil and political liberty. Out of the conference, which met in New York the following May, was born the National Association for the Advancement of Colored People. Its mission was declared to be "the completion of the work . . . the Great Emancipator began." In this circuitous fashion, the image of Lincoln the Emancipator was revived and renewed just as it seemed to fade from the memory of most Americans who were not his "stepchildren."[41]

The racist literature of the time offered two views on Lincoln and the Negro. In one he appeared as an abolitionist intent upon racial mongrelization. In the other and more respectable view he appeared as the true prophet of white supremacy, who saw that the two races could never live together as equals and sought to return the Negro to Africa. In 1890 a United States senator from South Carolina introduced legislation to support emigration to Africa, maintaining that this had been Lincoln's solution. Some Negro leaders, despairing of making emancipation work in the United States, also advocated separation and emigration. The American Colonization Society still stood ready to facilitate it.[42]

The most influential proponent of the white-supremacist Lincoln was Thomas Dixon, best remembered as the author of *The Clansman,* the novel and the play upon which D. W. Griffith based his sensational motion picture *The Birth of a Nation.* The son of a middling farmer in North Carolina, Dixon grew up during Reconstruction. He was a precocious student, and at the age of nineteen received a graduate fellowship to Johns Hopkins University, where he and another young southerner, Woodrow Wilson, became friends. An academic career did not suit Dixon's talents or ambition, however, and during the next three years he whirled through the theatrical world of New York City and the political world of the North Carolina legislature before settling on a career in the ministry. Dixon was a dynamic preacher, holding pulpits successively in Raleigh, Boston, and New York. Angered by a performance of *Uncle Tom's Cabin* in 1901, he decided to answer it in a novel depicting the horrors of Reconstruction from the southern point of view. The novel, *The Leopard's Spots*, set in the South Carolina Piedmont, actually revives some of Harriet Beecher Stowe's characters. Simon Legree, for instance, becomes a Scalawag leader of the Radical Republican legislature. Lincoln is only a spectral presence in this novel. His dream, it is said, was to restore the South as it was. With his death, the country descends into anarchy, race war, and barbarism. It is saved, finally, by the Ku Klux Klan.[43]

The book was a runaway best-seller. Dixon had found still another career. *The Clansman* took its place as the second of a trilogy of novels treating Reconstruction and the Klan. It opens with Lincoln, several days before his death, ordering the pardon of the rebel son of Mrs. Cameron. But Stanton refuses to execute it, and Dixon creates a little drama around the stereotypes of these men. Lincoln then argues vehemently with Austin Stoneman, the character modeled on Thaddeus Stevens, about the future of the Negro. He still advocates colonization. "The Nation cannot now exist half white and half black, any more than it could exist half slave and half free." The assassination at Ford's Theater is realistically depicted, as it would be later by Griffith on film, and Dixon draws upon the Easter sermons to conjure up a revengeful North. The story shifts to South Carolina, where it turns on the black Radical leader's lust for the white woman, Marion Cameron. Again, the Klan comes to the rescue of the South and civilization.[44]

Dixon plied his facile pen to construct spine-tingling plots that played upon the racial hatreds and fears of his white audience. He projected a hysterical new image of the Negro, no longer the harmless Uncle Tom or Sambo but the Black Beast. Amazingly, it received the kindly benediction of the Great Emancipator himself. Dixon professed to worship Lincoln, and to be no more than his humble disciple in advocating separation and deportation, which he did from pulpit and platform and in magazine articles as well as in his novels. In 1905 he turned *The Clansman* into a play. From its gala opening in Norfolk, then in cities throughout the South, it created a sensation. Packed with race, sex, and violence, the play was, as the first Norfolk reviewer predicted it would be, "like a runaway car loaded with dynamite." Negro leaders protested against this literary and theatrical demagoguery. "It is treason; it is criminal; it is untrue." But to no avail. Even John Hay, Lincoln's biographer and now Secretary of State, vouched for *The Clansman*.

The motion picture version, *The Birth of a Nation*, appeared a decade later, after collapse of the NAACP efforts to block its release. Griffith saw eye to eye with Dixon. He, too, loved Lincoln and hated the Negro. In several vignettes the President is portrayed as a compassionate father who is outwitted by the Radicals. The film ends with the magnanimous sentiments for the Second Inaugural Address. But, of course, as in the play, the message is in the plot, and it is an anti-Negro message. While the fate of the film was still in doubt, Dixon got his old friend Woodrow Wilson to view it in the White House. (*Birth of a Nation* was the first film ever screened there.) Enthralled, Wilson issued a ringing endorsement: "It is like writing history with lightning. My only regret is that it is all so terribly true."[45]

Dixon dedicated still another novel, *The Southerner*, to Wilson in 1913.* With the subtitle *A Romance of the Real Lincoln*, it was the author's main attempt to depict Lincoln as the advocate of racial discrimination and separatism. A prologue portrays the boy Abraham in the bosom of his family, where he acquires the poor whites' contempt for Negroes. The narrative begins in 1861, with the inauguration, and features a "hearts-divided" romance between the daughter of an abolitionist senator, Betty Winter, and a Confederate spy, complicated further by the fact that the spy's brother is a Union officer known to Betty. She ingratiates herself with the Lincoln family, and the President's opinions and actions are observed through her. He is portrayed as the soul of moderation. Dixon, who boasted of basing everything upon the authentic historical record, shows him with an assemblage of Negro leaders pleading for their support of racial separation and colonization for the good of both races. Only when this effort fails does he resort to the Emancipation Proclamation; even then, according to the author, he writes into it the principles of separatism, and at the end of his life he is planning with General Butler to effect removal of the freedmen. In advocating the inferiority and unassimilability of the Negro, Dixon had plenty of company; nor was he alone in ascribing this opinion to Lincoln and calling for a return to "Lincoln's solution."[46]

The Negroes' affectionate regard for Lincoln, though thinning with the passage of time, continued well into the twentieth century. He was their Moses, their deliverer, their savior. With one stroke of the pen he broke the chains of bondage.

---

* Dixon never wrote without political purpose, and obviously sought to associate Wilson with sectional reconciliation based upon the North's acceptance of the southern view of race relations, sanctioned by the name of Lincoln. Thus he wrote to Wilson's private secretary of the influence of *The Birth of a Nation:* "Every man who comes out of one of our theatres is a Southern partisan for life" (Richard Schickel, *D. W. Griffith: An American Life* [New York, 1984], 269).

Hallelujah broke out—
Abe Lincoln freed the nigger,
With the gun and the trigger
And I ain't going to get whipped no more.
I got my ticket,
Leaving the thicket,
And I'm heading for the Golden shore.[47]

Folk tales about Lincoln passed from generation to generation. One of the most common, of which there were many variations, expressed the conviction that Lincoln had actually visited the Negroes in their bondage, seen their distress, and returned to Washington to free them. Thus he came to Beaufort, South Carolina, sat down to dinner, and slept in master's bed. "He left his gold-headed walking cane dere and ain't nobody know de President of the United States been in Beaufort 'till he write back and tell um to look behind the door and send um his gold-headed walking cane." He came to Mississippi and made a speech. He stopped at a tavern in Tennessee, but nobody knew him. "He was just the raggedest man you ever saw." In Washington memories of Lincoln among Negroes were especially vivid. Children growing up imbibed the tradition that Ford's Theater was haunted by Lincoln's ghost; some of them had aged aunts and uncles who had served the President and his family—they came, it was said, "from the cream of Washington's colored society"—and they told stories about him and showed off their souvenirs. Negroes made pilgrimages to the tomb of "Lincoln, the God-like, the friend of our race," and thanked him for their freedom. Any Negro orator worth his salt had in his repertoire an oration on Lincoln, which he delivered countless times. Lincoln was more than a memory, the orators said; he was a living, vital force as the colored people struggled to realize the promise of emancipation.[48]

The annual commemoration of Emancipation Day traced back to 1864 and continued sporadically for a century to come. It was the Negroes' "natal day," "independence day," and "emancipation day" all in one. The fact that it was nowhere a legal holiday, or even recognized in the white community, hindered its development, and so did the inability of Negroes from one place to another to agree on a single calendar date for the celebration. January 1 was the most common choice, despite the conflict with New Year's. September 22, the date of the Preliminary Emancipation Proclamation, was often favored, however, and in Texas it was "Juneteenth," June 19, which was said to be the day in 1865 a slow-stepping mule brought word of the Emancipation Proclamation to

Emancipation Day, 1919. Color photogravure by F. G. Renesch, with likenesses of Lincoln, Paul Laurence Dunbar, and Frederick Douglass. (The Lincoln Museum, Fort Wayne, a part of Lincoln National Corporation)

Welcome Home, 1919, by F. G. Renesch. (The Lincoln Museum, Fort Wayne, a part of Lincoln National Corporation)

the faraway Confederate state. Ordinarily, the commemorations were under church auspices; but in Atlanta, by 1900, the Negro Literary and Historical Society stood as sponsor, and in Indianapolis, where the September event coincided with the State Fair, the Republican Club took the lead. The programs typically included a reading of the Emancipation Proclamation, an oration or address, and choral music. Sometimes a pageant, like the one William E. B. Du Bois wrote for the one hundredth anniversary in 1913, would be performed. Of course, the Negro press took notice of Emancipation Day. In time, it would be superseded by Negro History Week, which its initiator, Carter Woodson, placed on the calendar about the second week of February so as to encompass the birthdays of both Lincoln and Frederick Douglass.[49]

The Negroes' affection for Lincoln was rooted in a sense of kinship with his life and character as well as in gratitude for the gift of freedom. His birth, like theirs, was obscure; he was of lowly origins and had toiled up from poverty, as they had toiled up from slavery. He, too, had struggled to gain "a little learning." He, too, was "a man of sorrows acquainted with grief," as Kelly Miller said. His traits of character—simplicity, goodness, honesty—endeared him to blacks as well as to whites. He represented a moral ideal to which all should aspire. And if he was not, like Wendell Phillips, a lover of the Negro, he was a humanitarian to the core. In a reader for Negro children, Lincoln was the only person not of their race offered for instruction and emulation. "I confess," said Booker T. Washington in 1909, "that the more I learn of Lincoln's life the more I am disposed to look at him . . . not merely as a statesman, but as one to whom I can certainly turn for help and inspiration—as a great moral leader, in whose patience, tolerance, and broad human sympathy there is salvation for my race, and for all those who are down, but struggling to rise."[50]

Washington had succeeded Douglass as the foremost national leader of his people. In his autobiography, *Up from Slavery*, and in countless speeches, he recalled being awakened when a small boy by his mother's prayers for "Marse Lincoln and freedom." It was his earliest memory, and Lincoln's was the first name he heard spoken by anyone outside the Virginia plantation where he lived. Washington became famous as president of Tuskegee Institute, in Alabama, for advocating a beneficent work ethic for Negroes. In his "Sunday Evening Talks" to the students, Lincoln was held forth as a model of virtuous striving. He read the Bible regularly, thus directing his mind toward good; he earned the name "Honest Abe" for good deeds; self-denial, said Washington, was "the secret of Lincoln's success in life." Even his patience was a virtue for emulation by the race. The educator became a student of Lincoln, boasting that he had read every book and article written about him. And the more he read, the more impressed he became with the inspiration to be found in Lincoln's struggle up from poverty. "Like Lincoln, the Negro knows the meaning of the one-room cabin; he knows the bed of rags and hay; he knows what it is to be minus books and school-house; he has tasted the lowliest poverty, but through them all he is making his way to the top."

"The Wizard," as admirers called Washington after the *Oz* character, was much in demand as a speaker before audiences of both races. He often invoked Lincoln's name, perhaps nowhere more poignantly than in his speech before the National Republican Club on the centennial of the Emancipator's birth. He praised Lincoln for his patience and courage, and said that his proclamation was not for the Negro alone but for all oppressed peoples. "Today, throughout the world, because Lincoln lived, struggled, and triumphed, every boy who is ignorant, in poverty, despised, or discouraged, holds his head a little higher." Negroes should not be discouraged but should heed his example. "Like Lincoln, the Negro race should seek to be simple without bigotry and without ostentation. There is great power in simplicity."[51]

This was not the message that William E. B. Du Bois and the founders of the NAACP wished to communicate. Not that they were less enamored of Lincoln the Emancipator. Du Bois, a younger, more scholarly, as well as more radical educator, certainly venerated Lincoln. But he venerated him in a different way. Speaking at Hull House, Du Bois praised Lincoln most for his mental qualities, his clear-sightedness, and his capacity for growth. He shared Douglass's belief that, for all his greatness, Lincoln had been less than wholehearted in his commitment to Negro freedom and equality. Similar views had been put forth by George Washington Williams in his path-breaking *History of the Negro Race* in 1892, and they were held by Negro leaders like Archibald H. Grimké. Du Bois came to believe it was important to "demythicize" the Great Emancipator in order to improve the Negro's own self-confidence and clear-sightedness. "The picture of Lincoln looking out upon space with sad and loving eyes," he wrote, recalling the Freedman's Monument, "his right hand outstretched presenting the Emancipation Proclamation, his left hand resting tenderly upon the head of a newly freed and gratified slave kneeling at his feet, can no longer be satisfactory." Du Bois thought, too, that the Negro had been victimized by the Republican party. The idea that it was the party of Lincoln was part of the mythology that still enslaved him. Du Bois broke ranks and voted for Wilson, a liberal southerner and a Democrat, in 1912. Of course, he would be cruelly disappointed.

Another slap in the face was Congress's failure to enact legislation to commemorate the fiftieth anniversary of the Emancipation Proclamation in 1913, which was made all the more galling by an appropriation for the reunion of the Blue and the Gray at Gettysburg. The American people, it seemed, were in danger of forgetting what the Civil War was about. Keeping alive the image of the Great Emancipator was one way of remembering. It remained compelling among Negroes in that anniversary year. Albert Pillsbury, a Massachusetts leader, wrote a book in which he posed the question "Does lincoln deserve the title of Great Emancipator?" and answered with a resounding affirmative. Hatred of slavery was the dominant motif of his public life; he was neither reluctant, nor hesitant, nor contingent in his moves to abolish it. Still, among whites that image had been eclipsed by the more inclusive image of Lincoln the Man of the People, while some Negroes were beginning to wonder if it was not the image they needed emancipation from.[52]

The Centennial

The commemoration of the one hundredth anniversary of Lincoln's birth was one of those events that took up more space in the actual observance than it would in the historical record. As Americans again assessed Lincoln's place in history, they were guided by the themes of the apotheosis and by the images conveyed through reminiscence, biography, and politics.

Savior of the Union remained the favorite theme of the Grand Army of the Republic, the Union League, the Loyal Legion, and similar groups for whom the memory of Lincoln was entwined with memories of the war. Some hoped to consolidate the still alienated affections of many southerners to the national ideal, thus to finish at last the unfinished business of the war. Others were eager to forge the link between nationality and democracy in Lincoln's name. He had seen the Nation behind the Constitution: the vibrant democratic principle rather than the ancient legal compact. This added to the importance of the image of Man of the People. No conception had proven more protean. It encompassed the folk hero, the humanitarian, and the democratic leader. For reformers bent upon halting the march of predatory wealth and returning power to the people, no event of recent years, as one of them said, "was so pregnant with inspiration as the nation-wide anniversary of the birth of Abraham Lincoln." For Negroes and their friends Lincoln remained, above all, the Great Emancipator. They hoped to broadcast the message of racial justice and equality. But the commemoration, like everything else in American life, was Jim Crow, and to Negro leaders who had long worshiped at Lincoln's shrine it seemed that "the whites have outheralded Herod" on his anniversary. To an extent, of course, the NAACP could be considered a by-product of the occasion. The image of Lincoln as the First American, voiced by the poet Lowell, had prodigious connotations but was especially associated with western character and influence. Nature had taken "sweet clay from the breast of the unexhausted West" to mould this new hero. In 1909 Edwin Markham, already famous for verses in Lincoln, contributed to the commemoration a poem evocative of the West.

> So hidden in the West, God shaped His man,
> There in the unspoiled solitude he grew,
> Unwarped by culture and uncramped by creed;
> Keeping his course courageous and alone,
> As goes the Mississippi to the sea.

Lincoln the Self-made Man continued to instruct the young, in particular, and secured its own memorial at the great man's Kentucky birthplace.[53]

No issue had been more marked in the definition of Lincoln's character than that between the folk hero and the godlike statesman, and this tended to correlate, though not absolutely, with the division between the man of humor and the man of sorrow, the man of affection and the man of intellect, the Lincoln of romance and the Lincoln of work-a-day realities. It was, said an observer, as if two mental tickets were issued for Lincoln, one giving access to the almost superhuman savior of the Union and sad-eyed emancipator, the other to the droll humorist and the "great heart" who subsumed reason to sentiment. "Some of our countrymen pin their faith to one ticket, some to the other, and some—such is the delightful inconsistency of the human mind—accept both. They use one in the Sunday school and the other in the smoking car." The best minds of the afterwar generation—the generation that staged the Centennial—felt waning interest in the Lincoln of romance and reminiscence and, with Robinson, fixed their thoughts on the Olympian statesman. The viewpoint was well expressed in an anniversary editorial in the *Dial:* "The figure which was in the process of reconstruction from the time of Lowell's ode and Whitman's threnody to the time of the statue of Saint-Gaudens and which is still more definitely shaped in this centennial year, is far more the expression of our ideal than it is of our memory, and it speaks well for the national character in the twentieth century that this ideal is so pure and wholesome and altogether worthy of our devotion." As Herbert Croly wrote, there was "a kind of human excellence" in Lincoln which turned to moral and intellectual account every experience life offered, and which was more than nationalism or democracy or brotherhood or Americanism; and it was this that made him the supreme symbol of a worthy ideal.[54]

The first call for "The Lincoln Centenary" was published by the *New York Times* in 1905, four years in advance of the event. Because Lincoln was "the most representative, and most widely recognized National figure in our history," the anniversary offered an incomparable opportunity to appeal to the South, both white and black, and to strengthen national unity. It ought to provide as well for a great national monument in Washington. The further recommendation for a federal commission met with no response in Congress. As a result, the Centennial became an activity mainly of local governments, patriotic organizations, and the media. Things got underway in Illinois in 1907 when the legislature adopted a resolution calling upon the Governor to appoint a fifteen-member State Centennial Commission.

The year 1908 marked the fiftieth anniversary of the Lincoln-Douglas Debates. All seven of the cities where the debates had occurred celebrated the event. The State Historical Library published a new edition of the debates. Their modern fame dates from this time. Actually, it began in 1896 when Knox College, at Galesburg, unveiled a bronze tablet to memorialize the event. Robert Lincoln was present, and Chauncy Depew, who remembered the impact of the debates in the East, delivered the oration. Contrasting the contestants, he said, "Lincoln had humor and pathos, and Douglas possessed neither. . . . Unlike Douglas, Lincoln was weak unless he knew he was right. His whole nature must be stirred with the justice of his cause for him to rise above the commonplace." Needless to say, the celebration was of Lincoln rather than Douglas. Horace White, whose reports of the debates in the *Chicago Tribune* were said to have started "a new era in the history of journalism," spoke at the fiftieth anniversary commemoration in Alton, which, although a small place, turned out an audience of thirty thousand. It was another sign, White thought, of Lincoln's soaring fame as the Centennial approached.[55]

The simultaneous movement to memorialize Lincoln's birthplace was of unusual interest to the nation. In 1894 Alfred W. Dennett, a New York restaurateur, purchased the place called Sinking Spring Farm about three miles from Hodgenville. The 110 rolling acres were useless for farming, but Dennett thought to improve the landscape and turn the site into a memorial park. He also acquired from a neighbor, John A. Davenport, the beamed log cabin that, according to oral tradition, had stood on the site when Lincoln was born, but had been removed in 1861. The

From *McClure's*, December 1895.

Davenport cabin was reassembled on the original homestead—a photograph appeared in *McClure's* in 1895—but within three years it was again dismantled. The logs were marked and shipped to Nashville for exhibition, beginning a strange odyssey that would end back in Hodgenville nine years later. The authenticity of the cabin, and of its site as well, was disputed from the start. Locally, there were incompatible recollections about the history of the birthplace cabin; and the existence of other cabins, such as the one on the Knob Creek farm where the family went to live when the boy was two, simply added to the confusion. Public interest in log cabins associated with Lincoln seemed insatiable. Mention was earlier made of the Railsplitter's cabin from 1830. In 1891 the Abraham Lincoln Log Cabin Association purchased the dwelling at Goose Nest Prairie, in Cole County, which Lincoln had supposedly built for his father, dismantled it, and remounted it for exhibition in Chicago. For better or worse, the cabin at Pigeon Creek, in Indiana, disappeared without a trace.[56]

Dennett, having fallen into financial difficulties, failed to pay his taxes, and the sheriff of Larue County advertised the old Lincoln farm for auction in 1905. Meanwhile, a young newspaperman, Richard Lloyd Jones, had taken a deep interest in the property. One day, chatting in Louisville with "Marse Henry" Watterson on a favorite subject, Lincoln, the editor asked, "Have you ever been down to his birthplace?" When Jones confessed he had not, Watterson continued, "Well, go; you will find the spring broken down, the pigs and horses trampling upon sacred ground. You sit down on the bare knoll above the spring where the Lincoln cabin stood, the markers are still there. You sit there and there is something there that will move you." Jones went, was moved, and came away with the idea of making something of this "sacred ground."

Sometime later, in Chicago, he related his experience to his father, Jenkin Lloyd Jones, a well-known Unitarian minister and reformer and founder of the Abraham Lincoln Center in the city. He at once resolved on a pilgrimage to Hodgenville. Upon his return he wrote an editorial, "The Neglected Shrine," in his paper *Unity*, which brought the matter to the attention of a wider public for the first time. The younger Jones, when he learned that the property would be auctioned, interested Robert J. Collier, editor of *Collier's Weekly*, in bidding for it. Jones later regaled his friends with the story of how he spent the evening before the sale in nearby Elizabethtown getting two competitors hopelessly drunk. One of them represented a Louisville distillery which hoped to make and market "Lincoln Birthday Whiskey" from the spring. The next morning Jones was the only serious bidder. He bought the farm in Collier's name for $3,600. Immediately, the editor announced his intention of eventually conveying the place to the government as a Lincoln memorial.[57]

First, Collier and associates formed the Lincoln Farm Association to raise money for development. Under the presidency of Joseph W. Folk, a former governor of Missouri, it enlisted the support of famous men like Mark Twain and Cardinal Gibbons and built a membership of some seventy thousand people upon contributions of twenty-five cents to twenty-five dollars. *Collier's Weekly* promoted the cause, chiefly through the publication of Lincoln birthday numbers beginning in 1906. Mrs. Russell Sage gave $25,000, the only large gift received by the association. The design competition for the proposed memorial was won by young John Russell Pope, whose classic inspiration prefigured a distinguished career in American monumental architecture. How did one design a memorial to house and preserve a crude century-old log cabin and to honor the great man born in it? Polk's design, pure and serene and utterly unrelated to the historical Lincoln, called for encasing the birthplace cabin within a Greek temple. The oak logs from the cabin, meanwhile, were located on Long Island. To transport these sacred relics to Kentucky, the Pennsylvania Railroad furnished a special car, decorated by John Wanamaker, the Philadelphia department store baron—and a Lincoln enthusiast—which made well-advertised stops en route to allow children and adults to touch the logs.

In 1908 Congress gave a boost to the association's campaign with the appropriation of $50,000 toward the building of the memorial. The accompanying report lauded the objectives of preserving the birthplace cabin. "We believe American youths have too long turned longing eyes toward the holy places of Europe . . . while we have neglected to inspire them with the holy places at home." The memorial in Kentucky, the premier border state, the report continued, "will become the nation's commons, the patriotic mecca of North and South, East and West, the national symbol of peace and unity."[58]

The cornerstone of the memorial hall was laid on the one hundredth anniversary of Lincoln's birth. President Roosevelt spoke movingly before a huge rain-drenched audience, although nothing he said on this occasion cut closer to the heart of the matter than what he had written in *Collier's* the year before: "No more blessed thing could have happened to a great democratic republic like ours than to have had this man of the plain people, the rail-splitter, the country lawyer, develop into its hero and savior; for every feature of his career can be studied as a lesson by each of us, whatever his station, as we lead our several lives." The memorial would stand for all this, for it was here, on this quiet knoll in Kentucky, that that life began. The association made the decision not to place a statue before or within the temple housing the sacred relic lest it detract from it. The statue, a noble bronze by Adolph Weinman, was erected by the Commonwealth of Kentucky in Hodgenville and dedicated on Memorial Day of the centennial year. One November day in 1911, President Taft dedicated still another statue in Kentucky's new state capitol at Frankfort, and on the next day dedicated the birthplace memorial for the nation. It would serve as a constant reminder, he said, "of the unexplained and unexplainable growth and development, from the humblest and homeliest soil, of Lincoln's genius, intellect, heart, and character that have commanded the gratitude of his countrymen."[59]

The United States government assumed custody from the Lincoln Farm Association in 1916. Authentication of the logs of the templed cabin has remained a problem for the National Park Service, as it was earlier for the association. Today when as many as three hundred thousand of Lincoln's grateful posterity come to this somewhat isolated monument in Kentucky, ascend the fifty-six wide steps (each denoting a year of Lincoln's life), and pass through the columned entrance to gaze upon the cabin, slightly reduced from its original size of sixteen by eighteen feet, they are told it is thought to be reconstructed from *some* of the original logs. But the question scarcely matters to most visitors, since the cabin conveys the truth of Lincoln's beginnings, while the temple of gleaming pink granite endows it with the beauty of classic form.

Unique in several ways, it is perhaps most original in being the only monumental tribute of a nation to the birthplace of a great man. Presi-

Lincoln Birthplace Memorial, Hodgenville, Kentucky, 1909. By John Russell Pope. (U.S. Department of the Interior: NPS photo by Allan Rinehart)

Cartoon by H. T. Webster, Press Publishing Company, 1918.

dent Wilson, dedicating the memorial for the United States in 1916, called it an eloquent shrine to democracy. "Nature pays no tribute to aristocracy, subscribes to no creed or castle, renders fealty to no monarch or master of any name or kind. . . . Here is proof of it!" And so for some visitors the memorial became a civics lesson. "Every youngster in this country should be taken to the place of Lincoln's birth," said the Ohioan James M. Cox. "The combined lessons of all the classrooms in America do not give such an appreciation of the fact that in our country a humble beginning is not an impossible barrier to success." For others it invited ridicule, so incongruous was the architecture with the man it memorialized. Ida Tarbell confessed she first approached it with dread, yet found happily that in its extraordinary dignity and simplicity the memorial beautifully, serenely, stood for Lincoln.[60]

THE CENTENNIAL MEMORIALIZED Lincoln in many ways. President Roosevelt approved of placing Lincoln's head on one of the nation's coins, and he turned up on the penny—the commonest of all coins. It was the first United States coin to honor a president. An aspiring poet named Carl Sandburg, writing in the *Milwaukee Daily News*, hailed its arrival: "The common, homely face of Honest Abe looks good on the penny, the coin of the common folk from whom he came and to whom he belongs." So great was the demand for the Lincoln penny in 1909 that vendors sold it at a premium and enterprising tradesmen inserted it in metal casings advertising themselves. Eighty years later some 250 billion Lincoln pennies had been minted. Congress directed the Postmaster General to issue a commemorative stamp in the two-cent denomination that carried most first-class mail. Lincoln's head had previously appeared on special stamps—the first in 1866—but, of course, never on the first-class denomination, and some congressmen surmised that once ensconced he would evict Washington from that favored position. The carmine-colored stamp featured the profile bust (actually a reduced photograph of a plaster study) from Saint-Gauden's famous statue. Robert Hewitt, a collector, was responsible for the minting of a commemorative medal, which offered a striking frontal bust in high relief sculpted by Jules E. Roine.[61]

The events of the commemoration converged on February 12, which was a Friday, and in some states and cities a legal holiday. In major cities, like New York and Chicago, commissions were appointed to orchestrate the activities. Daily newspapers were the leading promoters of the commemoration. In Chicago the programs of schools, civic groups, patriotic organizations, and professional bodies spread over the entire week. It was ushered in by a blockbuster edition of the *Chicago Tribune* on Sunday, February 7. Proclaimed "The Greatest Issue of the World's Greatest Newspaper," it ran to 194 pages and weighed three and one-quarter pounds. A year in planning, over five weeks in printing on thirteen presses using five tons of ink, the sheets of the edition laid end to end would stretch eleven thousand miles, or all the way to Singapore. Nothing like it had ever appeared before. It marked, as the *Tribune* boasted, an epoch in newspaper publication. "Lincoln's Spirit Dominates City" ran the front-page headline, while the two-column editorial extolled the sixteenth President as both nationalist and democrat. The stories that filled the pages were, for the most part, lifted from extant writings about Lincoln. The center of the "picture section" featured a two-page spread of miniaturized portraits of Lincoln—ten across in twelve rows—which Osborn Oldroyd had assembled. (They resembled the end-papers of this volume.) Succeeding issues provided full coverage of the celebration in this city of two million people. In Saturday's climax, Woodrow Wilson spoke on "Abraham Lincoln: A Man of the People" at the Auditorium, and twenty thousand people later packed the Dexter Pavilion at Union Stockyards to hear an illustrated (stereopticon) lecture by the Reverend Jenkin Lloyd Jones and a choir of five hundred voices. The city's Negro community had its own great event at the Seventh Regiment Armory. Only here was attention paid to Lincoln the Emancipator.[62]

In Springfield the Jim Crow auspices of the commemoration threatened to become embarrassing. The capital city's program included an exhibit, with lectures, at the State Historical Society, services at the tomb, a reception at the homestead under the sponsorship of the Daughters of the American Revolution, an address by William Jennings Bryan at the Tabernacle, and a gala by-invitation-only banquet at the State Arsenal, where the British and French ambassadors headed the list of honored guests. Sponsored by the Lincoln Centennial Association, a private corporation created by the state commission, the banquet admitted the twenty-five-dollar subscribers to membership in the association. One of the invitations went to Edward W. Morris, a Negro lawyer in Chicago, because he was a member—the only Negro member—of the state legislature. Morris promptly subscribed, which was the cause of consternation, although he had no intention of showing up at the lily-white affair. The editor of the *Colored American Magazine* expressed amusement that the whites had suddenly grown so fond of Lincoln as to bar blacks from their meetings. At the Union League Club banquet in Brooklyn, colored waiters who wished to hear the speakers were sent to the kitchen, with the result, it was said, that dinner was served "very sullenly."[63]

The Lincoln Centennial Commission in New York City reported that

19 PARTS

The Chicago Sunday Tribune.

VOLUME LXVIII.—NO. 6. | 194 PAGES. | FEBRUARY 7, 1909. | 194 PAGES. | * * * PRICE FIVE CENTS.

"The Greatest Issue of the World's Greatest Newspaper."

CONCERT OF BLOWS AT THE PRESIDENT

Raymond Sees Conspiracy to Discredit the Roosevelt Administration as Its End Draws Near.

CITES THE PANAMA CASE.

Tennessee Coal and Iron Merger Another Instance of Hostile Unanimity Without Basis of Evidence.

BY RAYMOND.

Cowardice in the Attacks.

SUMMARY OF The Sunday Tribune

SUNDAY, FEBRUARY 7, 1909.

TELEPHONE CENTRAL 100.

POSTAGE ON TODAY'S TRIBUNE.

Readers who mail the special Lincoln edition of The Tribune should attach 15 cents postage for a domestic address or 30 cents for a foreign address. Newspapers short of postage are held up by the postoffice.

Summary of Features in Lincoln Sections.

HALF TONE SECTION.

SECTION A.

ATTACK SHERIFF'S PROFIT ON FEEDING

Citizens' Association Alleges Official Nets $30,000 in a Year on Poor Meals at the Jail.

'BLACKMAIL,' HIS RETORT

Strassheim Courts Inquiry and Secretary Singleton Says Complaint Will Be Pressed.

Charges Mailed Broadcast.

THE LINCOLN OF FORTY YEARS FROM NOW.

Somewhere in this country today there is an unknown boy who will be the country's greatest living man forty years from now.

LINCOLN'S SPIRIT DOMINATES CITY.

Week of Celebrating Centenary of the Birth of the Great Emancipator Begins.

PEOPLE JOIN IN HOMAGE.

Patriotism for a Time Blots Out Differences in the Politics and Religion of Citizens.

This is Lincoln week.

Greatest of Memorial Festivals.

Upper half of the front page of Centennial issue. *(The Chicago Tribune)*

Lincoln Centennial Association Banquet, Springfield, February 12, 1909. (Illinois State Historical Library)

one million people participated in programs in 562 schools and 624 meetings and related activities. The *New York Times*, which took the lead among newspapers in promoting the Centennial, sponsored an essay contest for schoolchildren. A series of seven biographical articles by Frederick T. Hill, subsequently gathered in a book, *Lincoln's Legacy of Inspiration*, provided the orientation. The aim, said the *Times*, was "to give Lincoln's message to those who feel they have not had a fair start in life," and who thus might find his career "an antidote to hopelessness and discontent." How that message could be delivered to slum children was a problem that weighed on the minds of many people. The *Boston Globe* conducted a forum on the question "Has the Poor City Boy the Chance that Lincoln Had?" Three of the four urbanologists said yes, provided, of course, the legacy of inspiration survived.

Judged by the response of ten thousand New York schoolchildren to Hill's earnest articles, America was still a country where the poorest and humblest might strive to greatness by adherence to the traditional virtues. The *Times* drew the conclusion: "It is neither Lincoln the President, nor Lincoln the Master of Men, nor Lincoln the Savior of the State who is winning the hearts of more and more Americans every year. All that history could tell of the President was told many years ago. It is Lincoln the Man who is inspiring his followers today—the man within touch of all the lowly of heart. This is he who of all Americans is 'leaving his imprint upon eternity,' " So, once again, it came down to character. As in Chicago, New York's commemoration extended through the week. The Lenox Library mounted a Lincoln exhibit. The New York Philharmonic Orchestra performed Fritz Stahlberg's "Abraham Lincoln: In Memoriam;" the piece was turbulent and mournful and, despite the presence on the stage of the Saint-Gaudens icon, about as suggestive of Lincoln as it was of Savonarola, in a critic's opinion.[64]

In the international reverberations of the Centennial—in London, Manchester, Paris, Berlin, Manila, and in Latin America—one tribute stood out from all the rest. Leo Tolstoy, perhaps the most famous man in the world in 1909, interviewed at his estate, Yasnaya Polyana, called Lincoln "a Christ in miniature, a saint of humanity." Of all the great national heroes and statesmen of history, he was the only true giant. Tolstoy had been amazed by the reach of Lincoln's fame. Once, traveling to the Caucusus, he met a Muslim chief who said of the American: "He was a hero. He spoke with a voice of thunder; he laughed like the sunrise and his deeds were as strong as the rock and as sweet as the fragrance of rose." The chief begged to learn more of Lincoln, and Tolstoy told him all he knew. Having himself become a saint, he naturally thought Lincoln was one. "Washington was a typical American"—a switch on the usual comparison—"Napoleon was a typical Frenchman, but Lincoln was a humanitarian as broad as the world. He was bigger than his country—bigger than all the Presidents together. Why? Because he loved his enemies as himself." Love was the foundation of his life, said the Russian, and he prayed that the Centennial would light the flame of righteousness among nations.

This emphasis on Lincoln's love had its other side in the love of the people for him. "One cannot read Abraham Lincoln without loving him," President-elect Taft declared in a notable tribute. In him there was more "inspiration for heroism" than in any other man in history. The *Times*, in its centennial editorial, said that if one name were to be given to the Martyr President a century after his birth it would be "Lincoln the Beloved." This love for him was founded on his love of humanity.[65]

The Centennial produced a bumper crop of books, essays, and poems on Lincoln. The only truly important historical contribution was the *Diary of Gideon Welles*, serialized in the *Atlantic Monthly* prior to publication. A centenary edition of Lincoln's writings appeared in nine volumes; Rice's *Reminiscences*, already a classic, was republished; and both Tarbell and Rothschild prepared second editions of their books. The most popular biography of the year was authored by the Progressive mayor of Cleveland, Brand Whitlock. In a compact two hundred pages, he added the elegiac tone of Vachel Lindsay to Herndon's characterization. With other Progressives, he emphasized Lincoln's philosophy of labor. The Civil War, in the President's eyes, was part of a world struggle for the rights of man over the rights of property. He was "essentially an idealist," said Whitlock, and his dream eluded him. Among volumes of special interest were J. Henry Lea's *The Ancestry of Lincoln*, which gave him a distinguished lineage, and Clara E. Laughlin's *The Death of Lincoln*, which reflected the public's continuing fascination with the assassination. No sooner had the first wave of reminiscences passed into folklore than a new wave followed. The *Memoirs of Gustave Koerner* made a significant addition to the literature. A political friend, Koerner had helped mobilize the Illinois Germans behind Lincoln and the Republicans. Heartwarming reminiscences filled the columns of newspapers. Those of William H. Crook, Lincoln's bodyguard, ran in the *Washington Post;* a four-page feature in the *Boston Globe* contained the recollections of one hundred living New Englanders who had seen Lincoln. Thus he remained an intimate presence in spite of the hero worship.[66]

The centennial year also saw the publication of a memoir that would attain the status of a cult classic, *The Valley of the Shadows*, by Francis Grierson. A prairie-born musical prodigy, Grierson achieved fame as a pianist in Europe. Later, after taking up the voguish spiritualism of the time, he launched a literary career with *Modern Mysticism* in 1899. *Valley of the Shadows*, written in London, was an attempt to recapture through a mystic veil the author's frontier boyhood. Elements of the supernatural—superstitions, premonitions and visions, the sense of being in the control of a higher power—had always figured in the Lincoln image. Grierson asserted that they were the key to understanding him. Agreeing with Carlyle that "all authority is mystical in origin," he viewed Lincoln as a prophet floated into power on waves of destiny. Not understanding this, politicians and lawyers and historians had not understood Lincoln. Grierson mentioned the unparalleled radiance of the comet Donati in the year 1858 and interpreted this as a heavenly benediction on Lincoln and the Republicans. As a boy he had heard the last of the great debates at Alton. His description of Lincoln as he "stood like some solitary pine on a lonely summit" and with wondrous power held the people breathless "under the natural magic of the most original personality known to the English-speaking world since Robert Burns" is almost worth the price of admission to the book.[67]

The Centennial spawned a quantity of verse. Percy MacKaye's *Centenary Ode*, recited at the Brooklyn Academy of Music, suffered the fate common to occasional poetry. It is contrived and ponderous. Yet it has its moments, as in

> Aesop and old Isaiah held in him
> Strange sessions, winked at by Artemus Ward.

And this incisive acknowledgment of Lincoln's complexity:

> The loving and the wise
> May seek—but seek in vain—to analyze
> The individual man, for having caught
> The mystic clue of thought
> Sudden they meet the controverting whim,
> And fumbling with the enchanted key,
> Lose it then utterly.

Finally, this transformation of the Great Emancipator into a prophetic world figure:

> He stands forth
> 'Mongst nations old—a new world Abraham,
> The patriarch of peoples still to be,
> Blending all visions of the promised land
> In one Apochalypse.

William Ellery Leonard, a distinguished professor at the University of Wisconsin, read his centennial poem at the dedication of a replica of the Weinman statue on the Madison campus. He speaks of Lincoln's "iron faith" and its mighty legacy to the nation. More important than that, however, is the man who lives in the hearts of the people.

> How often it seems we like to linger best
> Around the little things he did or said,
> The quaint and kindly shift, the homespun jest,
> Dear random memories of a father dead;
> His image is in the cottage and the hall,
> A tattered print perhaps, a bronze relief,
> One calm and holy influence over all,
> A household god that guards an old Belief.

A household god is an intimate god, one about whom loving sentiments gather. A Canadian-born writer, Edward William Thomson, was the author of several narrative poems about Lincoln which, as they were grounded in casual memories of him, belong as well to the literature of reminiscence. "Father Abraham Lincoln," published in *Collier's*, begins:

> *My private shrine. The Gettysburg Address*
> *Framed in with all authentic photographs*
> *Of him from whom the New Religion flows.*

The poet remembers how as a boy of fourteen munching a cheese-cake on Chestnut Street in Philadelphia he saw the President walk by.

> A sudden twinkle lit his downcast eyes,
> Marking the cheese-cake and the staring boy;
> Tickled to note the checked gastronomy,
> Passing, he asked, "Good, sonny?" in a tone
> Applausive more than questioning, full of fun,
> Yet half-embracive, as your mother's voice,
> And smiled so comrade-like the wondering lad
> Glowed with a sense of being chosen chum
> To Father Abraham Lincoln, President.

The boy became a soldier near the war's end and saw the President again at City Point, mounted, reviewing the troops, "a travesty of every point of horsemanship," and in black frock coat and antic stovepipe hat looking like "some old-time circuit preacher."

Too much by far for soldier gravity—
A breeze of laughter traveling as he passed,
Rose sudden to a gale that stormed his ear.
The President turned and gazed and understood
All in one moment, slightly shook his head,
Not warningly, but with a cheerful glee,
And sympathy and love, as if he spoke:
"You scalawags, you scamps, but have your fun!"
Pushed up the stovepipe hat, and all around
Bestowed his warming, right paternal smile,
As if his soul embraced us all at once.

Then strangely fell all laughter. Some men choked,
And some grew inarticulate with tears;
A thousand veteran children thrilled as one,
And not a man of all the throng knew why;
Some called his name, some blessed his holy heart,
And then, inspired with pentecostal tongues,
We cheered so wildly for Old Father Abe
That all the bearded generals flamed in joy!

The poem concludes with reflections on the miracle of Father Abraham and apprehensions for the republic's future as the money-mongers fouled the promise of his victory. Lindsay was not alone among poets in praying for the return of Lincoln's spirit to brighten the democratic promise.

The most popular poem of the season was Edwin Markham's "Lincoln, the Man of the People." Indeed, it was on its way to becoming the best-known verse ever written on the subject. Markham had rocketed to fame in 1899 with "The Man with the Hoe." Brought to eastern celebrity from his native Northwest, he was asked to read a poem on Lincoln at the annual dinner of the New York Republican Club in 1900. After some difficulty, he summoned the Muse, and the result, while it did not meet with the instantaneous acclaim of the earlier poem, attained comparable fame in the centennial year. Taking his cue from Lowell, Markham envisions Nature's moulding Lincoln from the common clay.

She took the tired clay of the common road—
Clay warm yet with the genial heat of Earth,
Dashed through it all a strain of prophecy,
Then mixed a laughter with the serious stuff.

Markham uses a number of natural metaphors—"patience of the rocks," "gladness of the wind," "justice of the rain that loves all leaves"—to evoke "the Captain with the mighty heart" who held up the ridgepole when the earthquake shook the house.

> He held his place—
> Held the long purpose like a growing tree—
> Held on through blame and faltered not at praise.
> And when he fell in whirlwind, he went down
> As when a kingly cedar green with boughs
> Goes down with a great shout upon the hills,
> And leaves a lonesome place against the sky.

No natural metaphor was more favored by poets than that of a tree—solitary and wind-beaten, typically "a gaunt, scraggly pine."[69]

The hopes of many that the Centennial would be crowned with a great monument to Lincoln in the nation's capital were unfulfilled. The movement in that direction was irresistible, however. Continuing dissatisfaction with the monument in Springfield had prompted efforts to revive the defunct National Lincoln Monument Association. In 1886 Senator Cullum, of Illinois, introduced legislation for an appropriation of $100,000 for each of ten years to erect a fitting national memorial. Cullom's idea was a columnar monument east of the Capitol, on the axis with the Washington Monument, and it assumed the erection of a similar monument to Ulysses S. Grant on a cardinal point north of the Capitol. Nothing came of this initiative. In 1901 the McMillan Park Commission, appointed by Congress, recommended the extension of the Mall one mile west to the Potomac and the erection of a Lincoln memorial at that point. (Another key to the McMillan Plan was the tunneling of railroad tracks under the Mall.) For seven years the plan lay dormant. Part of the problem was congressional indecision between competing plans for a memorial to Lincoln, one calling for a utilitarian monument in the form of a "Lincoln Way" connecting Washington and Gettysburg, another for a monument at or near the new Union Station north of the Capitol, and, of course, the Potomac Park monument. The upshot was the establishment of a Fine Arts Commission, which would offer expert advice on site and design. These questions were still unsettled two years later when Congress enacted Cullom's bill, a quarter-century after his original one, for a Lincoln Memorial Commission.[70]

ALTHOUGH THE CENTENNIAL aimed to be a festival of American patriotism, it was reluctantly observed, at best, in the former Confederate states. In Texas and Arkansas, though, Lincoln's birthday was observed for the first time; in Memphis and Little Rock commissions were appointed to plan commemorative events; in Birmingham schoolchildren were introduced to Lincoln; and in North Carolina the legislature adjourned for the anniversary. Ripples of praise ran across the South. The most impressive ceremony occurred in Atlanta. Union and Confederate veterans joined in services at the Trinity Methodist Episcopal Church, South. The commander of the United Confederate Veterans offered prayers, and a retired general of the United States Army, who had been with Lincoln at Gettysburg, read the immortal address. The Reverend James W. Lee, a son of the Confederacy, and the church's pastor, delivered an address in which he attributed the Union triumph to divine favor and thanked God that the soldiers both in blue and gray were now united on earth as in heaven and together regarded "the martyred president their commander-in-chief to all eternity." Such an event stands out because of its rarity, however. Across the South, Lincoln's one hundredth anniversary was better remembered among Negroes than in the dominant white community.[71]

The southern historian J. G. de Roulhac Hamilton, assessing "Lincoln and the South" in the centennial year, said that the "peculiar possessive affection" that had deified him in the North was to be found nowhere in the South, nor was it likely to arise, yet southerners increasingly recognized his greatness of heart and acknowledged he had never been their enemy. Hamilton distinguished three phases in the evolution of the southern attitude: first, Lincoln as "Black Republican"; second, from the moment of his assassination through Reconstruction and beyond, the Lincoln whose loss was the South's deepest regret, but for whom it still had no liking; third, the Lincoln of the New South generation, admired for his southern birth and blood, for his democracy, even for his nationalism, since it no longer threatened the South and its traditions. This third Lincoln found its sinister expression in Thomas Dixon's novels and its benevolent expression in Watterson's writings and addresses.

The Kentucky editor delivered another paean to the Doric hero of the Union in his oration dedicating the statue in Hodgenville. "All of us now are Unionists," he declared. The South as well as the North could participate in the exaltation of the log-cabin boy who became President and who was the divinely appointed savior of the Union. Lincoln found place as a southerner in the biographical volume of the encyclopedic *The South in the Building of the Nation*, which appeared in 1909. A non-slaveholding emigrant, he retained a sympathetic understanding of the country and its people, it was said, and despite emancipation never supposed the two races could be melded in common citizenship.[72]

Among the memorable southern tributes to Lincoln was Georgia-born

Maurice Thompson's Phi Beta Kappa poem, "Lincoln's Grave," at Harvard in 1894. He begged the privilege of one who fought for the Confederacy to sing at Lincoln's grave and hail his godlike humanity.

> He was the North, the South, the East and West,
> The thrall, the master, all of us in one.

He set freedom free, making real what had been a dream, and now his words are whispered by oppressed peoples everywhere.

> His was the tireless strength of native truth,
> The might of rugged, untaught earnestness;
> Deep-freezing poverty made brave his youth,
> And toned his manhood with its winter stress
> Up to the timbre of heroic worth,
> And wrought him to a crystal clear and pure,
> To mark how Nature in her highest mood
> Scorns at our pride of birth,
> And ever plants the life that must endure
> In the strong soil of wintry solitude.[73]

Some southerners, like the Populist and racist Tom Watson, were still trapped in the second of the three phases Hamilton described, while still others had yet to emerge from the shadow of the Black Republican image. Lyon G. Tyler, son of John Tyler and president of the College of William and Mary, was one of these. Responding to a *Baltimore Sun* feature that asked the question "What Do the Southern People Think of Lincoln Today?" Tyler replied that they were too polite to speak unkindly of him on his one hundredth anniversary. "But," he continued, "it is due to frankness to declare that, in spite of what a few enthusiastic Southerners may say, the mass of the Southern people can never be brought to see Lincoln in any other light than that of a representative of a section of the country." The South could not concur in the North's regard for him as a statesman or its estimate of Lincoln's moral character.

Similar opinions were expressed by George L. Christian in an address before the R. E. Lee Camp of the Confederate Veterans, at Richmond, and in the *Confederate Veteran* magazine. In the pages of the latter, Lincoln continued to be portrayed as a vulgar buffoon, a hypocrite in religion, and a purveyor of smutty jokes. He was guilty of crooked and duplicitous conduct in the provisioning of Fort Sumter; he was a wartime dictator; his Emancipation Proclamation violated his pledge not to interfere with slavery, and with it he sought to incite servile insurrection. Far from being a kind and generous ruler, he "Shermanized" the South. "What, then, has been the basis of all the fictitious greatness?" another Lincoln-hater asked. "We answer: Assassination. Assassination placed the crown of the martyr upon his brow. Henceforth, 'all things unclean become divine.' "

Such writing drew upon deep veins of bitterness and hatred. It assumed its most respectable form in Charles L. C. Minor's *The Real Lincoln,* published at Richmond in 1901, and in three revised editions over the next twenty-eight years. With chapter titles like "Was Lincoln Heroic?" and "Was Lincoln a Christian?" Minor educed an array of testimony, from Lamon, Herndon and Weik, Donn Piatt, and others, leading to a foregone conclusion. The book was deemed sufficiently dangerous to be removed from several public libraries in Massachusetts. A more poisonous rehash of the same materials was the pseudonymous *Facts and Falsehoods Concerning the War in the South,* published at Memphis in 1904. Here the apotheosis was seen as a staged affair managed by the Radical Republicans to consolidate their power. And sad to say, many in the South had fallen under the spell of this deified Lincoln.[74]

The South had its own heroes. The year 1907 marked the centennial of General Robert E. Lee. Thenceforth his birthday, January 19, was observed from Virginia to Texas. Whenever the North should join in that observance, challenged one unreconstructed rebel, it would be time enough for the South to honor Lincoln. A surprising advocate of the desouthernization of Lee stepped forward. He was Charles Francis Adams, a bred-in-the-bone Yankee of honorable lineage who had captained a Union cavalry regiment during the war and now presided over the Massachusetts Historical Society. In 1902 he addressed the question "Shall Cromwell Have a Statue?" and offered the example of the massive bronze erected to the great Puritan leader in front of Westminster after the passage of two centuries in a plea for a national monument to Lee. Lee had been great in war and great in peace. Although technically a traitor, he was entirely justified, Adams thought, in choosing to go with Virginia in 1861. If he was a traitor, so were Cromwell and Washington, and so, Adams hoped, would he have been if Massachusetts had been the seceding state. In closing, he extended his imagination to the distant future: "The bronze effigy of Robert E. Lee, mounted on his charger and with the insignia of Confederate rank, will from its pedestal in the nation's capital gaze across the Potomac at his old home at Arlington, even as that of Cromwell dominates the yard of Westminster upon which his skull once looked down." The perversity of the Adamses was legendary, and one scarcely knew whether to take this proposal seriously or whimsically.

Speaking on Lincoln at Baltimore in 1909, Woodrow Wilson remarked that not long before he had addressed a southern audience on Lee, and curiously, he observed, "there seemed to be no conceivable incompatibility between the two tasks now, in this generation." Perhaps, for General Lee was redeemed by his virtues, above all by his course of conduct from Appomattox to his death; but history was badly misread and the deserts of fame miscast by the suggestion that he belonged with Lincoln in the American pantheon. Better by far, certainly, that it would be Lincoln, from his memorial, who gazed across the Potomac to Arlington.[75]

5

Themes and Variations

THE IDENTIFICATION OF American ideals with the Allied cause in the First World War gave a new dimension to Lincoln's fame. It was internationalized. The image of the Great Emancipator as the secular Savior of Humanity, although its roots were in the world-wide outpouring of sympathy upon his death, now took definite form. The hero's national fame grew apace. The dedication of the Lincoln Memorial—temple and icon—in 1922 was a tribute to the place he had attained in American hearts and minds; and with such a memorial, his fame could only increase.

Year after year more printer's ink was spilled upon Lincoln than any other figure in history except Jesus Christ. A new bibliography in 1925, the work of the collector Joseph P. Oakleaf, added 1,576 titles—books and pamphlets only—to the number recorded by Daniel Fish in 1906. Lincoln was an inexhaustible subject, and in the quest for novelty every remote relation of his life became the theme of a book or essay. It was a joke in the publishing business that the title of a surefire best-seller would be "Lincoln's Doctor's Dog" because of the combination of favorite subjects. (To make the point, Christopher Morley later wrote a whimsical story with that title.) The stream of reminiscence subsided, though it was not yet dry. Julia Taft Bayne, who had played with the younger Lincoln boys in the White House and, of course, observed "that strong angel of the Lord," the President, wrote a charming memoir. In 1921 a reporter for the *New York Times* turned up in a pauper's home the man, Thomas Proctor, in whose bed, it was claimed, Lincoln died, and who with Robert Todd Lincoln was the lone survivor among the mourners at the martyr's deathbed. The front-page story was promptly disputed by Pauline Wenzing of Baltimore, daughter of William Peterson, the house's proprietor. She insisted the President had expired in her bed.[1]

The leading figures in Lincoln biography in the decade following the war were William E. Barton, Albert J. Beveridge, and Carl Sandburg—a preacher, a politician, and a poet. It remained a crowded field, however, and one into which academic historians ventured for the first time. Radio interpreted Lincoln to a mass audience, and his life was the subject of major motion pictures. Some of the old mysteries, such as Lincoln's ancestry, were unraveled, albeit not to everyone's satisfaction. Some of the old themes, such as Lincoln's religion, received unusual attention. In the "dry decade," prohibition was the foremost political issue drawing upon the power and authority of the Lincoln symbol. Southern feelings toward Lincoln continued to mellow, but the process was interrupted by outbursts that showed the persistence of the Confederate legacy. H. L. Mencken, the country's favorite cynic, reflected on the awesome power of the Lincoln legend: "He becomes the American solar myth, the chief butt of American credulity and sentimentality." Compared to him George Washington was a humanized hero. Although he joked a lot, every likeness showed Lincoln wearing the expression of a man about to be hanged. The hagiographers had pumped everything human out of him, leaving a plaster saint, "a mere moral apparition, a sort of amalgam of John Wesley and the Holy Ghost . . . fit for adoration in the chautauquas and Y.M.C.A.'s."[2]

The fixity of the image was amazing, yet to a degree it showed the chameleon-like tendency of all cultural symbols to reflect changes in the surrounding environment. In a decade when the President declared "the business of America is business," Lincoln offered a model for success in business as in life. He was the world's greatest Horatio Alger story. In the language of a college student's theme, "Lincoln was born in a log cabin that he built with his own hands." A little homily called "The Failure" circulated in popular magazines. In some 140 words it recounted the defeats Lincoln had suffered in politics: first beaten as a candidate for the state legislature; later denied a second term in Congress; rejected as his party's nominee, then defeated, in the race for the Senate; and so on. It concluded: "How would you stand in the face of such setbacks? Think it over!" A spurious apothegm, variously called "Lincoln's Prescription" and "Lincoln's Rule for Living," repeatedly turned up in the press. "Do not worry, eat three square meals a day, say your prayers, be courteous to your debtors, keep your digestion good, exercise, go slow and go easy," ran one version. Unsurprisingly, it was featured in advertisements for a medicine to cure indigestion.

The commercial exploitation of Lincoln's name began in earnest in 1905 with the founding of the Lincoln National Life Insurance Company in Fort Wayne, Indiana. Its founders requested and received from Robert Lincoln permission to use the name along with his father's likeness for the company's logo. The company tastefully featured Lincoln in its advertising. "How fast can you name a tunnel, a president, and the tenth largest life insurance company in America?" one ad asked. "Lincoln. It's a name you remember." Of course, the name conveyed *honesty* above all else. Its use was ubiquitous among banks. But several generations of children were initiated in the log-cabin tradition by Lincoln Logs, a building toy, while the automobile manufactured by the Lincoln Motor Company became "the car of the Presidents."[3]

In the crusade of Americanism after the Great War, Lincoln was often held up as the standard of civic education. "Let American High Schools teach at least one year of Lincoln," proposed a judge on the Ohio Supreme Court. "It is high time he became a staple of American education." His writings and speeches furnished the texts of Americanism. They had been edited and published for use in the schools since the 1890s, but the number of such volumes rapidly increased as the call to "Lincolnize America" sounded across the land. Charles Evans Hughes called upon the nation's colleges to offer courses of instruction on Lincoln. America's supreme need was "the reenforcement of character, the education of the soul," he declared when Secretary of State.

> For that purpose we all need a course in Lincoln. In making Lincoln the exemplar of the Nation, we are not merely recognizing heroic service but we are safeguarding our most vital resource. His love of the plain people, of whom he was the most distinguished representative, his tender sympathy and patience, his clarity of thought, tenacity of purpose and the essential fairness and understanding which dominated every action make him the embodiment of the democratic ideal. So long as we cherish Lincoln's principles and so long as his virtues inspire our youth, our security and progress are assured.

The only institution to make a point of incorporating the study of Lincoln into the curriculum was Lincoln Memorial University in the Tennessee mountains. There, it was said, Lincoln—his example, his words, his virtues and ideals—were made part of the conscious heritage of every student. The teaching apparently upheld the tenets of orthodox Republicanism. When in 1930, according to Chancellor John Wesley Hill, a professor "began unloading intellectual commodities incongruous with the trademark of Lincoln," he was fired forthwith. This provoked a student strike, followed by the firing of two additional professors, after which calm was restored "under the dominion of the principles of Lincoln." Immigrant children, in particular, might find inspiration in Lincoln's life. Michael Pupin, the Hungarian-born physicist, who at sixteen landed at Castle Island with five cents in his pocket, offered witness to that inspiration in his prizewinning autobiography in 1923. Pupin was among those, with John Dewey, Ida Tarbell, and others, who announced plans to found an international university in Lincoln's name in Westchester County, New York, with the stated mission to imbue the world's youth with "Lincoln-mindedness." The Wall Street Crash four months later doomed this audacious venture.[4]

The International Lincoln

The ghostly apparition, conjured by Vachel Lindsay, pacing up and down the streets of Springfield, carrying on "his shawl-wrapped shoulders" the bitterness, folly, and pain of the world, evoked the spirit of Lincoln as the apostle of brotherhood.

> He cannot rest until a spirit-dawn
> Shall come;—the shining hope of Europe free:
> A league of sober folk, the workers' earth,
> Bringing long peace to Cornland, Alp and Sea.

President Wilson shared "the shining hope"; increasingly, he experienced what he called "the holy and very terrible isolation" Lincoln had known in the wartime White House. Parallels between the Civil War and the World War were frequently drawn, both at home and abroad. In the leading paragraphs of the *London Spectator*, for instance, Lincoln's advocacy of the draft was urged in behalf of compulsory service (legislated in 1916) in Britain. The *Spectator* even printed a long and unpublished presidential defense of the draft in 1863. Lincoln's rejection of a negotiated peace at the Hampton Roads Conference was offered in partial justification of the Allied rebuff of the German overture in December 1916. The Vallandingham case became a leading precedent for Wilson's crackdown on dissenters and obstructionists. His proclamation of "a peace without victory" was compared to Lincoln's "charity for all" in 1865. The veteran Republican leader Joseph Choate believed it was the spirit of Abraham Lincoln that led the United States into the war. "If Lincoln were here today," Choate said in 1917, "his prayer [in the Gettysburg Address] would be verified and glorified into the prayer that all civilized nations shall have a new birth of freedom, and that government of the people, by the people, and for the people shall not perish from the earth."[5]

Allied leaders recognized the propaganda value of the Lincoln symbol. French Marshal Joffre, chairman of the Allied War Council, led a delegation to Springfield to render his country's homage to Lincoln just after the American declaration of war. For David Lloyd George, the British Prime Minister, Lincoln had been a personal hero long before he became a vehicle for propaganda. Lloyd George often told how his father, a Welsh shoemaker, acquired a portrait of Lincoln after his death and accorded it the place of honor in the family's humble home. Young David read and re-read the story of Lincoln's life; at every turn in his career, he took guidance from it. When the war came, he studied Lincoln's leadership in the American conflict. Rejecting the peace overture in 1916, the Prime Minister quoted Lincoln's own words: "We accepted the war for an object, a worthy object. The war will end when the object is attained. Under God, I hope it will never end till that time." In February 1917, Lloyd George sent a special Lincoln Day Message to the American people. Here he maintained that the issue at stake in the European war was basically the one that had divided the Union some fifty years before. "Has there not grown up in this continent a new form of slavery, a militarist slavery, which has not only been crushing out the freedom of the people under its control, but which in recent years has also been moving toward crushing out freedom and fraternity in Europe as well?" Regrettably, in the Civil War Britain's leaders had been blind to the issue. Now, inspired by Lincoln's example, the Allies fought to secure his democratic principles in Europe.

Around the conference table at Versailles in 1919, Lincoln was a favorite topic of conversation among the Big Three: Wilson, Lloyd George, and Clemenceau. Both the European leaders later made pilgrimages to Lincoln's tomb. Clemenceau, in a moving tribute, recalled his grief as a student in Paris upon hearing of Lincoln's death, and said the world followed him in the path he had so gloriously opened. No sooner did Lloyd George step off the boat in New York in October 1923 than he struck out for Springfield. Much as he wanted to see the country, he explained, Lincoln's home was the one spot he desired to see above all others. He had long been under Lincoln's spell, and having come through a great war, he felt an even stronger bond with him. He was, Lloyd George said, "the tenderest soul who ever ruled men, and the democracy for which he stood was now the hope of mankind."[6]

Lincoln's fame in Europe, consolidated during the World War, had been growing for a decade or longer. What the English thought of him was important to educated Americans, since many of them imported their opinions, like their china, from England. And so when a gentleman-historian, Sir Spencer Walpole, declared that Lincoln deserved the highest place among nineteenth-century statesmen, Henry Cabot Lodge, an American of the same stripe, was impressed. *Lincoln's Speeches and Letters*, with an introduction by James Bryce, the British Ambassador to the United States, appeared in the Everyman's Library in 1907. The same year saw the publication in London of Henry B. Binns's biography of Lincoln. (An Everyman's edition appeared in 1927.) Lincoln's life and writings became readily accessible in England for the first time. The Binns biography, otherwise insignificant, contained a critical assessment of the British government's sympathy for the Confederacy. Such books were harbingers of Lord Charnwood's *Abraham Lincoln*, published in London in 1916. It proved a milestone in Anglo-American relations as well as in Lincoln biography.[7]

Godfrey Rathbone Benson, Lord Charnwood, was born a commoner in 1864 and educated at Balliol College, Oxford. He grew up in Litchfield, Dr. Johnson's hometown, and later served as its mayor for a time. As a boy at school he had contracted an enthusiasm for Lincoln which, in retrospect, he found hard to account for, but supposed it had something to do with the ideal of manhood—honesty, humility, generosity, "the Christian graces"—represented by Lincoln. At any rate, he later read John T. Morse's biography, and when the war came, although he had never undertaken a literary work before, agreed to write a life of Lincoln for a series, "Makers of the Nineteenth Century," published by Constable and Company. According to Lady Charnwood, who presided over stately Stowe House in Litchfield, it was "conceived as a war service" and "written in the midst of a world's agony."[8]

The author acknowledged that his book added nothing previously unknown to the Lincoln story, and he apologized to his American readers for the large amount of general historical information he had included for the benefit of his English audience. He did not disguise his admiration for Lincoln; indeed he called the biography a "tribute" to him. Nor, on the other hand, did he gloss over his hero's faults and foibles. He believed that great men are more dignified off their pedestals than on. He accepted Herndon as Lincoln's Boswell and credited his veracity. Whatever may have been the faults of Lincoln's character, Charnwood philosophized, thanks to Herndon the world soon learned the worst of it. He was eccentric in viewing Lincoln's character as essentially English—a village type—rather than peculiarly American. Lincoln illustrated the paradoxes that genius is simple rather than complex and that the highest truths inhere in the commonplace. Charnwood was deeply impressed with the bravery, clarity, and power of Lincoln's utterance in the debates with Douglas and the Cooper Union Address. "How rare it is," he marveled, "for statesmen in times of crisis to grasp essential truth so simply."

In 1861 Lincoln saw clearly that secession not only broke the Union but threatened the American experiment in democratic government, which was, said the author, "the most hopeful agency for uplifting man everywhere." And this, of course, was Lincoln's highest claim to the interest of mankind. Charnwood imbibed the skepticism current in English opinion on Lincoln's military administration, yet concluded, on balance, that it was a brilliant achievement. The President sacked General McClellan after Antietam, Charnwood observed, less because he let Lee escape, which was forgivable, than because he had no regrets, which was not. On the matter of slavery and emancipation, Lincoln's course was marked by both candor and consistency. With many of his compatriots, Charnwood ranked Lincoln with Shakespeare in mastery of the English language. Of the Second Inaugural he wrote, "Here is one of the few speeches ever delivered by a great man at the crisis of his fate on the sort of occasion which a tragedian telling his story would have devised for him." In other words, it might have been written by Shakespeare.[9]

Charnwood's biography of 479 pages was the most readable that had ever been written on the subject. It was informative, thoughtful, and discerning; above all, it had literary grace and flair. In the *American Historical Review*, where its errors and thinness might have been expected to command attention, the book was hailed an instant classic. Reviewers were amazed that an English lord should understand Lincoln so well. William Dean Howells, who appraised the work in *Harper's* fifty-eight years after writing his campaign biography of Lincoln, thought that this must be because Charnwood became a titled aristocrat only in 1911. Howells, Morse, and others underscored the path-breaking significance of the book both in carrying American history beyond the nation's borders and in providing belated English recognition of the achievement of American democracy and its destiny in the world. Charnwood visited the United States in the fall of 1918 for the first time since his youth. He was genuinely astonished by the book's success in this country. His principal public appearance was in Springfield, where he delivered an address at the dedication of a statue of Lincoln on the State House grounds. After returning home the noble lord wrote some desultory notes on Lincoln but never followed up on his literary success. The book went through countless editions before it was finally superseded as the most-read biography by Benjamin P. Thomas's *Abraham Lincoln* in 1952.[10]

Charnwood's work directly influenced John Drinkwater's play, *Abraham Lincoln*, gratefully dedicated to him. This was the first successful dramatization of Lincoln's life for the stage. He had appeared incidentally in such plays as Thomas Dixon's *The Clansman*. Benjamin Chapin, six feet four inches tall and strikingly Lincolnesque in his stage makeup, had for over thirty years been impersonating Lincoln in dramatic monologues written by himself. Although welcomed as a wholesome antidote to *The Clansman*, Chapin's one-man show, *Lincoln at the White House*, seemed never to get under the surface of the makeup and struck at least one avid Lincolnian as "akin to sacrilege." Paramount Pictures, in 1917, released a series of ten "photodramas," *The Son of Democracy*, featuring Chapin.[11]

After its opening at the Birmingham Repertory Theatre in 1918, Drinkwater's drama was brought, not to London, which spurned it, but to a dingy playhouse in the suburb of Hammersmith. It was a spectacular success. "All Mayfair went to see it. Hammersmith became a nightly pilgrimage for the West End," it was said. Written as a chronicle play in six episodic scenes, it was far from great dramatic literature. But it was carefully wrought and stirred the audience. Drinkwater, a thirty-six-year-old author, sought to portray Lincoln under the duress of war. He employed a classic chorus speaking in irregular rhymed verse to enhance the tragic dimension of the drama.

In the first scene, at home in Springfield in 1860, a deputation calls upon Lincoln to urge him to seek the Republican nomination for the presidency. (This was one of many liberties the playwright took with the historical record: Lincoln aspired to the presidency and needed no encouragement to run.) Mary has greeted them, and she scolds her husband when he enters in a disheveled state. The talk is of the burgeoning sectional crisis. Recalling what he said about slavery when he first observed it in New Orleans ("If I ever get a chance to hit that thing, I'll hit it hard"), Lincoln is at once placed on the side of emancipation. Scene 2 focuses on the Sumter crisis and the showdown with Seward. Drinkwater could not resist the temptation to introduce into the dialogue the elevated language of Lincoln's public speech. Thus the close of the First Inaugural is spoken in conversation with Seward, which must have sounded doubly strange in the mouth of an actor, John Rea, with an Irish brogue. Between scenes the chorus intones:

> Two years of darkness, and this man but grows
> Greater in resolution, more constant in compassion.
> He goes
> The way of dominion in pitiful, high-hearted fashion.

Scene 3 occurs at the nadir of the Union cause. In it Lincoln bares his heartbreak to a group of female callers. In Scene 4 he presents the Emancipation Proclamation to the cabinet, after reading from Artemus Ward. Scene 5 places the President in a farmhouse near Appomattox where "Bob Lee" is about to surrender. While there he pardons a soldier sentenced to die for sleeping on duty. The last scene occurs in Ford's Theater. After delivering to that audience the immortal passage of the Second Inaugural Address, the President is shot; and Secretary of War Stanton, who is in the box with him, utters the last word: "Now he belongs to the ages."[12]

Such a hasty summary does the drama a disservice. It played well. Had Lincoln been thus exposed to an English audience before the war, one critic said, he would more likely have met with ridicule than with understanding. But in 1918–19 his tragedy was theirs.

> He hit us in our historical Puritan's wind. He seemed to incarnate our purpose, our usefulness, our sacrifice. . . . This nobility—was it not ours? This man of government of the people for the people by the people—was this not the "new order" promised by our politicians, nay actually being made in Paris by the peoples' representatives? And so Lincoln became the stuff our dreams are made of. . . . Lincoln caught the castigated soul of London at the hour of its release.

The play had the effect of a confessional. It was a national triumph and a national purification.[13]

Drinkwater's *Lincoln* opened at the Cort Theater on Broadway, December 15, 1919. (Simultaneously, lest there be any doubt about the play's message, the author published *Lincoln World Emancipator*, which portrayed the statesman as the ideal of the English-speaking race and appealed for Anglo-American union to extend his influence around the world.) The English triumph was repeated. *Abraham Lincoln*, Burns Mantle said, "is easily the most inspiring dramatic success of our time." Playing Lincoln, Frank McGlynn, heretofore unknown, had the role of a lifetime and made the most of it. He was awed by the responsibility. In his recollections, he said all the actors approached the play with "profound reverence," as though they were in a church instead of a theater. McGlynn became something of a student of Lincoln, pondering such questions as the authentic enunciation of the "of, by, and for the people" line of the Gettysburg Address. The only criticism he received on the accuracy of his portrayal, he said, was that he mistakenly "scarfed the shawl." Both Robert Lincoln and William E. Barton made the point.

Later, at the Chicago Historical Society, where the shawl was, Barton took pains to show McGlynn how it was worn.*[14]

Everybody—ambassadors, senators, professors, writers, and artists—saw *Abraham Lincoln*, and many were led to reflect more deeply than before upon the mystery of the man's character. Columbia University President Nicholas Murray Butler, recalling the playwright's line "Lonely is the man who understands," wondered if this was not the key to Lincoln's character. He had the gift to look deeply into the hearts and minds of his fellows. "Abraham Lincoln was lonely because he understood. Here is the secret to the pathos of the man."

Herbert Croly, editor of the *New Republic*, took as his point of departure Lincoln's chastisement, in the play, of a character named Mrs. Blow for speaking vengefully of rebels and pacifists. Paradoxically, Lincoln could fight and prevail without hatred and fanaticism. Contemporary statesmen had been unable to accomplish that. Clearly, the editor continued, it was not because Lincoln was simple or common. He was neither as these words were generally understood. His simplicity was an art, "an integrity of feeling, mind, and character which he himself elaborately achieved, and which he naturalized so completely that it wears the appearance of being simple and inevitable." Crusaders and ideologues who thought to find the truth of Lincoln in some particular cause or idea missed the point. He broke out of all the political breastworks. He was a spiritual and intellectual force. He was the hero of democracy, not from commonness, said Croly, but because he embodied "consummate personal nobility."[15]

The World War sparked renewed interest on both sides of the Atlantic in Lincoln as Commander in Chief. Whether or not he had burned the midnight oil studying Clausewitz's *On War*, it was often said that he had, and this led him to view warfare as an extreme form of political action, thereby requiring close collaboration of civil and military authorities. Nicolay and Hay had stressed the dominance of political considerations in Lincoln's strategic thinking, and, although unversed in military matters, the authors agreed with him. "War and politics, campaign and statecraft, are Siamese twins, inseparable and interdependent; to talk of military operations without the direction and interference of an Administration is as absurd as to plan a campaign without recruits, pay, or rations."

---

*Ten feet long, of a light brown color, the wool shawl was first laid over the left shoulder and dropped down below the left knee; the upper part was then draped across the back and thrown over the right shoulder, while the lower part was brought upward across the chest and the right shoulder; a pin held the shawl in place at the chest.

This was not the opinion of the Comte de Paris, who said Lincoln's interference brought "frightful disasters" on his armies.

The same opinion entered English military thinking through the influence of George F. R. Henderson's *Life of Stonewall Jackson*, in 1898, powerfully reinforced by British General Lord Wolseley's Introduction to the second edition of that work. One of the great lessons of the American Civil War, they maintained, was that civil authorities should keep hands off commanders in the field. Lincoln was offered as the bad example. Thus McClellan would have captured Richmond in 1862 had he been left alone. As for the picture of the President becoming a brilliant strategist by poring over the pages of Clausewitz and Jomini in the White House, Henderson sighed, "if it were not pathetic, it would be ludicrous." While Lincoln presumed to command, the Confederacy was victorious; when he abdicated to Grant, the Confederacy was defeated.[16]

This verdict was upset in the rethinking brought on by the World War. In 1917 Captain Arthur L. Conger, an expert in military science, delivered an address, "President Lincoln as War Statesman," before the Wisconsin Historical Society, that was deemed "revolutionary" in its estimate of Lincoln as well as in the place accorded to civil authorities in grand strategy. Conger saw Lincoln's hand everywhere, even in Grant's final campaign; it was, moreover, a brilliant hand. He predicted that Lincoln's fame as a warrior would, in time, match his fame as a statesman. The prediction seemed fulfilled in 1926 by the simultaneous publication of two books in England. In *The Military Genius of Abraham Lincoln*, Brigadier General Colin R. Ballard made a sweeping case. The blockade of the Confederacy in 1861 showed "true strategical foresight"; the President's decisions in the advance on Bull Run, in the Peninsular Campaign, in the West were unfailingly right. In fact, Ballard thought, he abdicated too much responsibility to Grant in 1864. Even the Emancipation Proclamation was a stroke of military genius. As a war measure it illustrated the "Higher Command," which Lincoln practically invented and which Britain and the Allies were slow to emulate partly because of the unfortunate influence of Henderson and Wolesley. Sir Frederick Maurice, who had been Director of Military Operations of the Imperial General Staff, made much the same point in his book *Statesmen and Soldiers of the Civil War*. Lincoln, the civilian, appeared as a model war executive, while Jefferson Davis, the soldier, was a failure. Lincoln showed how a democratic nation could conduct a great war. He held the nation to its purpose, and this was as great a victory as any on the battlefield.[17]

In 1926 the World Federation of Education Associations conducted a poll among students of high school age in the United States and Europe on the world's greatest heroes. Lincoln finished second behind Louis Pasteur. But Lincoln's international fame reached beyond the Western world. Sun Yat-sen, the Chinese revolutionary, was an ardent admirer of the famous American, and he made "government of the people, by the people, and for the people" the cardinal principles of his creed. His first rendering of these words in Chinese (Min yu, Min chih, Min hsiang), turned back into English, was "The people are to have, the people are to control, the people are to enjoy." After further consideration he came up with the formula of the "Three People's Principles: Nationalism, Democracy, Livelihood." A five-cent stamp issued in 1941 to commemorate the fifth year of Chinese resistance to Japan coupled the portraits of Lincoln and Sun Yat-sen and incorporated the three principles. The first book on Lincoln in the Japanese language appeared as early as 1890. In 1926, and for several years, the American-Japan Society sponsored a Lincoln essay contest for pupils at two levels of Japanese education. The judging was done by the Abraham Lincoln Association in Springfield. Lincoln's birthday seldom passed unnoticed in Japan. His story was taught in the schools. To the Japanese it was, quite simply, "the greatest story" in American history. Even the Emperor, it was said, worshiped Lincoln.[18]

Temple and Icon

On March 4, 1911, the Lincoln Memorial Commission, created by act of Congress two months earlier, held its first meeting in President Taft's office at the White House. The three senators and three representatives, including the two veteran Illinoisans Shelby Cullom and Joseph Cannon, Speaker of the House, elected Taft chairman and decided to seek the advice of the Fine Arts Commission on a site. That body, of course, was already committed to the site commonly known as Potomac Flats. In the daring McMillan Plan, it would be upgraded to a park and become the western terminus of the Mall, the third cardinal point on the line from the Capitol through the Washington Monument, in a composition potentially grander than the avenue between the Tuileries and the Arc de Triomphe in Paris or St. Paul's and Buckingham Palace in London. John Hay, before his death, had pleaded for a Lincoln monument on this site. He said in substance:

> As I understand it, the place of honor is on the main axis of the plan. Lincoln of all Americans next to Washington deserves this place of honor. He was of the Immortals. You must not approach too close to the Immor-

> tals. His monument should stand alone, remote from the common habitations of man, apart from the business and turmoil of the city; isolated, distinguished, and serene. Of all the sites, this one near the Potomac is most suited to the purpose.

Cannon favored a site in Arlington and vowed, "So long as I live I'll never let a memorial to Abraham Lincoln be erected in that God damned swamp." There was also support for the Soldiers' Home site in northwest Washington. And Champ Clark, another member of the commission, still favored making the memorial a work of transportation—a parkway to Gettysburg—rather than a work of art.[19]

In July the Fine Arts Commission formally recommended the Potomac site, and it was soon adopted. Thereafter, the two commissions worked closely together. The arts body recommended Henry Bacon as architect of the memorial. A native of Illinois who rose to prominence with his designs for the Chicago World's Fair in 1893, Bacon was a proponent of Greek classicism. Because of his friendship and association with Daniel Burnham, earlier head of the McMillan Park Commission and now chairman of the Fine Arts Commission, Bacon had the inside track for this brilliant opportunity. The Memorial Commission chose not to conduct a competition. The younger classicist, John Russell Pope, architect of the birthplace memorial, was invited, along with Bacon, to submit his design. Burnham gave the nod to Bacon, and in December 1912 the commission adopted and recommended his final plan to Congress. Highway and realty interests made a last stand for the Gettysburg road. "For Lincoln a finished memorial is not a fit memorial" was their slogan. Sundry bills had been introduced in Congress for alternate ways of commemorating Lincoln. One called for raising a memorial arch over Pennsylvania Avenue, while others looked to alternate "living memorials," such as a national vocational school. But without further ado Congress, on January 29, 1913, adopted the Memorial Commission's recommendation and voted start-up funds.[20]

It required a leap of faith in 1913 to envision the Lincoln Memorial on that site. No wonder there were skeptics. Nor was it surprising that Bacon's design should meet with criticism. "How Lincoln Would Have Laughed," ran the title of a story in the *Independent*. The planned memorial was both "a public confession of architectural insolvency" and "a bare-faced contradiction" to the life, character, and purposes of Abraham Lincoln. "Our national capital has Washington as a Roman general. Let us not add the more atrocious anachronism of Lincoln as Apollo." Instead of this course in compulsory Greek, a truly creative architect would design a monument both fitting and indigenous. "He would take the log cabin and rail fence of Lincoln's birthplace and transmute them into an edifice so glorious and beautiful that generations afterward men would admire and imitate it." Lincoln, the nation's tenderest memory, deserved better than an academic pile of marble. "In heaven's name," exclaimed Gutzon Borglum, the sculptor, "in Abraham Lincoln's name, don't ask the American people to associate a Greek temple with the first great American."

Grant LaFarge undertook to answer the critics. Greek *was* compulsory because nothing else remotely equaled it in memorializing immortals. What would you have? Perhaps "an experiment station for the uncultured commemoration of Lincoln's mere personal attributes." It is not for these—his poverty and pioneering and Rabelaisian humor—that he lives resplendent in the minds of men. He is a hero beyond time and place. Did Lincoln not embody tradition in his literary style? LaFarge asked. The Gettysburg Address is Doric. Bacon's design may recall a Greek temple, but that is only by implication, and because no other possesses the dignity, gravity, universality, and beauty of this form. And what if it does recall a temple? Could there be any place more fitting to revere a great soul? Perhaps not, but the answer would never be known because the commission barred the door to competition. The controversy passed quickly, but it touched an emerging issue in Lincoln iconography.[21]

In 1914, again without competition, the commission named Daniel Chester French to execute the gigantic work of sculpture planned for the memorial. Since the death of Augustus Saint-Gaudens, French had been the nation's foremost sculptor. He had already shown what he could do with Lincoln in a noble bronze, dedicated in 1912, before the Nebraska state capitol. Lincoln is standing, hands clasped in front, in a meditative posture. Behind him is a huge engraved tablet of the Gettysburg Address, which fixes the time and the occasion. The sculpture invited comparison with Saint-Gaudens's great work in Chicago; and while not its equal, it ranked very high. This, and the fact that Bacon was his friend and his collaborator on the *Lincoln*, made French's choice well-nigh inevitable. He submitted a one-quarter scale model for the commission's approval in 1916. It was instantly acclaimed, but Bacon realized that the statue as planned was too small for the space it would occupy and proposed to increase its size by about one-third. At the same time, French asked that marble be substituted for bronze, as originally planned. The commission approved both changes.[22]

AS WORK WENT forward, slowed by the war, on the Lincoln Memorial in Washington, a great fight raged over a statue to be erected in London's Parliament Square. The source of contention was George Grey Barnard's giant bronze dedicated in Cincinnati on March 31, 1917. A $100,000 gift to the city from Charles P. Taft, it was formally presented by former President Taft, the donor's half-brother. The poet of the occasion hailed the statue as "a symbol of democracy." Barnard, who worked in New York, had heretofore been acclaimed for his visionary and allegorical sculptures. Lincoln was quite a departure for him. Writing about his conception, he said his child's mind had been impressed with one image of Lincoln: "the mighty man who grew from our soil and the hardships of the earth." Beginning with that, he went on to study and restudy Leonard Volk's life mask and casts of Lincoln's hands. From these he imagined the bearing and features of the figure. "He must have stood as the republic should stand, strong, simple, carrying its weight unconsciously without pride in rank and culture."

Riveted to this Whitmanesque conception—"Lincoln is our song of democracy," Barnard said—he then searched two years for a model. Finally, in Louisville, he found a man forty years of age, 6′4½″ tall, who

Lincoln, Cincinnati, 1917. By George G. Barnard. (The Lincoln Museum, Fort Wayne, a part of Lincoln National Corporation)

had been born only fifteen miles from Lincoln's birthplace and on his own testimony had split rails all his life. He even brought to the sittings his father's old broadcloth suit, shirt, and tie. The finished sculpture was both original and arresting. Standing naturally, and rising straight up over twelve feet, one gnarly hand clasped above the wrist of the other, and in rumpled homely attire, Lincoln seems about to speak. The face is beardless, thin, and wrinkled; the eyes gaze forward with quiet earnestness. The aim, clearly, was to create a plain and earthy Lincoln. Alas, many people felt that Barnard had only succeeded in creating a coarse and vulgar effigy of the man.[23]

The statue might have received little attention but for the fact that arrangements were set in motion to place a replica in London. With that it became an international scandal comparable to the earlier one over Rodin's monumental *Balzac*. Just before the war an International Commission to Celebrate the Hundred Years Peace between Great Britain and the United States had been formed. The Americans, on their part, had proposed presentation of a replica of the Saint-Gaudens *Lincoln*. The British accepted and offered a splendid site in Parliament Square for it. But the war intervened and the commemorative effort collapsed. It was next heard of in 1917 when Mr. and Mrs. Taft offered a replica of the Barnard *Lincoln* for London. Without funds or prospects, the American committee of the Centenary Commission gratefully accepted the gift, as did its British counterpart.

Robert Lincoln learned of this plan at the time of the Cincinnati dedication. He penned an indignant protest to President Taft, asking his good offices to put a stop to what he considered an abomination. The statue, he said, "is a monstrous figure . . . grotesque as a likeness of President Lincoln . . . defamatory as an effigy," and to erect it in London would be a terrible insult to his memory. He enlisted the support of Joseph Choate, a former ambassador like himself, Nicholas Murray Butler, and other influential friends. Barnard attempted to reason with him. "Your father belongs to future ages, and all sculptors of this generation and those to come, must have as their birthright, as children of Democracy and Art, full liberty to express their interpretations of the life of Lincoln." He emphasized his reliance on the Volk hands and face, deducing the *real* Lincoln from these artifacts; and he suggested that the seventy-four-year-old son simply did not remember the young, rough, beardless man who was his father.

Robert's response to this was to elicit the testimony of Springfield relics, like Clinton L. Conkling, who knew his father in Illinois, on the accuracy of Barnard's portrait. In June, *Art World*, a journal under the editorship of F. Wellington Ruckstuhl, took up the cudgels against Barnard. The statue that was supposed to symbolize democracy, in fact, symbolized "hobo-cracy." *Art World* was a voice of the arts establishment. Some younger and more independent artists, like Borglum, approved of Barnard's work, while such a conservative sculptor as Frederick MacMonnies, seeing the question as one of artistic integrity, defended him. But the attack snowballed. In the end, the whole establishment—the American Academy of Design, the American Institute of Architects, even the Fine Arts Commission—was aligned against poor Barnard.[24]

Why were tempers so hot on this issue? In part, because it went to the heart of which Lincoln, the plain western politician or the grave and dignified statesman, should be the nation's, and the world's, icon. In sculpture, certainly, Barnard's Lincoln was a radical departure. As far as it was anticipated by anyone, it was by Borglum in his the youthful work. He, too, had become a deep student of Lincoln's features and been impressed by his compelling western presence. He, too, thought that the dualities of Lincoln's personality were disclosed in the opposite sides of his face: the right strong and masculine, the left soft and feminine. He, too, thought that Lincoln was divined through Volk's mask and casts. Both artists, with Truman Bartlett, the foremost authority on Lincoln's physiognomy, inferred much from the hands.* In 1908 Borglum had exhibited a gigantic marble head of Lincoln—four times life-size—which, although intended as a study, was sold, presented to Congress, and afterwards, placed in the Rotunda of the Capitol. Of the face he said: "You see half-smile, half-sadness; half-anger, half forgiveness;

---

* Bartlett was enamored of Edmund C. Stedman's poem "Hand of Lincoln" (1883; *Poems* [Boston, 1908], 435–36). Four of the twelve stanzas follow:

> Look on this cast, and know the hand
> That bore a nation in its hold:
> From this mute witness understand
> What Lincoln was,—how large of mould.
>
> This was the hand that knew to swing
> The axe—since thus would Freedom train
> Her son—and made the forest ring,
> And drove the wedge, and toiled amain.
>
> No courtier's, toying with a sword,
> Nor minstrel's, laid across a lute
> A chief's, uplifted to the Lord
> When all the kings of earth were mute.
>
> Lo, as I gaze, the statured man,
> Built up from yon large hand, appears:
> A type that Nature wills to plan
> But once in all a people's years.

half determination, half pause; a mixture of expression that drew accurately the middle course he would follow—read wrongly by both sides." Three years later the artist witnessed the unveiling of his seated Lincoln before the courthouse in Newark, New Jersey. The President sits wearily on a park bench, his hat resting next to his extended right hand, his left arm thrown over his knee, with the face of one pensive, tired, and alone. The conception was utterly original. Perhaps no sculpture of Lincoln, as F. Lauriston Bullard has said, better captured the man whom the people loved.*[25]

Regardless of the larger issue of interpretation, some argued that Barnard's Lincoln failed on its own terms, as the portrait of a homespun democrat. Rather, it depicted "a collosal clodhopper," one said. It was easy enough to see what Barnard intended, Kenyon Cox wrote, but the result was a caricature of the democratic Lincoln. People laughed at Barnard's model, the railsplitter, with his old clothes. Lincoln's face was wrinkled and ugly, like the clothes. The elongated neck and prominent Adam's apple contributed to the ungainly impression. Hands and feet were too big. Moreover, for all his efforts at authenticity, the sculptor erred in the choice of footwear. He showed Lincoln in shoes when, in fact, he wore boots and, according to his son, never even owned a pair of shoes. Nothing caused as much comment as the crossed hands below the waist. It suggested Lincoln was suffering an attack of indigestion. "The Tramp with the Colic" and "The Stomach Ache Statue" were two of the ribald names given to the work. A Massachusetts congressman, John Rogers, regaled his colleagues on the last day of the session in October by reciting these scathing criticisms. Ostensibly, he spoke in support of a resolution requesting the President to halt shipment of the Barnard replica to England.[26]

---

*Children loved it. They climbed over the statue, rode on Lincoln's knee, and rested in his arm. A poet, Leslie Pinckney Hill, thought he had seen no other monument so beautiful as this:

> Round all the high, horsed heroes I have known
> An undisturbed indifference has grown,
> Which neither time nor wonder can prevent.
> But he is on the ground, and children play
> Upon his knees, and stroke the earnest face
> That shines with their caresses, and all day
> He is their comfort in the public place:
> The rigid bronze itself cannot conceal
> That sheltering heart which little children feel.

Robert Lincoln disliked the statue partly because it was "used as a playground by the hoodlum children of the neighborhood." L. P. Hill, *Wings of Oppression* (Boston, 1921), 75; Lincoln to Clinton Conkling, November 24, 1917, Robert Todd Lincoln Papers.

Children at Work and Play about the Borglum Statue of Lincoln, Newark, New Jersey, 1917. (NYT Pictures)

This proved unnecessary. After the abuse heaped upon Barnard's Lincoln in the United States, British authorities could hardly be expected to take it even as a gift; besides, it was said, ship tonnage could not be spared to any statue during the war. The choice of a proper monument for Britain ceased to be a trifling matter, the *London Times* observed, once it became apparent that American pride and self-respect were involved. A Philadelphia daily conducted a referendum on the statue of Lincoln preferred by Americans. The Saint-Gaudens was overwhelmingly favored over Barnard's or any other. In January 1918, the American Centennial Committee polled its membership: of sixty replies, forty-one favored Saint-Gaudens, none favored Barnard, and seventeen expressed no preference. Eleven months later the committee formally endorsed the Saint-Gaudens *Lincoln*, thus returning to the original proposal. Robert Lincoln offered to pay for the replica, but this was avoided by the action of the Carnegie Endowment for International Peace, which made it a gift on behalf of the American people.

The Tafts promptly offered their gift to Manchester. No statue of Lincoln could be more appropriate for a working-class city, the *Guardian* observed. It was truthful to a man rough-hewn in every limb and lineament. The sculptor seemed to be saying, "Here is a man who needs no sentimental treatment." Manchester gratefully accepted. The Saint-

Gaudens replica was unveiled with fitting ceremony in the Channing Enclosure of Westminster on July 28, 1920. The Prime Minister, Lloyd George, accepting it for the British people, underscored the international significance of the dedication when he said of Lincoln: "In his life he was a great American. He is no longer so. He is one of those giant figures, of whom there are very few in history, who lose their nationality in death. They are no longer Greek or Hebrew, English or American; they belong to mankind."[27]

THE LINCOLN MEMORIAL in Washington was dedicated on Memorial Day 1922. Henry Bacon's white marble temple, a work of pure Hellenic beauty, magnificently completed a great vista and enshrined an American immortal who had become one of the world's gods. The exterior colonnade of thirty-six Doric columns, one for each of the states in 1865, stood for the Union Lincoln had preserved. Dominating the memorial hall was French's statue. No work of sculpture ever better fitted a grand space. Carved in Georgia white marble, the chaired figure measures nineteen feet from toe to head (on an eleven-foot pedestal); if standing, the figure would rise twenty-eight feet in the air. The linear space virtually dictated a seated Lincoln. The main problem French encountered with the hall was one of lighting. By design sunlight filtered through the marble slabs of the roof, as in the Parthenon, but it proved insufficient, leaving the head of the statue in darkness. Several years passed before the problem was analyzed and corrected by structural changes in the roof and use of floodlights. The inscription above the statue reads:

IN THIS TEMPLE
As In The Hearts Of The People
For Whom He Saved The Union
The Memory of Abraham Lincoln
Is Enshrined Forever

At the opposite ends, screened by rows of Ionic columns, are smaller halls dedicated to the Gettysburg Address and the Second Inaugural Address. Murals painted by Jules Guerin emblematic of Lincoln's principles decorate the halls.[28]

The dedication fell on a bright spring day in the capital. A multitude of people assembled before the grand flight of steps ascending the memorial and around the reflecting pool on the mall. Chief Justice Taft, in his capacity as chairman of the Lincoln Memorial Commission, presided. Robert Moton, successor to Booker T. Washington as president of Tuskegee Institute, was one of the featured speakers, voicing the senti-

Lincoln Memorial, Washington, D.C., 1922. By Henry Bacon. (U.S. Department of the Interior: NPS)

Statue by Daniel C. French. (The Lincoln Museum, Fort Wayne a part of the Lincoln National Corporation)

ments of twelve million black Americans. "As yet, no other name so warms the heart or stirs the depths of their gratitude as that of Abraham Lincoln. To him above all others we owe the privilege of sharing as fellow citizens in the dedication of this shrine." Edwin Markham delivered a revised and expanded version of his poem of 1900, "Lincoln, the Man of the People." With such lines as

> The grip that swung the ax in Illinois
> Was on the pen that set a people free

the white-haired poet added the western touch to the occasion. After the Marine Band played "The Battle Hymn of the Republic," Taft made the presentation speech. Noting that fifty-seven years had passed since Lincoln's death, yet, he said, "we feel a closer touch with him than with living men." And it was well, he thought, that this great tribute "should wait until a generation instinct with the growing and deepening impression of the real Lincoln has had time to develop an art adequate to the expression of his greatness." President Warren G. Harding, in accepting the Memorial, lauded Lincoln as "incomparably the greatest of our Presidents." In 1865 he gave his life that a nation might live. "Today, American gratitude, love, and appreciation, give to Abraham Lincoln this lone white temple, a pantheon for him alone."[29]

French's *Abraham Lincoln* was an instantaneous success. The contrast with the response to the Barnard statue could hardly have been greater. Of course, there were some, like Borglum, who did not like anything about the Lincoln Memorial; but seldom were artistic opinion and public opinion in such complete accord. The statue quickly became the nation's foremost sculptured icon. The reverence millions upon millions of viewers have felt before it is a function partly of the templed space, partly of its elevation as if upon an altar, and partly of the compelling figure itself. Lincoln sits gracefully, majestically, upon the august chair of state. His arms and hands rest upon its fasces-carved pillars. The hands are prominent. They are powerful, and they suggest the physical and mental determination that suffuses the entire figure. Although the body is in repose, there is tension in the limbs, the trunk, and the head. The face is strong, yet benign, with a trace of sorrow; the lips are closed, and the eyes gaze forward as if in calm but earnest prayer. People have thought to give a name to the statue, "Lincoln the Triumphant," for instance. But what some see as triumph other observers see as resignation; what some see as toughness, others see as tenderness; and what some see as majesty, others see as solemnity. In truth, the statue enshrined the man.

As Lincoln was consecrated in the hearts of the people, so was his grand memorial. Before long a million people a year were coming to visit it. Some came out of no more than curiosity. Some came in search of peace and strength. Some came with questions.

> O Father Abraham;
> O Liberator,
> What message for us now?[30]

Some, like Langston Hughes, came with mingled feelings of gratitude and hope.

> Let's go see old Abe
> Sitting in the marble and the moonlight
> Sitting lonely in the marble and the moonlight
> Quiet for ten thousand centuries, old Abe.
> Quiet for a million, million centuries.
> Quiet,—and yet a voice forever
> Against the timeless wall of time,
> Old Abe.[31]

And some, like the tourist of Charles Olson's verse, came to worship.

> Reverse of
> sic transit gloria, the
> Latin American whom the cab driver told me
> he picked up at Union Station had
> one word of english—link-
> cone. And drove him
> straight to the monument, the man
> went up the stairs and fell down on his knees
> where he could see the statue and stayed there
> in the attitude of prayer.[32]

Religion

Of the skirmishes over Lincoln's religion following his death, David Davis, his lawyer and friend, remarked, "I don't know anything about Lincoln's religion, nor do I think anybody else knows anything about it." The same opinion might be rendered after the millions of words since written on the subject, but that would be uncharitable and possibly untrue. The comparison of George Washington and religion is instructive. The literature is voluminous, but as a leading student has observed, nothing about Washington is more clouded by myth and legend and misinformation. Commonly portrayed as a devout Christian, and claimed by various denominations as one of their own, Washington was actually a lifelong Episcopalian whose practice was so lukewarm as to cast doubt upon his belief. Deists and agnostics took title to him. Yet, like Lincoln in the Civil War, Washington attributed his success in the Revolution to Divine Providence, and as President proclaimed days of public thanksgiving and prayer, as Lincoln would also. The parallel can be carried only so far, however. For whether or not he was a believing Christian, Lincoln plumbed depths of spirituality never touched by Washington. "Of all the Presidents of the United States," a leading clergyman has said, "Lincoln was one of the least orthodox, yet the most religious." The nub of the paradox was that for all the manifest religiousness, Lincoln was never baptized, never took communion, and never joined a church.[33]

From the hour of his death Protestant ministers portrayed the Martyr President as an exemplary Christian. Josiah Holland's biography fixed the conception. The next year, 1867, the Western Tract Society published William C. Gray's didactic life of Lincoln especially written for Sabbath schools. Despite the humiliating circumstances of his death in a theater—on Good Friday to boot—Lincoln rose to greatness by practicing the virtues of a true Christian: honesty, humility, charity, temperance, industry, and faith in God. Increasingly, by the end of the century Lincoln's life and teachings furnished the texts of sermons from Protestant pulpits, as well as in Jewish synagogues and, on occasion, Catholic churches. The Congregationalists, as earlier noted, introduced the annual Lincoln Memorial Sabbath. "If Lincoln was not a Christian," a Methodist preacher said, "it would make me feel like tearing myself away from the bonds of orthodox Christianity." Except for Jesus, according to the Unitarian Minot J. Savage, no one in the Bible was Lincoln's equal. Indeed, except for Jesus, no man in history was Lincoln's equal. John Hay may have been the first to say this, but the opinion was widespread in the twentieth century. "There is no need to preach the precepts of the Bible," declared Rabbi Joseph Silverman of Temple Emanu-El in New York, "when we have such a real Messiah, who lived in the flesh and never pretended to be more than a man."[34]

In *The Master and His Servant*, Jonathan T. Hobson developed an extended parallel between Jesus Christ and Abraham Lincoln. Both were born in forlorn hovels. Both Joseph and Thomas were simple carpenters. Lincoln, like Jesus, was unsure who his father was. Both were humble, kind, sorrowful, and loving of their fellow man. Both spoke in parables. Both were sent to fulfill divine missions and preceded by prophets who were executed: John the Baptist and John Brown. On Palm Sunday Jesus journeyed to Jerusalem, Lincoln to (or from) Richmond; one had his Last Supper, the other his last cabinet meeting; each met his death on Good Friday.

Bruce Barton picked up the parallel in *The Man Nobody Knows*, a popular portrait of Jesus as the personification of the modern business executive. Both men had fearful misgivings about their missions. Lincoln kept asking himself if he had read God's will right. "Was he, after all, only a common fellow—a fair country lawyer and a good teller of jokes?" Both Jesus and Lincoln were advertising geniuses. In communicating with the people, neither indulged in introductions but went directly to the point. Both were calm and lighthearted on the gravest occasions in contrast to humorless disciples, in one case, and to solemn cabinet officers in the other. But, of course, the parallel struck many orthodox Christians as blasphemous. It amounted to "The Belittling of Jesus" and "The Glorification of Abraham Lincoln," wrote a Southern Methodist of Barton's best-selling book.[35]

The conception of Lincoln as a second Christ—"a Christ in miniature," Tolstoy called him—was revealed as well in ecclesiastical art. The Church of the Ascension, in New York, incorporated a Lincoln effigy in the façade dedicated to those who had contributed most to Christianity. He was particularly favored in stained glass windows. A panel commemorating Lincoln the Emancipator was dedicated at Brooklyn's Plymouth Church in 1909; some years later a larger, full-size image of the Emancipator was incorporated in the cathedral of the Polish National Catholic Church in Scranton, Pennsylvania. Nothing of this kind was more impressive than the great window of the new Central Woodward Christian Church in Detroit, completed in 1929. The central panel, eighteen by seven feet, depicts the President striking the shackles from a slave boy; smaller side panels are emblematic of Justice and Mercy. The story is often told of a southern woman who attended the church while visiting in Detroit and, in the midst of the service, suddenly gasped aloud, "Oh my soul! Yes! Look!" nudging her son at her side, "Abe Lincoln in a church window!"[36]

For a number of preachers interest in Lincoln extended beyond the requirements of an occasional sermon or homily; they became students of his life and thought. Joseph Fort Newton, though ordained a Baptist, established an independent "Liberal Christian Church" in Cedar Rapids, Iowa, in 1908, and through a chance encounter with Frank B. Sanborn was led by way of the correspondence of Theodore Parker and William H. Herndon to write *Lincoln and Herndon*, the first study of their relationship. In this connection Newton interviewed old Henry B. Rankin of Springfield and got him to write his *Personal Recollections of Abraham*

Stained Glass Window, Central Woodward Christian Church, Detroit, 1929. (Burton Historical Collection, Detroit Public Library)

William E. Barton. (Illinois State Historical Library)

Lincoln, in 1916, the last truly important volume of reminiscences. For Newton, a leading liberal Protestant clergyman, Lincoln became a lifetime joy and inspiration. "What a life to study, what noble integrity, what high courage, what delicate justice and melting pity! What loyalty to the ideal . . . what heights of vision and valleys of melancholy, what tear-freighted humor!"

Another liberal Protestant, Jenkin Lloyd Jones, who came of age as a Union solider, founded the Lincoln Center in Chicago and, it may be recalled, had a part in the campaign to make Lincoln's birthplace a national shrine. A contemporary divine, Ervin S. Chapman, was among those who hoisted Lincoln's banner over the crusade of the Anti-Saloon League. He authored the compendious *Latest Light on Abraham Lincoln* in 1917. John Wesley Hill, a popular Methodist preacher in New York, became chancellor of Lincoln Memorial University in 1916. His most substantial book was *Abraham Lincoln, Man of God*, which appeared in 1920 with an Introduction by President-elect Warren G. Harding. The fact that the book went through several editions suggests its appeal, like that of *The Man Nobody Knows*, to the conservative political temper of the decade. William Allen White, the Emporia, Kansas, editor, for one, wondered if Hill had not confused his politics with his religion when he argued in his chapter "Lincoln and Labor," that the sainted hero, if alive, would enroll in the open shop movement.

Methodist Bishop Charles H. Fowler was a spellbinding orator on Lincoln over several decades. Edgar DeWitt Jones, who preached from the pulpit of the Central Woodward Church in Detroit, knew all the laborers in the Lincoln vineyard and wrote a book, *Lincoln and the Preachers*, in 1948. In it he said that during forty years he had spoken on Lincoln over a thousand times all over the world. Nobody, however, delivered more speeches on Lincoln than Louis A. Warren. Ordained a minister of the Christian Church in 1916, he became a student of Lincoln's Kentucky origins and in 1928 exchanged the pulpit for the directorship of the new Lincoln Historical Research Foundation in Fort Wayne, Indiana. Recognized as a leading authority on Lincoln's childhood and youth, Warren was an indefatigable lecturer.[37]

Foremost of the Lincoln clergy was William E. Barton, pastor of the First Congregational Church of Oak Park, Illinois. He was born in Sublette, a small town on the prairie, and his earliest memory was of his father draping the house in black muslin in mourning for the President's death. Stories about Lincoln abounded. The boy grew up thinking, "He and I were contemporaries." Young Barton went off to Berea College, in Kentucky, and after deciding on a ministerial career, he rode the circuit for several years preaching to people so much like those from whom Lincoln came that he felt he understood him. He attended Oberlin Theological Seminary, graduating in 1890, and held Congregational pulpits in Ohio and Massachusetts before returning to Illinois near the end of the century to lead the Oak Park church. As editor of a religious newspaper during the war years, he wrote occasional pieces on Lincoln; his interest thus aroused, he turned to serious collecting (he acquired a substantial part of the John E. Burton collection auctioned in 1915) and study. The first fruit of that study was *The Soul of Abraham Lincoln* in 1920, the year Barton turned fifty-nine. His next book followed but six months later, and they arrived as regularly as the equinoxes thereafter. In 1924 he retired from the pulpit in order to devote himself more assiduously to Lincoln. His pen was stilled only with his death in 1930.[38]

A tall man of distinguished bearing, with gray hair and a Vandyke beard, Barton was endowed with an ego that would crack the looking glass. If, as his son Bruce maintained, Jesus was an advertising genius, the elder Barton qualified as a true disciple. His entry in *Who's Who* was supposedly the longest in the volume. His blunt honesty and conceit grated on the feelings of many co-workers in the field. But if little loved, he nevertheless commanded their respect. A dogged researcher, he dared to question sacred truths. The whiff of his irreverent personality pervaded his writings. Although not a debunking biographer, he imbibed the debunking spirit of the age. His best work was a kind of historical sleuthing, for instance the monograph *A Beautiful Blunder: The True Story of Lincoln's Letter to Mrs. Lydia A. Bixby*. His faults were those of haste, prolixity, and random antiquarianism. As a reviewer observed of one of his books, "Dr. Barton belongs to the school of scholarship which indulges in unlimited draughts of minutiae, large mouthfuls of carefully prepared and discreetly served nothing." Professional historians scarcely noticed his work; at the same time, it was too dense both in style and substance to be truly popular.[39]

In the Preface to his first book, Barton wrote, "The fact that there are so many books on the religion of Abraham Lincoln is a chief reason why there should be one more." For previous works were either pietistic or patriotic or both. This was true, certainly, of William J. Johnson's *Abraham Lincoln, the Christian* in 1913, which interpreted his spiritual life as a pilgrimage ending in the embrace of the Savior, Jesus Christ. It was even truer of Hill's book published in the same year as Barton's. Lincoln's leadership was basically spiritual, according to Hill, and by becoming an orthodox Christian in the presidency he saved not only the Union, not only his own soul, but also the Christian soul of the nation. Contemptuous of such sentimentality, Barton professed to offer a more dispassionate view, one informed, moreover, by personal familiarity with Lincoln's religious background.

He began with a survey of Lincoln's religious environments, from his Kentucky boyhood to his life in Washington. Then in twelve succinct, episodic chapters he analyzed the conflicting evidence on the matter of Lincoln's religion. This was lawyer-like and, on balance, discreditable to the tradition of infidelity. Barton gave much weight to the influence upon Lincoln of the Reverend James Smith's *The Christian's Defense* in 1850, which had been earlier acknowledged, although Barton was apparently the first biographer to have looked at that book and examined its contents. Finally, *The Soul of Lincoln* offered a constructive argument for his religious faith as Barton understood it. First he disposed of false claims: Lincoln was not an infidel or atheist or materialistic theist, nor a spiritualist, nor Quaker, nor Methodist. To the question of "Why Did Lincoln Never Join a Church?" Barton was unable to go beyond the often quoted statement the President himself had given to Connecticut Congressman Henry C. Deming. It was because he was unable to assent without reservation to their diverse articles of belief and confessions of faith.

> When any church will inscribe over its altars, as its sole qualification for membership, the Saviour's condensed statement of the substance of both law and gospel, "Thou shalt love the Lord thy God with all thy heart, and with all thy soul, and with all thy mind, and thy neighbor as thyself," that church will I join with all my heart and all my soul.

Yet Lincoln was not creedless. In what a present-day theologian has described as "a classic *tour de force*," Barton assembled "The Creed of Abraham Lincoln in His Own Words." This creed of some three hundred words begins and ends with profession of belief in an Almighty God, in whose hands he is but a humble instrument, and to whose truth and justice he is ultimately responsible. It also contains the article: "I believe that the Bible is the best gift which God has given to men. All the good from the Saviour of the world is communicated to us through this book." This was as close to a profession of Christian faith as the creed came. And nowhere in Barton's book is it shown that Lincoln accepted the Revelation of Jesus Christ as set forth in the Gospels. Readers might well wonder if, after all the evidence had been sifted and the creed constructed, they had any clearer idea of Lincoln's religion than before. Regardless, *The Soul of Lincoln* was widely hailed as the "final authority" on the subject.[40]

Barton, with most writers on Lincoln's religion, recognized the need for a developmental approach. His religion evolved. The study of it was inseparable from his biography. Tradition defined three epochs, roughly identified with three of Barton's environments: New Salem, Springfield, and Washington. Lincoln was reared a Baptist, and though there was disagreement on just what kind of Baptist, Barton was not alone in thinking this was the source of his fatalism. William H. Herndon had established the New Salem period as one of strident infidelity. Lincoln read the works of Voltaire, Paine, and Volney and wrote a small treatise, unfortunately burned, advocating their doctrine. Of course, Herndon believed Lincoln never materially changed his views; in the tradition of Lincoln the Christian, however, they were dismissed as "youthful indiscretion." Barton said young Lincoln only imagined himself an infidel because all the Baptist preachers he had heard believed the world was flat and he rebelled against such absurdities. Barton also credited the testimony, in 1874, of the New Salem schoolmaster, Mentor Graham, that Lincoln's essay on religion, far from being an infidel production, was an argument for universal salvation through Christ's atonement. Thus the difference between the early and the final creed was much less than generally supposed.[41]

The Springfield period was considered transitional for Lincoln's evolving faith. When running for Congress in 1846, his Democratic opponent, Peter Cartwright, made charges of infidelity against Lincoln. No answer to the charges could be found. Henry B. Rankin, however, offered the testimony of his mother, who had asked Lincoln about them during a conversation in her home. He admitted to earlier skepticism but said that he was finding his way out of it as the truths of Scripture unfolded before him. He wished he might accept Christ and the Gospels; but he refused to discuss religious questions in a political campaign. Barton deigned to cite Rankin's secondhand testimony about a conversation seventy years past.

He noticed the claims of two ministers to the conversion of Lincoln in Springfield. One was advanced by the Reverend James F. Jacquess, a Methodist, at a reunion of his Illinois regiment in 1897. Having heard Jacquess preach in Springfield in 1839, Lincoln afterwards came to talk and pray with him for hours. "Now," said the preacher, "I have seen many persons converted . . . , and if ever a person was converted, Abraham Lincoln was converted that night in my house." Nothing in Lincoln's subsequent life tended to support this claim, and Barton dismissed it. However, he accepted Reverend Smith's conversion claim, as reported by his Presbyterian colleague James A. Reed in the controversy with Herndon in 1873. Hard upon Smith's arrival in Springfield came the death of Lincoln's second son, Edward, which profoundly disturbed his spiritual life, Barton thought, and this, combined with the persuasiveness of Smith's book in answering freethinkers and Mary's decision to unite with the Presbyterian Church, prepared the way for something that had the marks of a conversion. Lincoln also gave a kind of confession of faith to Newton Bateman in 1860, though the high gloss Holland put upon it helped Herndon to discredit the testimony.[42]

The third epoch offered the best evidence of Lincoln's Christian faith. Barton, who slighted the presidency in all his work, offered no systematic examination of this evidence. The tradition that the agonies of war, reinforced by family tragedy, transformed the President from a lukewarm theist to a devout Christian was supported by Lincoln's own words, such as those Barton used to construct his creed, and also by a rich stream of anecdote and reminiscence. The death of his dear son William in February 1862 was generally thought to have been a turning point in Lincoln's pilgrimage. In his grief he sought spiritual comfort. A visiting minister from New York, Dr. Francis Vinton, assured the President his son was alive in heaven. "Alive! Alive!" he sobbed uncontrollably. Most biographers concurred in Ida Tarbell's assessment: "It was the first experience of his [Lincoln's] life . . . which drove him to look outside his own mind and heart for help to endure a personal grief."

The President's appointment at Gettysburg may have been a more critical turning point. In sermons after his death, the story was told of a caller at the White House who had dared to ask him if he was a Christian. "The President buried his face in his handkerchief, turned away and wept." Then, composing himself, he said when he left home he was not a Christian, nor did he become one after Willie's death, though that was the severest trial he had known. "But when I went to Gettysburg, and looked upon the graves of our dead heroes . . . I then and there consecrated myself to Christ."

If reports were true, Lincoln prayed a lot. "I have been driven many times upon my knees in the overwhelming conviction I had nowhere else to go," he told Noah Brooks. And not only upon his knees: James E. Murdoch said that as a guest at the White House he went into the hallway one night because of a strange noise, and he found the President in an adjacent room face down on the floor, praying.[43]

The most often told story emanated from General Daniel E. Sickles, the colorful commander of Meade's Third Corps, who lost a leg at Gettysburg. When Lincoln visited the convalescent in Washington, he admitted to some alarm during the battle. In fact, though he would not wish it to be known, he went into his room, locked the door, and got down on his knees to pray to God for victory. "I told him this was His country, and our war was His war, and that we could not stand another

Fredericksburg or Chancellorsville. And then and there I made a solemn vow that if He would stand by our boys at Gettysburg, I would stand by him." And although he was not "a meeting man," Lincoln continued to Sickles, "somehow or other a sweet comfort crept into my soul that God Almighty had taken the whole business up there into His own hands, and that things would come out all right at Gettysburg."[44]

That the President read the Bible goes without saying. When the loyal Negroes of Baltimore presented him with an elegant $580 Bible in 1864, he said in reply, "All the Saviour gave to the world was communicated through this book"—words that Barton incorporated into Lincoln's creed. Of the 702 words of the Second Inaugural Address, according to Ervin Chapman, 266 of them were quoted verbatim from the Word of God. Remove them and only the shell of the address remained. Not long before his death, it was said, Lincoln sent for the Reverend Dr. Gurley, pastor of the New York Avenue Presbyterian Church where he rented a pew, and vowed to apply for admission at the next communion. Alas, that would be Easter Sunday. Sitting in his box at Ford's Theater on Good Friday, according to another story, the President talked to Mrs. Lincoln of his plan to journey to Jerusalem and walk in Jesus' footsteps once relieved of his great burden, and then the fatal shot pierced his brain. Hezekiah Butterworth evoked the memory of "Lincoln's Last Dream," closing with the lines:

> Lovely land of Palestina! he thy shores will never see,
> But, his dream fulfilled, he follows Him who walked in Galilee.[45]

THE DISSENTING TRADITION of Lincoln as a lifelong freethinker in religion had its apostles in William H. Herndon and Robert G. Ingersoll. Lincoln's soul answered to reason over love, to head over heart. His creed was the secular creed of the Enlightenment, as embodied in the thoughts of deists like Franklin, Paine, and Jefferson. Herndon had expounded Lincoln's freethought religion, popularly labeled atheist and infidel, not only in his famous lecture of 1874 but also in contributions to such obscure journals as the *Index*, in Toledo, and *Truth Seeker*, in New York; and they entered into both the Lamon and the Herndon and Weik biographies. Ingersoll stated his views in orations and polemical writings. In 1893 he engaged in an epistolary debate, subsequently published, with General Charles Collis. Each was able to show the weakness and inconclusiveness of the other's case. Most of the evidence on the Christian side consisted of pious falsehoods, Ingersoll maintained. Invocations of the favor of Almighty God, as in the Emancipation Proclamation, should not be read as professions of religious belief but, rather, as concessions to public opinion. Ingersoll cited the testimony of Jesse Fell, John G. Nicolay, and Mrs. Lincoln that he experienced no change of religious belief after leaving Springfield. If he had become a Christian, the great agnostic declared, Lincoln was too honest a man not to have avowed it.[46]

The freethought argument was essentially negative. Its fullest statement was in John E. Remsburg's book of over three hundred pages, *Abraham Lincoln: Was He a Christian?* Remsburg, a Kansas "village atheist," had been Herndon's friend; he inherited some of the Illinoisan's Lincoln relics along with his convictions. He sought testimony from Lincoln's contemporaries about his religion, using what he agreed with and discarding the rest. He believed there was a clerical conspiracy to suppress the truth about Lincoln's infidelity. Unwelcome evidence that he could not refute Remsburg branded lies, for instance Lincoln's statement on accepting the Baltimore Bible in 1864: "but for this book we could not know right from wrong," "the best gift of God to man," and so on. How could Lincoln have said things so unreasonable? How could he have blessed a book that justified slavery? Ingersoll supposed he must have been "temporarily insane." Remsburg echoed Herndon in denouncing "this speech—this supposed, this fraudulent speech—a lie." It was like the caption, "Lincoln Reading the Bible to His Son," the press usually put on Mathew Brady's famous photograph, when in fact the volume in the President's hands was a picture album.[47]

In the next generation, Joseph Lewis continued the freethought tradition. His address before the annual banquet of the Freethinkers Society of New York in 1924 revealed the sterility, even the dishonesty, of this line of thinking. Except for fragmentary evidence from the New Salem years, there was little basis for maintaining that Lincoln was a freethinker in religion, and the case was fatally compromised when advocates invented evidence to support their position. Lewis held that Lincoln's infidel opinions were unaltered in the White House, and introduced as "irrefutable" proof a letter written to Judge J. A. Wakefield, described as an old friend, in which Lincoln said, "My earlier views of the unsoundness of the Christian scheme of salvation and the human origin of the scriptures, have *become clearer* and stronger with advancing years and I see no reason for thinking *I shall ever change them.*" But if Lincoln wrote such a letter, it has not been produced, nor is J. A. Wakefield known to Lincoln's *Collected Works.*[48]

Less dogmatic and more creditable were the efforts of Ethical Culturists, Unitarians, and similar modernists to enroll Lincoln among their patron saints. Horace Traubel, editor of an Ethical Culture journal in 1890, named the quartet of Lincoln, Emerson, Whitman, and Ingersoll to represent the standard to which the movement aspired. In the same spirit, the Reverend Charles F. Potter of New York's West Side Unitarian Church, in 1924, announced plans for "a new all-American Bible" to be made up of texts from such democratic prophets as Lincoln, Jefferson, Jane Addams, and the like. Potter felt confident it would make a spiritual appeal to the average American as great as, or greater than, the Gospels of Jesus. On Lincoln Sunday in 1927 the independent Community Church on Park Avenue unveiled, at either side of the altar, bronze busts of Lincoln and Emerson. In his sermon, John Haynes Holmes said they were the two greatest Americans. Holmes denied that Lincoln was a Christian, and knew not what to name a religious outlook that combined rationalism and mysticism in a unique way. Lincoln's faith in the rule of God in history was annexed to a secular faith in the ongoing progress of humanity. David S. Muzzey, better known as a historian, was prominent in the movement to humanize and ethicize religion, and he placed Lincoln in its vanguard. "Lincoln was a prophet," he declared, "like Socrates and Bruno and Jefferson and Emerson, of the day when a religion founded on . . . the vision of human brotherhood and moral rectitude should dawn."[49]

To one marginal group, the spiritualists, the key to Lincoln's spiritual life was not to be found in conventional religion, rational or revealed, but in the belief that the human personality exists after death and communicates with the living through psychic mediums. He was, in the words of a recent spokesman, "too spiritual to be religious," and this explained why he never joined a church. Spiritualism had found many followers in the United States since the modern craze began in 1848. It was different from mysticism, which is inextricably linked to religion, and to which Lincoln admitted a strong predisposition. As President he spoke often of the consciousness of being in the control of a higher power; his visions, as in the dream of a phantom ship hurrying toward a dark shore on the night before his assassination, were of a mystical nature. Mysticism informed the writings of Francis Grierson and may be seen as well in Denton J. Snider's curious biography of 1908 wherein Lincoln appears as the mediator and harmonizer between the nation's Folk-Soul and a kind of Hegelian World-Spirit.

The President's interest in spiritualism was stirred by Willie's death. It manifested itself most obviously in seances he attended in Georgetown and, on at least one occasion, in the White House. Mrs. Lincoln was the prime mover in this matter; the President may simply have been indulging her whim. Newspaper reports of these events caused some merriment at Lincoln's expense. Obviously, if the President took the counsel of mediums at critical times during the war, as a New York daily opined, "it would be the literary event of the nineteenth century." Vague reports continued after Lincoln's death. However, the first authentic account came from the celebrated medium Nettie Colburn Maynard, in the book *Was Abraham Lincoln a Spiritualist? or, Curious Revelations from the Life of a Trance Medium*, published in 1891 at the time of her death.[50]

Nettie Colburn was introduced to the President at the White House in December 1862. A frail and timid young woman from White Plains, New York, she was already known for her psychic powers. After some polite conversation among the small group gathered in the Red Parlor, she walked directly in front of the President, fell into a trance, and commenced an hour-long oration on the nation's history and destiny, closing with a majestic plea to sign the Emancipation Proclamation on January 1, as scheduled, despite the political pressures upon him to postpone it. The act would be, she assured him, "the crowning event" of his administration and would "fulfill the mission for which he had been raised up by an over-ruling Providence." After the medium regained consciousness and the guests shook off the spell, one of them asked the President, "Did you notice anything peculiar in the method of address?" "Yes, it is very singular, very!" he replied, and his eyes fell upon a full-length portrait of Daniel Webster hanging on the wall, as if to say it was he who had spoken through Nettie Colburn. Hers was a singular gift, he told her, and he did not doubt it was from God. Lincoln met the young woman on other occasions, once at a Georgetown home when a piano on which he sat levitated, and he found her a job in the Department of the Interior.

The incident concerning the Emancipation Proclamation, in particular, was verified by a competent witness, Simon P. Kase. However, Kase's story differed in significant details. A Pennsylvania ironmaker with a faddish interest in spiritualism, he wrote his account of the incident in an undated pamphlet, *The Emancipation Proclamation: How, and by Whom It Was Given to Abraham Lincoln in 1861*. Maynard lectured Lincoln on emancipation in the same Georgetown home of the Laurie family where the piano moved—Kase sat on it with the President!—and it occurred in 1861. So the Emancipation Proclamation itself was a gift of spiritualism. In 1926 Sir Arthur Conan Doyle, in his two-volume *History of Spiritualism*, called it one of the most important instances ever recorded of spirit intervention in earthly events, yet one might search the histories and biographies and find no mention of it.[51]

Spiritualists gained respectability by claiming Lincoln one of their own. A large portrait of him hung in the hall where a convention of spiritualists met in 1896. They sought to strengthen the association during the

Lincoln Centennial. Mediums attested to conversing with Lincoln in the invisible world. One received a message from him written on a slate: "One cannot help being born in a log cabin, but one can avoid living in one forever." A scandalous forgery, still to be discussed, in the pages of the *Atlantic Monthly* in 1928 turned out to be a spiritualist hoax. Writing in this vein has continued to the present day. In *Lincoln Returns* (1957), Harriet W. Shelton, an elderly psychic, gave an account of seances over a period of four years in which she was in communication with Lincoln. He it was, she alleged, who urged her to write the book that he dictated "clairaudiently" to her inner ear. More recently, a "vindication" of Lincoln's spiritualism, *Willie Speaks Out!* was written as Willie's own story of his "Pa" from the "other side."[52]

The fact that Lincoln never united with any Christian church undoubtedly contributed to the love American Jews felt for him. The Jewish population at the time of the Civil War was about two hundred thousand. A significant number lived in the Confederacy; and Jews in the North were not solidly Republican. Several leading rabbis in 1861 opposed the war and voiced pro-slavery sentiments, citing Old Testament warrant for the institution. Nevertheless, by the time of his death Lincoln had won a special place in Jewish affections. He had had several Jewish friends in Illinois. He appointed one of them, Abraham Jonas, a lawyer, postmaster of Quincy. He met Abraham Kohn, a devout Jew who was city clerk of Chicago, in 1860, and Kohn worked for his election to the presidency. Upon Lincoln's departure from Springfield, Kohn presented him with a large American flag inscribed in Hebrew with verses from the first chapter of the Book of Joshua ending: "Have I not commanded thee? Be strong and of good courage; be not afraid, neither be thou dismayed: for the Lord thy God is with thee whithersoever thou goest."

In Washington the President had repeated occasion to call upon his chiropodist, Isachor Zacharie, to "put him on his feet," and he employed him in a peace mission to New Orleans. Simon Wolf, a Washington lawyer who became a Lincoln convert in 1858, had easy access to the White House; the President patronized the bookseller Adolphus Solomans on Pennsylvania Avenue; and, of course, he received delegations of rabbis on matters of special concern to Jews. One of these was their appointment as chaplains. The law provided for the appointment of chaplains of "some Christian denomination"; after the rabbis intervened, this blatant discrimination was removed, and Lincoln immediately appointed three Jewish hospital chaplains. Even more important in Jewish eyes was the President's revocation of General Grant's General Order No. 11 prohibiting Jews as a class from trading with the Army of

Tennessee. This was the litmus test of Lincoln's freedom from prejudice against Jews.[53]

Upon his death it was perhaps inevitable that Jews should look upon Lincoln as a modern Moses who had brought them within sight of the Promised Land, alternately as a second patriarch—was his name not Abraham?—of their people. Thus Lewis N. Dembitz, a Louisville lawyer, one of three Jewish delegates in the Republican National Convention of 1860, and uncle of the later great Justice Louis D. Brandeis, addressed the congregation of his synagogue mourning the President's death:

> You often called him, jocosely, Rabbi Abraham, as if he were one of our nation—of the seed of Israel. But, in truth, you might have called him "Abraham, the child of our father Abraham." For indeed, of all the Israelites throughout the United States, there was none who more thoroughly fitted the ideal of what a true descendant of Abraham ought to be than Abraham Lincoln. And, if he was uncircumcised, we are told, "all the nations are uncircumcised in flesh, but all they of Israel are circumcised in heart."

Rabbi Isaac M. Wise of Cincinnati, who became the father of American Reform Judaism, even suggested that Lincoln was of the flesh as well. Wise had been but a half-hearted Unionist until he met with Lincoln to protest Grant's infamous order. Henceforth a fervent advocate, he lauded the martyr in his sermon preached one week after the assassination as "the highest jewel, the greatest hero, and the noblest son of the nation." Then he said, "Brethren, the lamented Abraham Lincoln believed [himself] to be bone to our bone, and flesh to our flesh. He is supposed to be a descendant of Hebrew parentage. He said so in my presence, and, indeed, he preserved numerous features of the Hebrew race, both in countenance and in character." In all the curious byways of Lincoln genealogy, no one, and surely not he, had ever hinted at Hebrew parentage. Wise himself seems never to have returned to this aberrant memory, though he continued to believe, as he wrote in his magazine, *The Israelite,* "Old Honest Abe was the greatest man that ever sprung from mortal loins."[54]

As more and more Jews immigrated to the United States, Lincoln became in their sight the symbol of all America was meant to be. "Lincoln," it has been said, "was the first American folk hero among Jews." In his person, as in his ideals, he was the hero with whom they could most closely identify. Common, honest, and upright, man of sorrows and man of laughter, he had a sense of kinship with the poor and downtrodden, for he was of them. Spiritually, he shared Judaism's faith in an Almighty God and passion for truth and justice. He was, said a New York rabbi, "the realization of the true spirit of Judaism." Rabbi J. Leonard Levy of Pittsburgh, perhaps the closest student of Lincoln among Jews early in the twentieth century, called him "America's suffering Messiah" whose mission was freedom and brotherhood. It became customary in some temples, as in some churches, to set aside the Sabbath nearest Lincoln's birthday to remember him. When a Reform rabbi from the North took the practice to Fort Worth, he was told by the president of the congregation, "Young man this is Texas, not Illinois." Still, he kept the practice.[55]

In the centennial year, Isaac Markens, a New York cotton broker and Lincoln enthusiast, contributed a path-breaking study of Lincoln and the Jews to the journal of the American Jewish Historical Society. After Markens, the leading student was Emanuel Hertz, another New Yorker, who became an avid collector and propagator of Lincolniana, and in this connection sought to recover the record, as far as it survived, of Jewish thought about Lincoln. In 1927 Hertz, who believed Lincoln a prophet divinely appointed to his task, published *Abraham Lincoln: The Tribute of the Synagogue*, consisting of sixty-six sermons and speeches. All comparisons of Lincoln to ancient Hebrew seers failed to satisfy. He was variously linked to Saul the patriot, Elijah the messenger, Hillel the teacher; Hertz, however, favored the first and oldest conception of Lincoln as a modern Moses. His oratorical tour de force, *Abraham Lincoln—The Seer*, originally published in the *American Hebrew*, was widely reprinted and rapidly translated into Hebrew, Polish, Russian, Czech, Bohemian, Hungarian, and German.[56]

Ancestry

In the autobiographical sketch penned for Jesse Fell in 1859, Lincoln wrote, "My parents were both born in Virginia, of undistinguished families—second families, perhaps I should say." Without being ashamed of his ancestry, he called attention to his commonness, thereby shaping his own myth. He did not care to talk about his family. "I don't know who my grandfather was," Herndon quoted him as saying, "and I am much more concerned what his grandson will be." Actually, he did know who his paternal grandfather was: Abraham, who came to Kentucky from Virginia in 1782 and was killed by Indians two years later. Thomas, the youngest of three sons, was then six, and grew up "a wondering laboring boy . . . litterally *[sic]* without education." His son, the President, at the time of his death could trace no ancestor beyond this grandfather. A death notice in the *New England Historical and Genealogical Register* conjectured that the family descended from the Lincolns of Massachusetts. Of the family of his mother, Nancy Hanks, he knew nothing but, according to Herndon, suspected she was illegitimate and, indeed, had no certain knowledge he was born in honest wedlock. Lincoln's genealogy was a mystery inside the enigma of the man. "The career of Abraham Lincoln," a Harvard historian wrote, "casts doubt on all our ideas of heredity and education." Quite simply, there was no satisfactory explanation of that career in the light of Lincoln's ancestry, birth, and early life.[57]

The problem called forth a range of responses. One deduced Lincoln from "special providence" or a kind of immaculate conception. He had no ancestors. One mystery—the old mystery of Christ—explained another. And so the poet-chronicler Denton Snider had Lincoln say:

> Fathering myself I am the rightful son
> Not of Tom Lincoln, but of God the Father
>
> . . .
>
> I shall take up my race's life in mine,
> Abbreviate in me humanity,
> The people's dateless representative,
> The incarnation of all ages' folk.

A second response was to supply Lincoln with respectable parents and ancestors. Unfortunately, there was no way to change his mother, though the Hanks family might be elevated above the sordid "white trash" status to which Herndon and Lincoln had consigned them, or perhaps the best blood of Virginia might be discovered in Nancy's veins by virtue of a first-family father. Nothing, on the other hand, seemed to preclude giving Lincoln a proper father. A third response was to say, basically, that Lincoln was sui generis, either a product of atavism or the ultimate proof that heredity does not matter. The latter went along with the notion that Lincoln invented himself.[58]

The first difficulty to be overcome was that of Lincoln's legitimacy. Herndon had unloosed the notion, got from Dennis Hanks, himself a bastard, that Lincoln was "a spurious child," probably the son of Abraham Inlow (or Enlow) of Bourbon County, Kentucky. However, the discovery of crucial documents restored the marriage record of Thomas and Nancy in 1806, and Herndon repented his error before he died. But nothing deterred believers in folk tales. One of these was James H. Cathey of North Carolina. In a substantial book, *Truth is Stranger Than Fiction* (1899), he maintained that Lincoln was the offspring of an illicit union between Nancy Hanks and a different Abraham Enlow, an esteemed citizen of Rutherford County in the shadow of the Blue Ridge. Nancy, the daughter of an indigent mother, had been "bound out" in the Enlow family. When she became enceinte, Enlow shipped her off to Kentucky, where the son Abraham was born, presumably before Nancy's marriage, which would make Lincoln several years older than he claimed to be. Cathey based his "truth" on local tradition and striking physical resemblances, especially in the instance of Wesley Enlow, an alleged half-brother, who was "a walking epistle of Abraham Lincoln."

Many years later, James C. Coggins, a college president, produced a variant of Cathey's work in *Abraham Lincoln a North Carolinian*. Lincoln, it seems, was Tarheel-born. Coggins located his birthplace, Lincoln Hill, as it was long known in the county. After he was elected President, many of the local ladies boasted they had bounced him on their knees as a baby. Unfortunately, he and the mother proved an embarrassment to Mrs. Enlow, who drove this poor Hagar and her brat to Kentucky. By 1940 Coggins had discovered eugenics and realized that Lincoln should have good blood on both sides. So he said that Nancy, instead of being low-born as Cathey had supposed, was the daughter of Michael Tanner, sire of a fine German family that came to Virginia about 1750. Little Abe's mother was Nancy Tanner, not Nancy Hanks. Furthermore, he was a guest at her wedding to Thomas Lincoln.[59]

The woods were full of fanatics who continued to believe such stories well into the twentieth century. The derivation of Lincoln's genius from a union so ordinary as that of Nancy and Thomas strained credulity. If not divine, it must have been the spawn of another genius, just as Beethoven was sometimes thought to be the natural son of Frederick the Great. While traveling in South Carolina in 1896, Clarence King encountered a similar legend and wrote to John Hay about it. He viewed John C. Calhoun's statue in Charleston, and could not help but notice a certain likeness to Lincoln, especially in the face. Mentioning this to a native, King learned that some quite respectable people believed Calhoun was Lincoln's father. A state judge was gathering the evidence for publication. According to the story, Calhoun, a young lawyer touring the western circuit, stopped at a lonesome tavern and got maid Nancy in a family way, then spirited her away to a farm in the mountains, the Abraham Enlow farm, as it happened, and the story unfolded from there.

Curious details were added in published accounts. A North Carolina judge, Felix E. Alley, cited the testimony of James L. Orr, an Anderson County friend of Calhoun's. Orr met Lincoln in Congress in 1849 and, upon learning his mother's name was Hanks, inquired among members of that family who were his constituents, and they told him of the old contretemps in the Craytonville tavern kept by widow Ann Hanks, adding that Calhoun put up $500 to get Nancy to leave the district.* Reflecting on what he had heard in Charleston, King supposed it was all moonshine, but he could not help but marvel at the dramatic possibilities of a wayward son rising up to demolish his father, for Calhoun became the political father of the Confederacy. "That would be nearer the Greek than anything I ever came across in modern life," he told Hay, "and unless you know to the contrary it might be worth while to trace up the threads." Robert Lincoln would not be pleased, yet even he "might be willing to sacrifice a grandmother in exchange for an Aeschylus situation." Some southerners took a perverse delight in this theory, as if only a Lincoln endowed with Calhoun's brains could have licked the South. Ought that not admit Lincoln to the southern aristocracy? John Temple Graves, recalling his father's pride of ancestry, said he waxed enthusiastic upon learning that Calhoun might be Lincoln's father. Not that he approved of Lincoln. "The point is, that it made Abraham Lincoln our cousin."[60]

The paternity suit was gradually resolved, then closed in rousing fashion by William E. Barton in 1920. The Lincoln family pedigree was established. Solomon Lincoln, of Hingham, Massachusetts, had begun to assemble it and had actually corresponded with the congressman in 1849. The American line traced from Samuel Lincoln—Abraham's great-great-great-great-grandfather—who arrived in 1637, and it included governors, congressmen, cabinet officers, and other dignitaries. Abraham descended from the pioneer branch that homesteaded in western Pennsylvania, thence to Virginia and Kentucky. For a time Pennsylvania Germans contested this lineage, claiming that Lincoln, the name being derived from Linkhorn, was of German descent. But this was exploded by the work of Marion Dexter Learned, professor of German at the University of Pennsylvania, in 1909. The centennial year also saw publication of *The Ancestry of Lincoln*, by James H. Lea and J. R. Hutchinson, which ended on a note of genealogical triumph: "There is not a trait in the broad and lovable character of Abraham Lincoln that we may not find foreshadowed in one or many of his ancestors." It was, moreover, as others insisted, a peculiarly Yankee inheritance. The genealogical investigation culminated in Waldo Lincoln's *History of the Lincoln Family* in 1923. Meanwhile, a bust of the President was unveiled in the English

*On October 21, 1941, *Time* magazine (p. 21) reported that a judge of the North Carolina Supreme Court handed down an opinion asserting that Lincoln was Calhoun's natural son, born of the natural daughter of a genteel Virginia planter. He cited Judge Alley's book published three decades earlier.

parish church of his first American ancestor's birthplace in Hingham, Norfolk. The gentrification of the Lincoln lineage undercut the rustic image. Honest Abe began to look more and more like George Washington; and his achievement, which had seemed a miracle, appeared the logical outcome of the marriage of good stock and pioneer spirit.[61]

In his book *The Paternity of Lincoln*, Barton set up seven putative sires and knocked them down one by one. It was easy, for they were straw men by 1920. Barton showed uncharacteristic restraint in stopping at seven. A North Carolina professor, observing that nowhere else in history had such a multiplicity of fathers been ascribed to any man, identified thirteen sires. Why was such an inquiry necessary? Barton asked in his four-hundred-page book. Every biographer of Lincoln, he answered, must "skin this skunk" lying in the way at the outset. All the delicate maneuvering around it had not disposed of the problem. It was time to get rid of the skunk. He expressed surprise at the poverty of imagination shown by those who sought to furnish Lincoln a worthy father. Why could they not find better sires than a trio of Enlows? To be sure, there was Calhoun. But why not Henry Clay? He was, after all, a Virginian removed to Kentucky, virtually the same age as Thomas Lincoln, and with a reputation for philandering unknown to Calhoun. Oddly, too, the inventors of these sires all thought they were doing Lincoln a service. In his case being born a bastard was counted a blessing. Only among certain hostile southerners was it a canard. Even they had not suspected it until Lincoln's death. The lie, the lore, the tradition was all posthumous. In attacking it Barton understood, as the subtitle of his book indicated, that the issue was less one of Lincoln's paternity than it was one of the chastity of his mother. This, in turn, led him to the still vexed question of her own legitimacy. For the rest of his career Barton was preoccupied with working out the maternal line of Lincoln's ancestry.[62]

In this matter, two distinct traditions had emerged by 1900. The first may be named the Herndon tradition, since it stemmed from his gleanings of Lincoln's talk and from his garrulous sixty-seven-year-old informant, Dennis Hanks. Nancy was the natural child of Lucy Hanks, the daughter of Joseph Hanks; she later married Henry Sparrow and raised a large family. Nicolay and Hay, in one concise sentence, had said as much, while tactfully omitting any mention of the father. Who he was no one knew. But as in the paternity legend, it was sometimes supposed that Lincoln's other grandfather was a first-family Virginian. *The Sorrows of Nancy*, the fiction of a Virginia woman, Lucinda Boyd, in 1899, argued that Nancy's father was a son of John Marshall, the great Chief

Justice. She became an orphan when he was killed in border warfare, and subsequently immigrated to Kentucky.*

The other tradition may be named after the first genealogist of the Hanks family in America, Caroline Hanks Hitchcock. Her little book *Nancy Hanks*, published in 1899, was a by-product of her genealogical investigation. It was far from complete—indeed, she never would complete it—but she placed the first of the family in Massachusetts in 1699 and traced one line to Virginia, thence to Kentucky, in a migration resembling the Lincolns'. Meanwhile, Mrs. Hitchcock, who lived in Boston, had made an important discovery about Nancy Hanks, and at the behest of Ida Tarbell rushed into print with it. In Bardstown, Kentucky, she had uncovered the will of Joseph Hanks, who died in 1793 bequeathing property to eight children, the youngest being Nancy, aged nine. "I give and bequeath unto my daughter Nancy one heifer yearling called Peidy," read the laconic testament. Further evidence showed that Nancy went to live with Richard Berry, her uncle, whose name was on the marriage bond of Nancy Hanks and Thomas Lincoln in 1806. Mrs. Hitchcock had established, she thought, not only that this was the Nancy Hanks who gave birth to Abraham Lincoln but also that she was herself of legitimate birth.

Many agreed with her, among them Lea and Hutchinson in their study of Lincoln's ancestry. Tarbell, who wrote the Preface for the book, was her champion. She had an aversion to genealogy; "it makes me crazy," she said. So she was grateful to Hitchcock for her work. It exploded Herndon. Moreover, it smoothed the way for a much happier view of the Hanks legacy to Lincoln. Tarbell's hope that the old slurs on Lincoln's ancestry might soon be forgotten proved premature, however. Caroline Hitchcock, instead of settling the question, had woven a crazy quilt. There was too much guesswork, too many evasive phrases—"it is related," "tradition says"—in her work. She failed to take account of the ubiquity of the name Nancy among families interwoven with the Lin-

---

*A poet, George Lawrence Andrews, might have used this scenario for his tribute to "dark, dream-eyed Lucy" who one moonlight night yielded her body to a dashing cavalier.

> Out through the gap one rode; there is no record of his name.
> He loved, betrayed and left as time on time has many another
> But Lucy faced her neighbors smiling. No amount of blame
> Could daunt the dark-haired girl who lived to be our Abe's grandmother.

Andrews's poem appeared in *Plain Talk*, February 1928; a clipping is in the Lincoln Museum.

colns, not only Hanks but Sparrow, Shipley, and others. No lawyer, as Jesse Weik, who was one, told Tarbell, doubts the right of a man to legitimize children by his will. So what did the legacy of a heifer yearling to a girl named Nancy prove?[63]

In distant retrospect, whether or not Nancy Hanks was illegitimate had no bearing on the greatness of Abraham Lincoln. But to many who worshiped at his shrine in generations past, it truly mattered. A cult gathered around Nancy Hanks almost simultaneous with Lincoln's canonization. "A great man never drew his infant life from a purer or more womanly bosom than her own," Holland wrote. Herndon's line, attributed to Lincoln, "All I am or hope to be I owe to my sainted mother" was more often quoted than any other from his pen. Napoleon had said, "The future destiny of a child may be learned from the mother," and a theory of heredity agreed with him.[64]

Nancy was characterized as gentle, delicate, sensitive—qualities which offset the rude and coarse image of Thomas. She was, according to Nicolay and Hay, "of appearance and intellect superior to her lowly fortunes." The inference that she was of "distinguished ancestry" came easily. As for her intellect, it was commonly said that she could read and write, though evidence of either was lacking. A romantic novel in 1920 portrayed her not only as a lover but as a young businesswoman and "the first citizen of Elizabethtown." Nancy was usually described as fair and petite, but even those who remembered her could not agree on her physical appearance. She was slender, she was plump; her hair was light, her hair was dark; her eyes were blue, her eyes were gray. Lincoln himself, only nine when his mother died, left no clear recollections of her. He had fond memories of his stepmother, Sarah; and these were sometimes mistaken for memories of Nancy. The mystery surrounding her was inseparable from the greater mystery surrounding him.[65]

A simple monument was erected over the grave of this "Madonna in the wilderness" in 1879; it gave way to a more elaborate monument in 1892. The memory of Nancy Hanks captivated the poet Harriet Monroe.

> Prairie child.
>   Brief as dew,
> What winds of wonder
>   Nourished you?

After reflections upon the obscure death that extinguished her "heat of fire," the poet continues:

> Wilding lady,
>   Still and true,

Who gave us Lincoln
  And never knew.

To you at last
  Our praise, our tears,
Love and a son
  Through a nation's tears!

Mother of Lincoln,
  Our tears, our praise,
A battle flag
  And the victor's bays![66]

The conflict between the rival traditions on Lincoln's grandmother was brought to a head by Barton in 1925. After resigning his pulpit, he had relocated himself in the woods of Foxboro, Massachusetts, and in the library annex he called "The Wigwam" immersed himself in Lincolniana. He wrote profusely. Any reader of popular magazines, to say nothing of Lincolnologists, recognized Barton as the preeminent authority. He carried on a lively correspondence with students and collectors. They, or most of them, admired his industry but disliked him personally. He was smug, impudent, conceited, and intolerant of criticism. The reverend gentleman's morals were not above reproach. When Daniel Fish declined the offer of a Lincoln title he had admittedly "swiped" from Lane Theological Seminary, Barton blithely replied, "Believe me, if you are going to have any scruples about stealing from a Theological Seminary, you will never make a Collector." As earlier noted, his book on Lincoln's paternity led him into the Hanks genealogy. Here he held tentatively to Caroline Hanks's thesis that the Nancy named in Joseph Hanks's will was the mother of Lincoln. But as he waded into "the slough of the Hanks genealogy," he came to doubt it. Fish did his best to steer him away from old Herndon, whom he dismissed as a drunk and fanatic, but to no avail.[67]

Discrepancies between Hitchcock's story and the evidence Herndon got from Dennis Hanks aroused Barton's suspicions. He made a pilgrimage to the Lincoln country in Kentucky, where his escort was young Louis A. Warren. In 1909 Warren had come from Massachusetts to attend Transylvania University. Seven years later he was ordained a minister of the Disciples of Christ; and his first pastorate, providentially, was in Hodgenville. Here, where he became editor of the local newspaper as well, Warren was indoctrinated in the Lincoln tradition. Learning his way around courthouse records, he became convinced that too much attention had been paid to reminiscences and not enough to documentary research. Tarbell, who discovered Warren after Barton, was amazed by his close study of Lincoln's origins. "He has done ten times over more than anybody else," she wrote. Tarbell had returned to the Lincoln field after a long absence and was then embarked on a series for the McClure Newspaper Syndicate, published as a book, *In the Footsteps of Lincoln*, in 1924. On most questions, over the years, she saw eye to eye with Warren.

But a document he turned up in court records, a summons to Lucy Hanks to answer the charge of fornication, proved Tarbell's and Hitchcock's undoing and delighted Barton. Like a birdhound on the track of game, he pursued this Lucy in the surviving records. He found discrepancies with the story told by Hitchcock, and this sent him to Virginia to search through the Hanks family history. He discovered what Caroline Hitchcock had overlooked: Joseph Hanks had a ninth child named Lucy, unacknowledged in his will. Barton jumped to the conclusion Lucy Hanks was Nancy's mother. Herndon, and Dennis Hanks, had been right after all. Weik had never doubted it. Without the benefit of Barton's discovery, he repeated the old aspersion on Nancy in his book *The Real Lincoln*, in 1922.[68]

As word of Barton's findings spread in Lincoln circles, anticipation for their publication soared. On February 3, 1923, he offered a preview before a select audience at the Chicago Historical Society. He was prepared to produce, he said, not just one grandmother—the one he had set out to find—but two, for on a detour from the main investigation he turned up the paternal grandmother, wrongly identified before, Bathsheba Herring, who actually lived until Lincoln was twenty-seven years of age, though he apparently never saw her. But interest centered on the Hanks lineage. Caroline Hitchcock was beside herself with rage at a minister of the gospel reviving salacious slander she had laid to rest, she thought, a quarter-century ago. Ida Tarbell tried to calm her friend. It was all very repellant, she agreed, but no lover of Lincoln would believe Barton. In her new book she adhered to the Hitchcock thesis, and while she acknowledged that Barton had evidence supposedly "shattering" this tradition, she pointed out that the proof had not yet been adduced. Meanwhile, she called upon Hitchcock to finish the long-awaited documented genealogy, which was the only thing that would silence scoffers like Barton.[69]

Barton's proof was laid out in his two-volume *Life of Abraham Lincoln* in 1925. An article in the *New York Times Magazine* prior to publication called national attention to the author's revelations of the "Lost Grandmothers." Adding Lucy as the ninth child, and fourth daughter, of Joseph Hanks, Barton inferred that she had been disinherited in the will because of misbehavior. Nancy, who inherited the calf and who

Hitchcock made into Lincoln's mother, was actually his great-aunt. Among the documents Barton produced was Lucy's certificate of age and consent to marry Henry Sparrow in 1790. Written and signed by her, it was "her sole literary monument," which was one more than her daughter Nancy left. The Sparrows had eight children; numbers of their descendants lived in a place called "the cut-off" in Anderson County. Nancy grew up among Sparrows, thus accounting for the not uncommon belief during and after her lifetime that she was one. Barton had no proof regarding her paternity, but backed by the Hanks-Herndon tradition he assumed she was illegitimate.

The biography was well received and praised for its "realism," though as more than one reviewer pointed out the background overwhelmed the man. The meat of the book was in the first one hundred pages. Ida Tarbell reviewed it with some trepidation in the *Christian Science Monitor.* She was gracious, as always, but of Barton's pièce de résistance she thought he had proven much less than he assumed. Joseph Hanks's will said he made provision for "all his children," yet Barton tried to accommodate it to the fact of another child, Lucy. He did not prove that the Lucy indicted for fornication and the Lucy who filed a marriage certificate were the same person. Nor, of course, did he prove that Nancy was born a bastard. Waldo Lincoln agreed with the criticism, adding in a personal letter to Barton that, right or wrong, "it was unwise to stir up an old scandal."[70]

In the end, Barton's chief adversary was Louis Warren. Encouraged by Albert Beveridge, who employed him as a researcher, he had moved to Indiana. He filled the pulpit of a small church near Indianapolis and became a lecturer for the Indiana Lincoln Union dedicated to memorializing Lincoln's youth in Spencer County. No publisher would take Warren's thousand-page manuscript in two volumes: one devoted to documents, the other to interpretation under the title "Lincoln's Old Kentucky Home." In 1926 he was satisfied to publish a compressed version of the latter by itself as *Lincoln's Parentage and Childhood.* The book was at once recognized by serious students as a serious contribution. Warren's scholarly attitude was a refreshing change from Barton's idiosyncratic manner. Generally, the book offered solid documentation for raising the Lincoln family in the social scale. Part of the evidence was proof of Nancy Hanks's legitimacy. In Warren's view, both the established traditions were in error. His heart lay with Hitchcock. He had to concede, however, that her work was badly flawed. Having no confidence in the testimony of Dennis Hanks, skeptical of Herndon as well, he could not agree with Barton. In Warren's genealogy, Nancy's mother was born Lucy Shipley—one of five sisters—and she married a Hanks, though there was no positive proof of this, by whom she had her daughter. Early widowed, she "farmed out" orphan Nancy to Mrs. Berry, Lucy's sister, and in 1790 married Henry Sparrow.

The critical document for Warren was Lucy Hanks's marriage certificate. Barton had included a photographic reproduction of it in his *Life of Lincoln*. As Warren read it, Lucy plainly stated she was a *widow*, thereby legitimating Nancy. But in his transcription of the document, Barton ignored this evidence.* The gentleman from Foxboro did not take kindly to Warren's interpretation. Reviewing his book in the *Monitor*, he said Warren had allowed his feelings to override his judgment, and he discredited the book as a whole because of the disagreement over Lincoln's grandmother.[71]

Unfortunately for Warren, the crucial document had disappeared. He had not seen the original but hoped to prove his case by the testimony of a handwriting expert. Warren suspected Barton had "snitched" the document from the Mercer County Courthouse. Barton *did* have it; he had borrowed it, he said, and such was his pride of discovery he had had the precious "literary monument" framed. Warren pleaded with him to return it to the archives, but it remained with his papers when they were sold to the University of Chicago after his death. There Ida Tarbell saw it in 1932. She had wanted to believe Warren was right, but had to confess her disappointment when she could see no sign of the word "widoy" or any evidence of tampering. Others more expert than she, Worthington C. Ford for instance, had already reached this conclusion. Both Beveridge and Sandburg, in their major biographies, adopted Barton's view of the matter. He summed up his genealogical studies in *The Lineage of Lincoln*, published in 1929. Typically, Barton could not refrain from sensationalism, in this instance the contention that Lincoln

---

*Warren was justified in his reading. The concluding words and lines on the four-by-five-inch piece of paper were as follows:

. . . given under my hand
this day
Apriel 26, 1790
[wi] doy
Lucey
Hanks

The bracketed letters are barely visible, and "doy" might be "day," though that was redundant. Barton omitted it altogether from his transcription (*Life of Lincoln*, I, 61-62):

. . . given under my hand this day
Apriel 26th, 1790
Lucey Hanks

and Robert E. Lee were distant cousins. The *jeu d'esprit* did not set well with the latter-day Virginia chivalry.[72]

But in the genealogical wars of the time Barton prevailed. When someone requested Ida Tarbell's considered judgment on where the truth lay on the vexed question of Nancy Hanks, she replied, "I am not sure though I think the weight of evidence at the moment is on the side of illegitimacy." And so it remained until a genealogist of the rising generation, Adin Baber, took up the work Caroline Hitchcock had left unfinished, cleared Nancy Hanks of the old scandal, and consigned to the junkheap of history the long tradition started by Hanks and Herndon and closed by Barton.[73]

Controversies Old and New

The disputes over Lincoln's ancestry and religion were obviously important both for his biography and, in varying degrees, for his place in the American mind. They were not easily settled. Smaller controversies on smaller matters arose from time to time. They, too, intermingled the Lincoln of history and the Lincoln of memory and imagination. As a general rule, the reverence surrounding the symbol protected it from the heats of political debate; however, there were exceptions to the rule, and on the issue of temperance and prohibition, in particular, opposite forces competed for the favor of Lincoln's name and authority. One of the oldest controversies concerned the Lincoln image in the South. Old animosities flared up in the 1920s, but as in other losing battles the smoke and the noise masked a retreat.

William E. Barton figured prominently in two little literary skirmishes. In his biography he tried to explode the legend of Lincoln and the sleeping sentinel. Rummaging in War Department records, he could find no evidence that Lincoln knew of Private William Scott or the last-minute pardon that saved him from the firing squad. An editorial in the *Boston Herald*, "One Question Settled," written by Lincoln expert F. Lauriston Bullard, had a note of finality. But it was not settled. An eighty-two-year-old Union veteran in Vermont produced a faded clipping from the *New York Times* the day after the event which reported the President's intervention. Another Vermont veteran, Luke E. Ferriter, said he was the soldier Scott had relieved from duty at midnight before his near-fatal sleep and described in detail the marshaling of the firing squad and Lincoln's miraculous appearance with Scott's reprieve. The daughter of Lucius Chittenden, the Treasury officer whose account of Lincoln's actions was a principal source, vouched for its substantial accuracy and said he had dictated it to her.

Barton was forced to acknowledge his error in not consulting the newspaper record. Yes, it seemed, Lincoln did know about the Scott case and ask that the petition of pardon be honored by General McClellan. But this was the extent of his involvement. Prompted to research the matter further, Barton learned that the War Department had finally located the pardon. It was signed by McClellan, just as he had expected, and the President's intervention had been pro forma. The point, at any rate, was that Lincoln, instead of the merciful saint portrayed in this story, was simply a man of judgment and discretion. Barton entitled his article in the Sunday *New York Times* "An Old Myth Dispelled," ending with a mock sigh of distress over the loss of another beautiful story. This was the reverend scholar's last irreverent word on the subject.[74]

Far from being dispelled, the myth grew under the glare of publicity. In 1927 Allen Clark, a Washington historian, published *Abraham Lincoln: The Merciful President*, which focused on the Scott case. Two years later a radio dramatization premiered on Lincoln's birthday. The people of Scott's hometown, Groton, Vermont, erected a monument to him with the inscription "Pardoned by Abraham Lincoln." The revival culminated in Vermont historian Waldo F. Glover's monograph *Abraham Lincoln and the Sleeping Sentinel*, in 1936. After a thorough review, Glover reached several conclusions. First, although McClellan signed the pardon, he had been led to act only because of the President's interest in the case. Second, Lincoln did not make a personal appearance at the last moment to save Scott's life. That was a dramatic embellishment by the poet Janvier, and somehow it had displaced reality in the minds of old soldiers like Luke Ferriter. Third, Scott did die heroically at Lee's Mill with a prayer for the President on his lips. Glover cited the contemporaneous report of a war correspondent in the *Philadelphia Inquirer*, widely copied in the Vermont press. Bullard, who had earlier been persuaded by Barton, changed his mind. Glover's book "clinches the case," he said.[75]

Barton's biographical notice of Lincoln's celebrated letter to Mrs. Bixby coincided with widespread newspaper discussion set off by Bullard's investigation into the whereabouts of that letter. As in the former case, the public interest caused Barton to pursue the matter further, but here the result was "an expanded footnote" of 135 pages entitled *A Beautiful Blunder*. The point of this lapidary exercise was that although the President's letter of condolence to Mrs. Bixby assumed the truth of information later proven to be false, and she had sacrificed not five but only two of her sons on the field of battle, nevertheless it was a beautiful blunder since it called forth a letter now recognized as a priceless legacy.* [76]

The letter had a curious history. Written upon information furnished by Lydia Bixby to the Adjutant General of Massachusetts, transmitted to the Governor, thence to the War Department, finally to the President, it was such a deeply felt expression of the nation's sorrow for "so costly a sacrifice upon the altar of Freedom" that it would become, in Sandburg's words, "a piece of the American Bible." Lincoln probably meant for it to be published, and it was, first in the *Boston Transcript*, immediately after Mrs. Bixby received it. The letter found its place in several books, among them Holland's biography, where it stood without comment, published just after Lincoln's death. It did not become instantly famous. Nicolay and Hay made no mention of it in their *History*, nor Herndon in his biography, nor Tarbell in hers. Engraved reproductions of the letter made their appearance in 1891. They purported to be facsimiles of the original. One of these was the printed keepsake for guests at the annual Lincoln Day Dinner of the National Republican Club in 1906. Increasingly recognized as a literary masterpiece, it was included in the Everyman Library volume of Lincoln's writings in 1907, repeatedly placed in the forefront of the Lincoln canon, and esteemed one of the treasures of world literature.

That the letter was a brilliant stroke of statesmanship was also recognized. According to Barton, President Wilson, whose own artistry with the pen was well known, despaired of equaling Lincoln in the expression of condolence, and so decided to send the parents of four United States Marines killed at Vera Cruz in 1914 copies of the former President's letter to Mrs. Bixby with a covering note in his own name. Alas, this was one of Barton's peccadilloes. Nothing in the voluminous Wilson

---

*The letter to Mrs. Lydia Bixby, November 21, 1864, is as follows:

> Dear Madam,—I have been shown in the files of the War Department a statement of the Adjutant General of Massachusetts, that you are the mother of five sons who have died gloriously on the field of battle.
>
> I feel how weak and fruitless must be any words of mine which should attempt to beguile you from the grief of a loss so overwhelming. But I cannot refrain from tendering to you the consolation that may be found in the thanks of the Republic they died to save.
>
> I pray that our Heavenly Father may assuage the anguish of your bereavement, and leave you only the cherished memory of the loved and lost, and the solemn pride that must be yours, to have laid so costly a sacrifice upon the altar of Freedom.
> Yours, very sincerely and respectfully,
>
> A. Lincoln

papers, at least, supports him. A front-page story in the *New York Times* in 1918 focused national attention on the Bixby letter. Frau Meter, a German mother who had sacrificed nine sons in defense of the Fatherland, received this cursory note from the Kaiser: "His Majesty is immensely grateful at the fact, and in recognition is pleased to send you his photograph." The contrast to Lincoln's compassionate letter was manifest.[77]

But where, if anywhere, was the original of the Bixby letter? Until it was found no one could be absolutely certain that Lincoln wrote it. One suspects that the original ended up in the *Boston Transcript*'s trash after it was set in type, but antiquarians and scholars faithfully pursued the document for decades. In 1904 one of the latter, Ellis Oberholtzer, in a perfunctory biography, suggested that the letter was on display at one of the Oxford colleges. Other reports said that the college was Brasenose. With such hints Bullard, then a forty-year-old novice in Lincolnology, launched a search he pursued off and on for the next forty years. He wrote to Brasenose College, and from the Principal learned that nobody there had ever heard of Lincoln's Bixby letter. Yet many Americans believed the letter could be found at Oxford, and some even swore they had seen it. "There it hangs in the place of honor, above the trained scholarship of the ages," declaimed an orator at the National Republican Club. In the summer of 1925, Bullard, a busy Boston newspaperman with an insatiable appetite for Lincoln study, found the opportunity to investigate the matter for himself at Oxford. Unhappily, not even a facsimile could be found, he wrote in a letter to the London *Times*, republished in his own newspaper, the *Herald*, and elsewhere in the United States. The great American libraries searched high and low for it to no avail.[78]

What a mystery! No one doubted that Lincoln, or someone, probably John Hay, wrote the celebrated letter. Many suspected the facsimile was a forgery, executed by someone—again, the finger pointed to Hay—who had mastered Lincoln's handwriting. And some kept the faith that the original would turn up. Should it turn up, Barton guessed the letter would be worth more than the $10,400 recently paid at auction for Lincoln's astonishing epistle of January 26, 1863, to General Hooker. Four decades later its worth was reliably estimated at $100,000. Bullard brought his search to a close with the publication in 1946 of *Abraham Lincoln and the Widow Bixby*, the most authoritative study. While he never found the Holy Grail, he ended the quest in the sure belief that Lincoln authored the letter ascribed to him.[79]

LINCOLN'S NAME WAS early and constantly invoked in the legions that marched in the crusade against Demon Rum and triumphed with the adoption of the Eighteenth Amendment in 1919. Opponents of prohibition—"wets," as they were called—protested that the appeal was fraudulent and answered with fraudulent claims of their own in which Lincoln sometimes sounded like a lobbyist for the whiskey distillers.

Both sides agreed that Lincoln practiced temperance himself and advocated it publicly. The record was clear on that. Leonard Swett, his lawyer-friend, remembered Lincoln saying to him sometime before the election of 1860 that he had never even tasted liquor. "What!" Swett exclaimed. "Do you mean to say you never tasted it?" "Yes, I never tasted it," was Honest Abe's unequivocal reply. His insistence on serving "Adam's ale"—water—to the delegation that informed him of his presidential nomination was widely reported. Campaign biographies portrayed him as a temperance candidate. The total abstinence Lincoln practiced in his own home he maintained within the family in the White House. Gifts of wine and spirits were routinely shipped to hospitals. Of course, Mrs. Lincoln served wine to guests on formal occasions, and it was often remarked that the President, despite his moral support for temperance in the military, chose not to enforce such a policy. Nothing he said in this regard was as well known as his joke about sending Grant's brand of whiskey to other generals in the field. Obviously, if he was a total abstainer himself, he was not eager to impose that standard on everybody else. So there was room for debate on his public position toward prohibition.[80]

The debate touched upon Lincoln's words and deeds in three periods of his life. During the New Salem years, the liquor interests maintained, he had been a saloon keeper. In 1908 Adolphus Busch, the St. Louis brewery baron, announced that Lincoln had been granted a liquor license in 1833, and he was issuing a reproduction of the license for public distribution. The National Liquor Dealers Association and similar groups reproduced the license, and it graced the walls of saloons. A liquor, or tavern, license had, in fact, been issued to the firm of Berry and Lincoln. Twenty-five years later Lincoln was confronted with the fact by Stephen A. Douglas in the debate at Ottawa. Saying that when they first became acquainted he was a poor schoolmaster and Lincoln "a flourishing grocery-keeper"in New Salem, Douglas meant that his opponent sold liquor, for that was the common practice of groceries on the frontier. He did not stop there but went on to say that Lincoln could "ruin more liquor than all the boys of the town together." Daniel Burner, one of Tarbell's informants for her "Early Life," had said there was

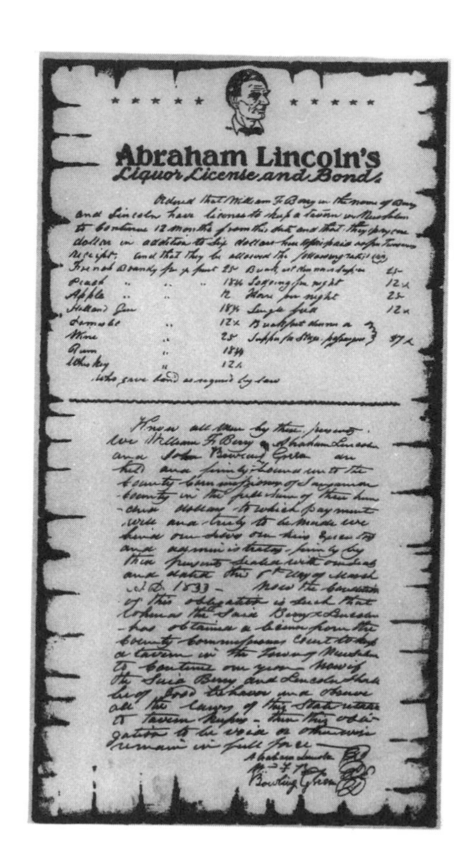

Abraham Lincoln's
Liquor License and Bonds

Ordered that William F. Berry in the name of Berry and Lincoln have license to keep a tavern in New Salem to continue 12 months from this date and that they pay one dollar in addition to six dollars heretofore paid as per Treasurer's receipt, and that they be allowed the following rates (viz)

French Brandy per ½ pint 25 — Breakfast, dinner or supper 25
Peach " " " 18¾ — Lodging per night 12½
Apple " " 12 — Horse per night 25
Holland Gin 18¾ — Single feed 12½
Domestic " 12½ — Breakfast dinner or supper for stage passengers 37½
Wine " 25
Rum " 18¾
Whiskey " 12½

who gave bond as required by law

Know all men by these presents. We William F. Berry, Abraham Lincoln and John Bowling Green are held and firmly bound unto the County Commissioners of Sangamon County in the full sum of three hundred dollars to which payment well and truly to be made we bind ourselves our heirs executors and administrators firmly by these presents sealed with our seals and dated this 6th day of March A.D. 1833 — Now the condition of this obligation is such that whereas the said Berry & Lincoln has obtained a license from the County Commissioners Court to keep a tavern in the town of New Salem to continue one year. Now if the said Berry and Lincoln shall be of good behavior and observe all the laws of this State relative to tavern keepers — then this obligation to be void or otherwise remain in full force —

Abraham Lincoln (Seal)
Wm F. Berry (Seal)
Bowling Green (Seal)

"nothing for sale but liquor" in the Berry and Lincoln store and described how "Big Abe"—the strongest man he ever knew—would take a full barrel of whiskey, containing forty-four gallons, and, gripping each end with his two hands, raise it to his mouth and drink from the bunghole. (D. W. Griffith recorded the feat in his motion picture *Abraham Lincoln*.) So much for liquor never passing his lips! The credibility of Burner's testimony is another matter; but Lincoln, in 1858, ignored Douglas's tribute to his capacity for strong drink, while flatly denying he had ever kept a grocery.

William H. Townsend, a Kentucky lawyer who, like Bullard and Markens among others, made Lincoln his avocation, and who in 1932 was embarked on a study entitled *Lincoln and Liquor*, wrote to Tarbell: how could she reconcile Lincoln's denial with the evidence that the Berry and Lincoln store was a grocery and sold liquor by the drink? She could not, but thought that Lincoln was reacting against the bad connotations of the word "grocery." After all, as she had said before, and as Townsend would say in his little book, the tavern license authorizing sale of liquor by the drink was issued only after the Berry and Lincoln partnership "winked out," in part because of the partners' disagreement over this very issue, and although Lincoln's name was on the bond, it was not his signature.[81]

The temperance forces made much of Lincoln's advocacy in the legislature and on the platform during the score of years before he was elected President. The evidence was not unmixed, however. Near the close of the 1840 legislative session, radicals moved to amend a bill that would further restrict liquor sales by mandating total prohibition, but the move was squelched by Lincoln's motion to table. Like Saul among the prophets, declared the anti-prohibitionists in later years, Lincoln rose to smite the enemy. In 1842 he made an address before the Washington Temperance Society of Springfield. In it he appealed for reform through reason and persuasion rather than denunciation and coercion. "It is an old and true maxim, that a drop of honey catches more flies than a gallon of gall." The address was published; temperance moderates always found it to their liking. In 1855, according to the prohibitionists, Lincoln actually drafted a stiff Maine Law and stumped the state for its adoption. Basically, this story rested upon the testimony of one man, the Reverend James B. Merwin. Beginning with an address at the Lincoln Monument in Springfield in 1904, Merwin presented himself as a close associate during this campaign and later in Washington. Unfortunately,

---

TOP LEFT: The Wets Publicize Lincoln the Tavern Keeper, *c.* 1910. (Illinois State Historical Library) BOTTOM LEFT: Lincoln Pledging Cleopas Breckenridge to Total Abstinence. Drawing by Arthur I. Keller, in Louis A. Banks, *The Lincoln Legion*, 1903.

he said, all his correspondence was consumed in the Chicago fire. (None may be found in Lincoln's *Collected Works.*) The Anti-Saloon League, in which Merwin was a leader, took his word as gospel. After his death, Charles T. White, in his book *Lincoln and Prohibition,* made Merwin and Lincoln the heroes of the movement.[82]

During the war Lincoln commissioned Merwin a chaplain with a mission of temperance to the troops. The President seems to have been pleased with his work. But he would surely have been astonished by the words Major Merwin put into his mouth on the day of his death.

> "Merwin, we have cleaned up a colossal job. We have abolished slavery. The next great movement will be the overthrow of the legalized liquor traffic, and you know my heart and my hand, my purse and my life will be given to that great movement. I prophesied twenty-five years ago that the day would come when there would be neither a slave or a drunkard in the land. I have seen the first part come true."

Prohibition was thus Lincoln's testament to posterity.[83]

Wets and moderates repeatedly challenged the prohibitionists with their own verbal concoctions. The most fascinating of these originated in Atlanta in 1887. The question of local option in the sale of alcoholic beverages was before the electorate. Its advocates circulated a handbill which invoked Lincoln's name and quoted him at length. "Prohibition will work great injury to the cause of temperance. It is a species of intemperance itself . . . ," said the handbill. "A prohibitory law strikes a blow at the very principle on which our Government was founded. I have always been found laboring to protect the weaker classes. . . . Until my tongue be silenced in death I will continue to fight for the rights of man." When the quotation kept popping up, Nicolay and Hay were asked about it, and they said it was fabricated. There is, in fact, no source for the statement in Lincoln's speeches and writings; users sometimes traced it to his supposed speech in the legislature in 1840, for which there is no record. This remarkable piece of spurious Lincolniana continued to be employed in public controversy into the middle of the twentieth century.[84]

Lincoln came to the temperance cause through the Washingtonian movement, and he remained faithful to its spirit. It was non-sectarian; it employed methods of persuasion and self-improvement; and in its work it used such devices as the pledge of abstinence. The Lincoln Legion, organized in 1903 under the umbrella of the Anti-Saloon League, professed to be guided by this spirit. Lincoln was the hero of total abstinence. The book that introduced the Lincoln Legion to the public told how its founder, Dr. Howard Russell, stopped at the Springfield drugstore Lincoln had once patronized and learned from its current proprietor about a man still living in Sangamon County whom Lincoln had pledged to total abstinence as a boy. This was Cleopas Breckenridge. He came to Springfield and told his story to Russell. At a temperance meeting in the schoolhouse during the summer of 1846 or 1847, the speaker, who Breckenridge later realized was Abraham Lincoln, was mighty persuasive, and after finishing distributed a pledge he had written himself. As ten-year-old Cleopas signed, Lincoln said to him, "Now, sonny, you keep that pledge, and it will be the best act of your life." And he did. It was supposedly this pledge that the Lincoln Legion adopted, prefaced by the statement, "That we accept the inspiration of the life and teaching of Abraham Lincoln upon the question of Abstinence from Intoxicants; that we recognize his temperate example as a model for our practice, and that we honor his sacred memory by organizing a host of pledged abstainers." The untold number of youths who signed this pledge presumably associated their vows with all the rewards of a life lived in Lincoln's footsteps.[85]

AS THE TRADITION of the Lost Cause waned in the South, so did the old southern animus toward Lincoln. That tradition was amazingly resilient, however. In 1911 a professor would be dismissed from the University of Florida for telling his students the North had been right in the Civil War and Lincoln was a greater man than Jefferson Davis. Books full of the old slang and sarcasm continued to be published, for instance *A True Vindication of the South*, in 1917, by a Georgia lawyer, recently deceased, Thomas N. Norwood. Lincoln was a gawk, a boor, and a monster. "Caesar had his Brutus! Charles the First his Cromwell, and Abraham Lincoln his John Wilkes Booth!" Donald Davidson, one of the younger generation of tradition-scarred writers, persisted in viewing the Lincoln Memorial, staring across the river at the Lee mansion, as an affront to southerners.

The Lincoln myth endangered the Confederate myth. The latter was sustained, in large part, by the United Confederate Veterans, the United Daughters of the Confederacy, and similar organizations which shrank and dwindled with the passage of years. In 1922 the veterans held their reunion in the old capital of the Confederacy, Richmond, Virginia. They heard the report of the Historical Commission headed by Mildred ("Miss Milly") Rutherford of Athens, Georgia. "The War Between the States," it affirmed, "was deliberately and personally conceived and its inauguration made by Abraham Lincoln, and he was personally responsible for forcing the war upon the South." Cheers greeted the report. It was adopted by acclamation. New winds were blowing across the Southland, however. Leading newspapers condemned the action. Respect for the Confederacy did not require making a villain of Lincoln, one said. "His name is as high in the Southern States today as anywhere in the nation."[86]

Another incident in Richmond six years later offered a second gauge of changing southern opinion. The Virginia General Assembly adopted a resolution to adjourn in honor of Lincoln's birthday for the first time in its history. The author of the resolution, R. Lindsay Gordon, said that "every southern gentleman now agrees with Lincoln on the slavery question" and that his murder by a "Southern lunatic" was a grievous blow to the South. Keepers of the Confederate flame rose in protest. Major Giles B. Cooke, the lone surviving officer of General Lee's staff, said the resolution dishonored the Confederate hero; the UDC demanded the General Assembly expunge this blot on the historical record. A sympathetic northern Democrat, Claude G. Bowers, said the resolution was "simply a cold-blooded, vote-getting proposition of the Republicans," and warned "that the Southern Democrats who line up in these hallelujahs to Lincoln" were the dupes of their political foes. But the resolution stuck. Even the Richmond newspapers, the *Times-Dispatch* and the *News Leader*, commended it. "However clear or beclouded Lincoln's title to greatness may be," the former editorialized, "it is conceded by all that his fame is second to none in the annals of this country," while the latter declared flatly, "Whoever abuses Lincoln sinks himself."

Although some northerners held their noses at comparisons between Lincoln and Lee, the Plutarchian parallel as developed by historians like Woodrow Wilson and William E. Dodd lifted Lincoln in the estimation of southerners. An Emory University professor, speaking at the dedication of the heroic relief sculpture of Lee at Stone Mountain, Georgia, astonishingly broke into an encomium of Lincoln.

> I, son and grandson of Confederate officers . . . stand uncovered at the name of Lincoln. Let us thank God that in the holy of holies of America's heart sleep such great ashes. Let us thank God that in the morning stars of the flag above us shines the gentle and immortal light of his soul. Son of the cabin, child of the wilderness, we salute you!

Schoolboys in Alabama cities, a scholar has said, cited Lincoln more often than Lee as their ideal historical character. Schools across the South set up busts of Lincoln, and children lisped the Gettysburg Address on Lincoln's birthday, as had generations of children in the North.[87]

The most recalcitrant of the resisters to the Lincoln myth invading the South was Lyon Gardiner Tyler. Born in 1853, the son of President John Tyler and his second wife, Julia Gardiner, he taught at the College of William and Mary before becoming its president in 1888, the office from which he finally retired in 1919. Most of this time, indeed until his death fourteen years later, Tyler edited a historical magazine to which he was the principal contributor. His public crusade to denigrate and, as he saw it, demythicize Lincoln was set off by an editorial in the *New York Times*, "The Hohenzollerns and the Slave Power," in 1917. The identification of the planter aristocracy of the Old South with Prussian militarism was false and slanderous. Rather it was President Lincoln who had behaved as a Hohenzollern, said Tyler. "Indeed, no two men ever stood further apart in principle than Wilson and Lincoln." For in the World War the United States was fighting for the rights of small nations, for self-determination, for democracy—the very foundations of the Confederacy in 1861. Other professional southerners made the same associations. Had not Lincoln "Shermanized" the South, just as the Kaiser had devastated Belgium? Were not "the unspeakable inhumanities" of the German government analogous to those unleashed by Lincoln with the Emancipation Proclamation?[88]

Once started, Tyler never stopped. He fought Demon Lincoln with the same tenacity that prohibitionists fought Demon Rum. In an article, "Propaganda and History," Tyler placed the Lincoln myth in the larger context of Yankee domination of American history and culture. Turning the myth on its head, he said Lincoln was coarse, inhumane, undemocratic, weak and indecisive, and irreligious. His unprovoked aggression started the war. There were no mitigations of his character or his statesmanship. For too long southerners had believed that but for his assassination the horrors of Reconstruction would have been averted. This was part of the myth. "Lincoln," in fact, "by his abolition policy . . . was the true parent of reconstruction, legislative robbery, negro supremacy over their masters, cheating at polls, rape of white women, lynching and the acts of the Ku Klux Klan." Almost every issue of *Tyler's Quarterly* contained something derogatory to Lincoln, if not by the editor then by one of the corps of literary gladiators—Landon C. Bell, Paul S. Whitcome, David R. Barbee—he recruited for the crusade. They provided the texts of speeches in the United States Senate by the South Carolina demagogue Cole Blease.[89]

Tyler, along with Mildred Rutherford, led the fight against the subversive influence of northern textbooks in southern schools. The fight antedated the Civil War, heated up in the 1890s, and came to a head thirty years later. History textbooks were the main concern. Too many of them propagated the Lincoln myth and the northern view of the War Between the States. Tyler welcomed publication of a new and revised edition of R. G. Horton's *Youth's History of the Civil War*, a Copperhead work from 1866, since it was one of the few such books written from a southern point of view. He believed that northern propaganda, by the resentments it engendered in the South, kept up the old division in the nation. The Lincoln myth, instead of promoting harmony and reconciliation, as misguided southerners like Henry Watterson had believed, promoted discord. In 1931 the Virginia Board of Education, shockingly, replaced a history textbook written by a native Virginian, John H. Latané, by one written by David S. Muzzey of Columbia University. It held that slavery was the sole cause of the war and exalted Lincoln, "the Savior of our Country," to the eminence of Washington. Muzzey had agreed to certain changes, for instance the substitution of "inexcusable" for "unworthy" in describing the southern cause, and this had apparently sanitized the book for the educators. But it was an atrocity, Tyler declared, and he fought against the book's adoption.[90]

All in vain. Tyler could take some satisfaction in the rise of revisionist thinking on the war among northern writers like Claude G. Bowers and Edgar Lee Masters. But the Lincoln myth was unshaken. The national spotlight fell upon Tyler in 1928 because of his part in the protest against the birthday resolution. *Time* published a story, "Tyler v. Lincoln," in which it not only ridiculed his opinions but dismissed his father, President Tyler, as "historically a dwarf" beside Lincoln. This provoked a pamphlet, *John Tyler and Abraham Lincoln: Who Was the Dwarf?* wherein the two were compared to the latter's great disadvantage. *Barton and the Lineage of Lincoln* followed. It assailed the biographer for claiming a blood relationship between General Lee and "the blackguard Lincoln."

Did the gods laugh when Tyler died on Lincoln's birthday in 1935? An editorial in the *New York Times*, the newspaper that started him on his mad career, called Tyler perverse and wrong-headed but paid respect to his tenacity. Another Virginian of the time, James Branch Cabell, the novelist, liked to play with the ironies of the southern tradition. What if John Wilkes Booth had succeeded in his original plan to abduct the President? Cabell wondered. For it was the assassination that deified Lincoln. Poor Booth, his masterwork went unrecognized and unacclaimed. There ought to be a monument to him. "You gave us," said the author in an imaginary letter, "in place of the dingy truth, a demi-god and a national messiah." Cabell, with most southerners in the second quarter of the twentieth century, subscribed to the Lincoln myth. In 1937, the historian-general of the UDC, Dolly Blount Lamar, addressing the Georgia convention, ended her peroration thus: "Let the world know of the wisdom, the kindness, and the justice of the great President of the Confederate States of America, Abraham Lincoln!" It was just one of those slips, she said embarrassingly. But *Time* took it more seriously. Here, finally and definitively, the South's subconscious belief in the superiority and righteousness of Lincoln was acknowledged.[91]

6

From Memory to History

AT THE FIFTIETH ANNIVERSARY meeting of the American Historical Association on December 28, 1934, James G. Randall, a fifty-three-year-old professor of history at the University of Illinois, read a paper which in retrospect would be considered a landmark in the scholarly study of Abraham Lincoln. The occasion fittingly called forth learned reassessments of the historiography on many familiar subjects. Randall's paper was entitled "Has the Lincoln Theme Been Exhausted?" Answering that question, he disparaged the vast literature that had piled up around Lincoln and noted that professional historians had rarely ventured into the field. Lincoln was an old subject, but the study of Lincoln under the discipline of scholarly inquiry was young, indeed it had scarcely begun. Randall's essay proved "a bugle call" to the profession. The bulk of Lincoln scholarship for the next twenty years was a response to it.[1]

Randall had staked out his claim a year earlier with a long, brilliant, and authoritative article on Lincoln in the *Dictionary of American Biography*. Now, before his academic audience in Washington, he described the rich opportunities that awaited researchers in the field. Lincoln's writings had yet to be properly edited. The research in newspapers, as they might document Lincoln's movements, thoughts, and activities, had been meager. His record as a lawyer was woefully incomplete, even for his cases before the Illinois Supreme Court. Every other year brought announcements of "new light" shed upon Lincoln; more often than not, however, it turned out to be "the light that failed." Thus it had been with the alleged discovery of the Lost Speech. Randall mentioned several doctoral dissertations emanating from his seminar at the University of

Illinois, for instance Harry E. Pratt's on David Davis and William A. Baringer's on Lincoln's rise to power between 1858 and 1861; and he listed a range of topics that awaited close monographic study: Lincoln in the Illinois legislature, Lincoln as President-elect, Lincoln and foreign affairs, among others. So much that had been written about Lincoln had been written from the point of view of the Republican party, said Randall, that "the historian must turn revisionist." This was particularly true about Lincoln and the Civil War, of course, but Randall challenged the historian to study the American past strictly on its own terms and hew to the line of disinterested inquiry wherever it might lead. "Not only must he be free from party and sectional bias; he must be *innocent of the hero tradition*." How that might be possible after all that had gone before, and in view of Lincoln's place in the American pantheon, the professor did not say.[2]

Almost seventy years after Lincoln's death, Randall could declare there was still no acceptable biography of him. Thus were the recent efforts of William E. Barton, Nathaniel Wright Stephenson, and others, to say nothing of older books, implicitly dismissed. The last several years had seen an explosion of Lincoln biography. Unfortunately, the most impressive works were incomplete. Only the first part, *The Prairie Years*, of the poet Carl Sandburg's biography had appeared; and Albert J. Beveridge's *Abraham Lincoln* stopped in 1858, when the author's pen expired with him, and so it remained a great truncated monument of American biography.

One of the most important developments of the time, coincident with the rise of historical scholarship, was the organization of the Lincolniana enterprise. Associations in Illinois and Indiana, particularly, provided skillful direction to the multifaceted activity of researching, preserving, and commemorating the Lincoln heritage; and although its main thrust was "the hero tradition" of which scholars were counseled to be innocent, it submitted itself increasingly to exacting standards. The scandal that erupted in the sedate pages of the *Atlantic Monthly* in December 1928, concerning the Ann Rutledge romance, proved a caution to every Lincolnian. Some, like Ida Tarbell and Sandburg, were embarrassed by it, while others, like Paul M. Angle, director of the Abraham Lincoln Association in Springfield, exalted in the crushing triumph of fact over fiction, scholarship over reminiscence and romance. If Randall's paper in 1934 was "a bugle call" to the profession, then the so-called Minor Affair in the *Atlantic Monthly* stood as the watershed between those for whom Lincoln was an affectionate memory and those for whom he was a great historical figure, still raised upon an altar, to be sure, but an altar that neither needed nor desired mortal polishings.

Organizing the Lincoln Enterprise

Robert Todd Lincoln's lock on his father's papers remained a deterrent to scholarship. "The Prince of Rails," to give him his popular title, did not fit the Lincoln legend. The double entendre of the title acknowledged that he was son of the Railsplitter as well as the president of a great railroad company. Easily dismissed as but a pale shadow of his father, Robert Lincoln was, in truth, that rarity among sons of Presidents, "A Man in His Own Right," as maintained by his biographer. Among Republican leaders he had long commanded widespread support among Negroes, and as president of the Pullman Company—the nation's largest employer of Negroes—he had earned a reputation for benevolence, though this was disputed by the Brotherhood of Sleeping Car Porters.* The awkwardness of being Abraham Lincoln's son contributed to Robert's reticence on the public stage. He loved to hear his father extolled, but only once, at Knox College, chose to speak at an occasion honoring him. After President McKinley's assassination in 1901, Robert Lincoln became something of a recluse because of what he perceived as "a certain fatality about the presidential function when I am present." He had, in fact, been present at the assassinations of both Garfield and McKinley; to this was added the lifelong burden of guilt over his absence from the presidential box at Ford's Theater on the night of April 14, 1865.[3]

Robert Lincoln remained a jealous guardian of his father's papers. After Nicolay and Hay had finished with them, he persisted in the policy of denying all access, though on occasion, as with Ida Tarbell, he would furnish tidbits to quiet the importunate. He could be alternately petty and furious in the kind of proprietorship he assumed over Lincoln's fame, the latter having been demonstrated in his crusade against Barnard's statue, while the former was shown in his tedious effort to force the publisher of Alonzo Rothschild's posthumous book to drop the title, "*Honest Abe*," on grounds of vulgarity.[4]

But unlike the literary trustees of some other famous men, Robert Lincoln apparently did not purge his father's papers, nor was he anything less than responsible in his custody of them. After Nicolay's death in

---

*In its labor relations, the Pullman Company made much of the fact that its president and chairman of the board of directors was the son of the Great Emancipator. In 1905 the porters responded with a pamphlet entitled *Freemen Yet Slaves Under Abraham Lincoln's Son*, and A. Philip Randolph later said of the company's appeal: "It is a most unhappy and pathetic question, for Abraham Lincoln freed Negroes from economic exploitation as chattel slaves, whereas his son . . . has bent his influence and name to the notorious exploitation of Negroes as *Pullman slaves*."

1901, the Lincoln Papers passed briefly into Hay's care at the State Department, then returned to Robert, who placed them in a bank vault in Chicago. At that time Herbert Putnam requested the collection for the Library of Congress, which he headed, where they might be opened to qualified researchers. Robert Lincoln thought this premature, however. After retiring from the Pullman Company, he sold his Lake Shore Drive mansion and moved to Washington, thereafter dividing his residence between the capital and his country home, Hildene, near Manchester, Vermont, always keeping the Lincoln Papers with him. Putnam, Tarbell, and others turned up the pressure on him to donate this treasury to the Library of Congress, or at least deposit it there.[5]

On May 6, 1919, quite out of the blue, seven trunks of Lincoln Papers were delivered to the Library on deposit from the owner and under an injunction of secrecy. Less than four years later Robert Lincoln deeded the papers to the Library on the condition that they remain closed for twenty-one years after the date of his death. The story was later told, and widely believed after Nicholas Murray Butler vouched for it in a personal memoir, that the papers were miraculously saved from the flames to which Robert Lincoln had consigned them. In August 1923, according to Butler, while he was vacationing in Vermont, a mutual friend alerted him to Lincoln's plan to burn the papers. Rushing to Hildene, he found the octogenarian sitting by the fire, with the papers in an open trunk by his side. "Burn your family papers!" Butler exclaimed. "Why, Robert Lincoln, those papers do not belong to you. Your father has been the

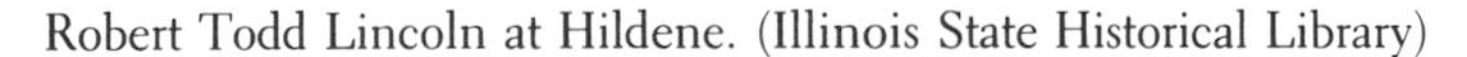
Robert Todd Lincoln at Hildene. (Illinois State Historical Library)

property of the nation for fifty years, and the papers belong to the nation. . . . For heaven's sake, do not do anything like that!" The hero of this little drama, Butler, then persuaded Robert to give the papers to the nation. In fact, of course, they had already been given to the nation and, with minor exceptions, had been in the Library of Congress for four years prior to the university president's visit to Hildene.

Robert Lincoln died in 1926. For the next twenty-one years scholars could rest assured that the Lincoln Papers were in good hands, though they had no access to them. The widow, Mary Harlan Lincoln, had access and might grant it to others, but with one minor exception she chose not to do so. Her death in 1937 extinguished the Lincoln family name. Two daughters married and had children, but the children left no heirs. That melancholy genealogy further enhances Robert Lincoln's place as the solitary descendant history is likely to remember.[6]

As long as the great collection of unpublished manuscripts remained under lock and key, it was useless to think of a new edition of Lincoln's writings. The first edition by Nicolay and Hay in 1894 had purported to be complete. But, as it may be recalled, previously unknown or uncollected letters, speeches, and documents kept turning up as researchers in Shandyite fashion traveled the odd byways and peeked into the dark corners for Lincolniana. First were the 225 new manuscripts plus 500 telegrams Ida Tarbell collected for the appendix to her biography. In 1905 came two new editions, again purporting to be complete, one by G. P. Putnam's Sons in eight volumes, the other by the Francis D. Tandy Company in twelve volumes. (Both contained supplementary material.) The latter of these, which was superior, added 518 items to the 1,736 contained in Nicolay and Hay's *Complete Works*. In 1917 an eighty-year-old Connecticut Yankee, Gilbert Tracy, who had clerked in the War Department in Lincoln's administration, published the *Uncollected Letters* he had painstakingly assembled over many years. In 1926 John D. Rockefeller, Jr., made an extraordinary gift of 485 Lincoln manuscripts to the collection at Brown University, and forty-three of these were published for the first time during that year.

Paul Angle, a name omnipresent in Lincoln studies for a quarter-century or longer, brought out *New Letters and Papers* in 1930. The book added 430 items to the growing catalogue, including Lincoln's eulogy of President Zachary Taylor, which Barton had recovered from oblivion. At this time Angle in Springfield, Louis Warren in Fort Wayne, and Emanuel Hertz in New York City engaged in a not always friendly rivalry to expand the catalogue. In the second volume of his 1931 biography, Hertz made the absurd claim of adding 3,500 documents, of which a thousand appeared in print for the first time. This was false. Besides, as Angle pointed out, the Hertz collection was indiscriminate, undocumented as to sources, and full of fraudulent material. It was, in short, like much of that enthusiastic Lincolnian's work, worse than useless. Warren reported his findings from time to time and in 1938 published his master list.[7]

A new generation of private collectors succeeded the old. Joseph B. Oakleaf, the last of the Big Five, published his long-awaited *Lincoln Bibliography* in 1925. It proved invaluable to collectors. The premier collector of the second generation was Oliver R. Barrett. Born eight years after Lincoln's death in Pittsfield, Illinois, where Nicolay had edited the newspaper, he started his hobby at the age of thirteen. After earning a law degree at the University of Michigan, Barrett settled in Chicago and commenced a successful practice. When it came to Lincolniana he was like a bloodhound in pursuit of game. The paucity of published letters from Lincoln to Horace Greeley sent him on the hunt. He approached the famous editor's surviving daughter in New York and was told that all

Oliver R. Barrett. (Illinois State Historical Library)

but one of Lincoln's letters in the Greeley files had been burned. Barrett finally got that one after her death. He went door to door in Springfield searching for Lincoln's "old house desk." When he finally found it the pigeonholes were still stuffed with letters to the President-elect. And there were other collecting coups, for instance, the "Bonfire Letters," which survived Mrs. Lincoln's housecleaning in February 1861, and the "Carpetbag Papers" discovered in a forgotten repository. Once described as "a veritable Croesus among manuscript collectors," Barrett gathered treasures from many authors and statesmen; but Lincoln was his main interest. No collector knew more about his subject, nor was any more generous toward others in the Lincoln guild.[8]

As in the past, Chicagoans were at the forefront of the Lincoln enterprise. Henry Horner, for many years a Cook County judge, became the New Deal governor of Illinois in 1933. He installed his Lincoln collection of nearly two thousand books and four thousand pamphlets and broadsides in the governor's mansion. The acorn from which the collection had grown was Henry Clay Whitney's *Life on the Circuit with Abraham Lincoln*, a gift of the author's son, Horner's first law partner in Chicago. The collector's zeal carried over to his governorship, where he did what he could to preserve and perpetuate Lincoln's memory. "No man can know Lincoln too well," Horner averred, "and no man can know him at all without becoming better for that knowledge." Alfred Whital Stern, another Chicagoan, long a friend of Horner's began his great collection after a chance encounter with Lincoln's writings. It, too, was essentially of printed Lincolniana. In 1941, however, Stern paid $15,000 for the famous epistle to Major General Joseph Hooker, the most ever recorded for an original Lincoln letter.* Several prominent Lincoln authors—Barton, Hertz, Bullard—also built large collections. Most of them remained intact and went eventually to research libraries. Oakleaf's collection was acquired by Indiana University, Barton's by the University of Chicago, Bullard's by Boston University.

Governor Horner generously donated his collection to the Illinois State Historical Library just before his death in 1940. It became the nucleus of that library's Lincolniana. Not to be outdone by his old friend, Stern

---

*In his letter of some 370 words, dated January 26, 1863, the President implored his new commander of the Army of the Potomac to go forward with vigilance and said: "I have heard, in such way as to believe it, of your recently saying that both the Army and the Government needed a Dictator. Of course it was not *for* this, but in spite of it, that I have given you the command. Only those generals who gain successes, can set up dictators. What I now ask of you is military success, and I will risk the dictatorship." *The Collected Works of Abraham Lincoln*, Roy P. Basler, ed. (New Brunswick, 1953), VI, 78.

presented his collection to the Library of Congress in 1950. The catalogue published a decade later annotated 5,201 items. Barrett's assemblage of Lincolniana went under the auctioneer's hammer in 1952. The estate had offered it whole to the Illinois State Library for $220,000. The campaign to raise the purchase fund fell short of the goal. Nevertheless, with the funds in hand, the library managed to purchase at auction a substantial part of the collection. The prize acquisition was a set of fourteen autograph letters written between 1841 and 1848 to Joshua Speed, for which it paid $35,000. The Barrett sale at Parke-Bernet's, New York, yielded $273,610.[9]

LEADERSHIP IN PURSUIT of Lincoln in Illinois was divided between the State Library, where it was only part of a larger mission, and the Abraham Lincoln Association, which had but one mission. The latter began life as the Lincoln Centennial Association, the corporate spinoff of the state commission in 1908. For fifteen years after the Centennial it continued a kind of half-life, devoted to honoring Lincoln on his birthday and entertaining visiting dignitaries like Lord Charnwood and David Lloyd George. Membership was by invitation only. Dinner dress was black-tie. Most of the members belonged to Springfield's social and business elite; a considerable number were sons and grandsons of men who had known Lincoln. In 1924–25, however, the association was transformed under the presidency of Logan Hay, a local lawyer, who was himself the grandson of Stephen Logan, Lincoln's second law partner, and a cousin of John Hay. The association opened itself to dues-paying members and also hired an executive secretary to set it on a scholarly course.

Hay and his friends were fortunate in the choice of young Paul Angle, who had just finished an M.A. degree at the University of Illinois. Given a makeshift office at the downtown Sangamo Club (exchanged two years later for improved quarters in the National Bank Building), Angle quickly turned the association into a force for the advancement of Lincoln study. Two serials, the quarterly *Bulletin* and the annual *Lincoln Centennial Association Papers*, published speeches and articles and reported work in progress. Better to express its new character, the association changed its name to the Abraham Lincoln Association in 1929. Membership by then had risen to over seven hundred. The avowed purposes of the association, as stated on its letterhead, were "to observe each anniversary of the birth of Abraham Lincoln; to preserve and make more readily accessible the landmarks associated with his life; and to actively encourage, promote and aid the collection and dissemination of authentic information regarding all phases of his life and career."[10]

Paul M. Angle. (Illinois State Historical Library)

A fast learner, Angle soon laid out several promising avenues of research. All concerned Lincoln's Illinois years. Touring the counties of Lincoln's old circuit, Angle stopped in courthouses and ransacked records. "I hunted in basements, in vaults, and in files covered with the dust of three-quarters of a century," he wrote. Basic research on Lincoln's legal career had simply not been done. Until it was, it would be difficult to measure his stature at the bar. Herndon and Whitney, Angle noted, had rated him a notch above mediocre, while others had expressed exalted opinions of his talents. Preliminary investigation of his trials before juries, Angle reported in 1928, did not sustain the common view that he was the best jury lawyer in the state.[11]

Angle was more interested in another project to document Lincoln's movements day to day in Springfield and elsewhere. He began with a critical year, 1858, and published the chronological record in a pamphlet of fifty-six pages in 1926. And he went on year by year. In 1932 the association published *Lincoln 1854–1861. Being the Day-by-Day Activities of Abraham Lincoln from January 1, 1854, to March 4, 1861.* Carried forward, and backward, under Angle's successors, the work would be brought to triumphant conclusion during Lincoln's sesquicentenary. This was fundamental research. Nothing in the annals of biography equaled it. The books of the series, Angle later said, "are to Lincoln study what the steel frame of a skyscraper is to the finished structure." In 1932, worried for the future of the Abraham Lincoln Association during the Great Depression, Angle accepted the position of historian in the Illinois State Library. He was succeeded at the association by a Ph.D. from Johns Hopkins University, Benjamin P. Thomas, whose field of expertise was Russo-American relations in the nineteenth century. But like Angle before him, Thomas quickly turned himself into a Lincoln scholar.[12]

As researchers sought to locate Lincoln precisely in time, so did they seek to locate him precisely in place. The activity was not confined to Illinois, of course. The place where he was born, the places where he had lived and died, the pews where he had worshiped, the platforms he had occupied—every place he had touched became a shrine. But nowhere was this topographical remembrance quite so complete as in Illinois. Henry B. Rankin, in 1915, made an urgent appeal to the citizens of Springfield, heretofore indifferent, to preserve its heritage from Lincoln. Pew Number 20 of the First Presbyterian Church should be set aside; Lincoln's law offices should be appropriately memorialized; a bronze tablet inscribed with his Farewell should be mounted at the Wabash Station. These things would be done. The Lincoln home at Eighth and Jackson was already a state-maintained museum. Angle's book *"Here I Have Lived": A History of Lincoln's Springfield, 1821–1865*, in 1935, was the literary monument of this effort. Over the years the Lincoln Circuit Marking Association erected markers at most of the places Lincoln stopped or stayed or spoke in central Illinois.

> The little towns, the country roads,
> The wood, the prairies, the abodes
> Of humble men where malice fails
> And charity for all avails—
> These are the shrines that still enfold
> The heart of Lincoln as of old
> Whose living legend runneth thus:
> We loved him; he was one of us.

During the World War the Sucker State joined with Indiana and Kentucky to chart Lincoln's route from Knob Creek Farm to Pigeon Creek in Indiana and thence to Macon County, Illinois. This Lincoln Way, dotted with shrines and markers, evolved into the Lincoln Heritage Trail.

If patriotic devotion faltered, the tourist industry hastened to revive it with a liberal dose of nostalgia.[13]

In 1932 Lincoln's New Salem, twenty miles from Springfield, was dedicated as a State Historic Site. Citizens in nearby Petersburg took the first steps to restore the pioneer village, abandoned for some sixty years, in 1906. William Randolph Hearst, the newspaper publisher, purchased the site and conveyed title to a local association. Just over a decade later, the Old Salem Lincoln League commenced research on the ground and erected the first cabins. The restorers relied upon the work of local historians and the memories passed on to the third generation. Thomas P. Reep, the principal researcher, gathered the information thus turned up in *Lincoln and New Salem*. And so the folk tradition was built into the reconstructed village. The state took over the project in 1931 and appropriated $50,000 for construction. Under the direction of Joseph F. Booton, the historical investigation was refined and backed up by archaeological evidence. The cabins erected by the Old Salem League, failing the test of authenticity, were dismantled. Booton's reconstruction, when completed, numbered twenty-three structures. It included the Rutledge Tavern, the Denton Offutt store, and the two Berry-Lincoln stores among the familiar places of the Lincoln story. This village on a bluff above the Sangamon was "Lincoln's Alma Mater," as Barton said. The historic site may be visited simply as a reconstructed pioneer village, without much regard to its most famous resident; but for him, however, it would have remained a deserted clearing in the forest. To the attentive visitor, strolling the thoroughfares, the whole village passes before the eyes as a kind of poem, with Lincoln the hero.[14]

COMPARED WITH THE New Salem period, Lincoln's growing-up in Indiana from 1816 to 1830 remained an obscure chapter in his biography. Herndon had visited Spencer County and interviewed several of Lincoln's boyhood acquaintances, but had dismissed the Indiana years as the "dark age" of his life. Hoosier patriots protested and sought to establish the importance of these years in the shaping of Lincoln's mind and character. The early researches of William Fortune of Boonville, Anna O'Flynn, employed by Ida Tarbell, and J. Edward Murr, a Methodist preacher, were succeeded by an organized campaign under the name "The Lincoln Inquiry." It was spearheaded by John E. Inglehart, the founding president of the Southwestern Indiana Historical Society. Other groups—the Spencer County Historical Society, the Lincoln Trail Club, the Indiana Lincoln Union—joined in the effort to uncover Lincoln's Hoosier heritage. In addition to the assemblage of reminiscences, almost three hundred capsule biographies were written of citizens who had been among Lincoln's friends and neighbors. The aim was to document Indiana's formative influence. "His environment emphasized human equality, ambition, nationalism, shrewdness and native wit." There was no mystery about the man Lincoln became. He was a true product of Indiana pioneering.[15]

In their attack on the "dunghill" thesis, the Hoosiers set forth a positive interpretation—happy, propitious, hopeful—of Lincoln's boyhood. It has been labeled the "chin fly" thesis. The Pigeon Creek farm was not isolated from the great world. The youth, after all, had gone down Anderson Creek to the Ohio, thence to New Orleans. Moreover, he told Leonard Swett "that he had got hold of and read through every book he had ever heard of in that country for a circuit of fifty miles." Tradition had him attending sessions of the district court at Boonville and borrowing books from a prominent attorney, John A. Brackenridge. The boy took lessons from three Hoosier schoolmasters; the Gentryville grocery, where he held forth, was better than any schoolroom; and Robert Owen's experiment at New Harmony was part of his intellectual universe, though unhappily out of reach physically. Visiting his old home in 1844, Lincoln remarked it was "an unpoetical as any spot on earth," yet he was stirred to write the sentimental verses beginning "My childhood's home I see again."[16]

Louis A. Warren became the leading authority on Lincoln's youth. After he moved to Indiana, "the Lincoln Ministry," as Warren called it, crowded out the ministry of Jesus Christ in his life. While working for the Indiana Lincoln Union, which aimed to build a memorial to the "angel mother," Warren met Arthur F. Hall, president of Lincoln National Life Insurance Company in Fort Wayne. Hall had decided that the company, having prospered under the motto "Its Name Indicates Its Character," owed the public a larger service in that name. In 1928, impressed by Warren's talents as a scholar and lecturer, Hall offered him the position of director of a new historical department, named and endowed the Lincoln Historical Research Foundation. Beginning with a blank slate, Warren committed the foundation to two main purposes: first, to build a library of Lincolniana, and second, to serve as a medium of communication and education about Lincoln. The acquisition in 1930 of Judge Fish's collection put the foundation in business as a research library. Warren had already started a single-page weekly bulletin, *Lincoln Lore;* and he would soon start a one-man speaker's bureau, through which he made as many as two hundred speeches a season—the Lincoln season running from January to March—in all parts of the United States.[17]

At the dedication of the foundation on February 11, 1931, President Hall declared:

Louis A. Warren. (The Lincoln Museum, Fort Wayne, a part of Lincoln National Corporation)

The Hoosier Youth, by Paul Manship, Fort Wayne, Indiana, 1932. (The Lincoln Museum, Fort Wayne, a part of Lincoln National Corporation)

> No motive of commercialism or profit entered into our plans to assemble this wealth of Lincolniana. We seek merely to provide the means and the channel through which there may continue to flow an ever increasing volume of information concerning Lincoln, especially to the youth of our land, that they may be influenced to think and to live as Lincoln did—"with malice towards none and charity for all."

In this spirit the company had commissioned Paul Manship to carve in marble a heroic statue of the Hoosier youth. Warren worked closely with the sculptor to resolve ticklish questions of pioneer dress and accouterment. Did the lad wear suspenders? Did he wear boots or moccasins? What did his ax look like? And assuming he had a dog, what was the breed? The statue was dedicated before the company's headquarters building in 1932. Named *The Hoosier Youth,* it portrayed Lincoln twice life-size sitting on a stump at the age of twenty-one. The conception was idealized, clean-cut, like most of Manship's work. It represented the dreamer and poet, rather than the railsplitter, though an ax was one of three properties the sculptor used, the others being a book, held in one hand, and a hound, under the other hand. Speaking at the dedication, Ida Tarbell acclaimed the statue as an idealization of beauty, health, and aspiration, and predicted it would do much good to present-day youth. The editor of an Iowa newspaper read the statue thus: "Success in life will come for any boy who rightly divides his time, as Lincoln did, between his dog, his book, and his axe." For the next half-century, troops of Boy Scouts massed before *The Hoosier Youth* on Lincoln's birthday.[18]

Although Warren had begun research on *Lincoln's Youth* before he became director of the Lincoln National Life Foundation, it was not finally published until 1959. But he never left anyone in doubt as to his views. The Hoosier heritage was entirely consistent with the gentle heritage Warren had established in Lincoln's *Parentage and Childhood.* He flatly denied that Thomas Lincoln and his family had spent the first winter in a crude "half-faced camp" open on one side, and that the eighty-acre farm was located a mile from water and provided but a miserable existence. Thomas was "a worthy parent," said Warren, not only an industrious farmer but a skilled carpenter and cabinetmaker, some of whose works survived to become antiques. In the "dunghill" tradition the father discouraged his son's efforts at education, even cuffing him for reading a book rather than feeding the hogs. Warren, on the other hand, insisted that Thomas Lincoln encouraged his son's education, just as his mother and stepmother did. What was encouraged and what was discouraged may have contributed to the prevailing confusion on this issue. According to Swett, Lincoln told him, "My father had suffered greatly for the want of an education, and he determined at an early day that I should be well educated. And what do you think," Lincoln continued, "his ideas of a good education were?" The answer: elementary reading, writing, and arithmetic. At any rate, said Warren, when the family pulled up stakes to go to Illinois it was comfortably situated in Spencer County. Thomas was an upstanding member of the Little Pigeon Baptist Church as well as a respected carpenter and cabinetmaker. He owned a hundred-acre farm, with forty in cultivation; he had just sold one hundred hogs at market and had begun building a new house. In the light of Warren's account, the dumbest thing Thomas Lincoln ever did was to move to Illinois.[19]

The Lincoln Inquiry came to fruition in the 1930s. Memorials sprouted on the southern Indiana landscape in and around Lincoln City. Today the National Park Service administers the Lincoln Boyhood National Memorial, which incorporates the Nancy Hanks Memorial and the Lincoln Living Historical Farm. Historic Rockport on the Ohio River boasts the Lincoln Pioneer Village, a piece of make-believe, which includes among its fifteen cabins an alleged replica of the Pigeon Creek homestead. The epoch found its fictional chronicler in Bruce Lancaster, whose *For Us the Living* was published in 1940. Young Hugh Brace and his family are neighbors of the Lincolns and follow the same migratory trail. Abe is portrayed as a great talker and hell-bent for learning. "Thet Abe!" someone says. "Read in the loft till nigh sunup. Some book he got an' he kean't stop till he's gobbled it down whole." The fictional Abe actually visits New Harmony. The novel's title, of course, came from the Gettysburg Address, but Lancaster has the Hoosier youth saying in a little speech it was "for us, the livin' " that the Founding Fathers risked their lives in the Revolution; and it is on this note that the book ends with Lincoln's decision to run for the legislature from Sangamon County.[20]

The Lincoln enterprise spanned the continent. In 1932 Robert and Alma Watchorn presented to the city of Redlands, east of Los Angeles, the Lincoln Memorial Shrine—the only memorial to the sixteenth President west of the Mississippi. Watchorn, an American rags-to-riches story, devoted much of his fortune to building the octagonal marble shrine and endowing it as a research center. The Henry E. Huntington Library, in nearby San Marino, already had substantial holdings in the Lincoln field and would continue to add to them. Another remote outpost of the enterprise was at Lincoln Memorial University, where R. Gerald McMurtry, who had apprenticed under Warren in Fort Wayne, became director of the Department of Lincolniana in 1937.[21]

Second Culmination: Sandburg, Beveridge, and Others

The cycle of Lincoln biography dominated by William E. Barton, the preacher, was distinguished by the masterly works of a poet and a politician, Carl Sandburg and Albert J. Beveridge respectively. Yet it would be a mistake to overlook the earlier work of Nathanial Wright Stephenson. His 1922 biography, *Lincoln: An Account of His Personal Life,* found many admirers and offered a readable alternative to Lord Charnwood's single volume. A fifty-five-year-old professor at the College of Charleston, Stephenson had come to history by way of journalism. As the title of his book suggested—the full title ran on, *Especially of Its Springs of Action as Revealed and Deepened by the Ordeal of War*—Stephenson anticipated the later vogue of psychological interpretation of Lincoln. He postulated "a double life," inwardly solitary and mysterious and outwardly coarse and companionable, the influence of his mother and father respectively. In the ordeal of the war, the two sides were fused into a single coherent and masterful personality. This was provocative but unsophisticated, and more than one reader wondered, with Allan Nevins, if it did not say more about the author than the subject. Stephenson's account of Lincoln's early life was marred by his use of questionable sources, for instance J. Rogers Gore's *The Boyhood of Abraham Lincoln,* in 1921, which wove from the reminiscences of loquacious ninety-year-old Austin Gollaher a kind of Huck Finn and Tom Sawyer story of childhood. Stephenson was more perceptive on Lincoln the President, where he anticipated the historical revisionists in portraying his political struggle as one fundamentally with "the vindictives"—the Radicals—of his own party.

Both Sandburg and Beveridge began their books in the same year, 1922; both were on a generous scale, yet neither reached the period of Lincoln's presidency. Sandburg's *Abraham Lincoln: The Prairie Years* appeared in two volumes in 1926. Beveridge's biography carried Lincoln into October 1858, at which point death stilled the author's pen. Thus left incomplete, it was published in 1928, also in two volumes, though they packed twice as many words as the rival work.[22]

That Carl Sandburg, troubadour of the common man, should write a biography of Lincoln was doubtless decreed by fate. Born of Swedish stock in 1878 in Galesburg, Illinois, he would dedicate his big book to his parents, August and Clara Sandburg, "Workers on the Illinois Prairie," like Lincoln. It was, he liked to say, the biography of the son of an illiterate mother written by the son of an illiterate father. As a boy growing up, he heard the stories of persons who had known Lincoln and participated in their feelings of familiarity. Although he dropped out of school after the eighth grade, his insatiable curiosity led him to devour the books in the public library. Galesburg was the home of Knox College, where the great debate of 1858 was impressively commemorated in 1896. Sandburg, a hustling eighteen-year-old, stole away from his milk wagon to hear the speeches. He roamed the country and roughed it for several years, even enlisted in the Spanish-American War, before returning home and finishing his education at Galesburg's other college, Lombard, at the turn of the century. After graduating, he went to work as a newspaperman, first in Milwaukee, then in Chicago.[23]

A drifter and a dreamer with a gift for self-dramatization, young Sandburg wrote poetry and lectured from the lyceum platform. His first subject, his first hero and inspiration, was Walt Whitman, "Poet of Democracy." In 1904 he made a pilgrimage to the poet's tomb in Camden, New Jersey. Whitman's *Song of Myself* became Sandburg's song, along with his vagabondage of the "open roads" and his "democratic vistas."

> I AM the people—the mob—the crowd—the mass.
> Do you know that all the great work of the world is done through me?
> I am the workingman, the inventor, the maker of the world's food and clothes.
> I am the audience that witnesses history. The Napoleons come from me and the Lincolns. They die. And then I send forth more Napoleons and Lincolns.
> I am the seed ground. I am a prairie that will stand for much plowing.

In 1914 Harriet Monroe's magazine, *Poetry*, published a group of Sandburg's poems, naming them "Chicago Poems." His landmark book of that title came two years later. Sandburg became the heart and soul of Chicago's literary renaissance. *Cornhuskers* and *Smoke and Steel* followed in rapid succession. Not only in books of verse but from the platform, Sandburg sought the "great audiences" Whitman had called for. Broad-shouldered, blond, with a weathered face, a mischievous gleam in his eyes, and a shock of unruly hair over his forehead, Sandburg strummed the guitar as he chanted folk songs and read his poems to popular audiences. Ben Hecht, his colleague on the *Chicago Daily News*, crowned him "the Peepul's Poet." And the Emporia editor, William Allen White, wrote in a fan letter that of all the modern poets "you have put more of America in your verses than any other."[24]

Whitman led him back to Lincoln. He alone was large enough to embody Sandburg's vision of the common man ennobled and enthroned. Lincoln had never been far from his thoughts. A simple quatrain in the *Chicago Poems* posed the question behind all the people's memory:

> Remembrance for a great man is this.
> The newsies are pitching pennies.
> And on the copper disk is the man's face.
> Dear lover of boys, what do you ask for now?

For some time he had been collecting stories and compiling notes about Lincoln. In the Preface to *The Prairie Years* he said he had been planning such a portrait for thirty years. But this was a piece of retrospective license. In between books of poetry, Sandburg had written the *Rootabaga Stories* for children. Its commercial success led him to propose to his publisher, Alfred Harcourt, a children's biography of Lincoln. Thus the work began. Harcourt was dumbfounded three years later when he received the manuscript of *The Prairie Years*.[25]

Sandburg was a novice, unapprenticed in any way, in the writing of history and biography. Nor did he seek help or guidance. The biography was as solitary a creation as one of his poems, though he acknowledged

Carl Sandburg. (Illinois State Historical Library)

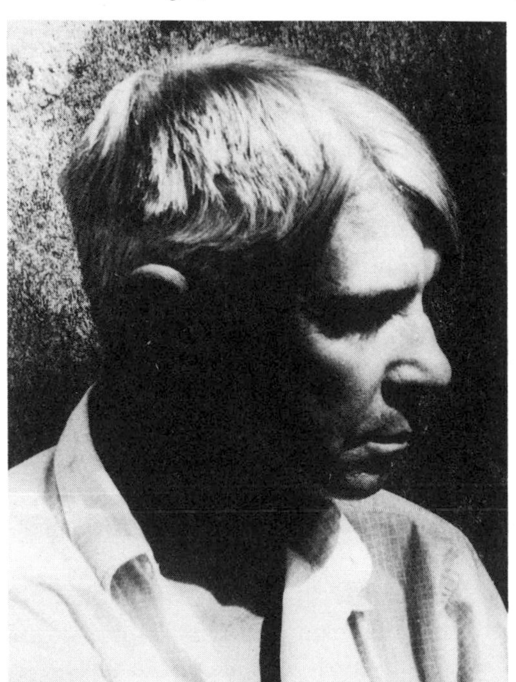

the influence of Ida Tarbell upon his thinking about Lincoln and paid respects to Oliver Barrett and Frederick Meserve for their collections. (Tarbell and Barrett alone received advance sheets of the book.) Sandburg, nonetheless, brought to the undertaking unique qualifications. He felt in his bones that he knew Lincoln. Was he not a rough-hewn dreamer like himself? Were they not both prairie philosophers? "Sandburg's portrait of Lincoln," according to his own biographer, "is a mirror of his own interior life." In this sense, not unlike some other eminent biographies, it was as much a work of self-discovery as it was of discovering another self. While this added to the insight and interest of the work, some readers distrusted "the blueprint seemliness," in Max Lerner's phrase, of the author and the subject.

Sandburg also brought to *The Prairie Years* a prose style commonly more poetic than his verse. His penchant, especially in the early chapters, for rhapsodic passages lush with mawkish sentiment made the glory of the work for some readers while it sickened others. A reporter as well as a poet, Sandburg had a sponge-like capacity to absorb vast amounts of raw data—facts, images, reminiscences—and lay them before the reader in more or less coherent fashion. Whether or not they were well digested is another matter. Edmund Wilson, who hated the book, called it "an album of Lincoln clippings." This was unkind, but it pointed to the tendency of the biography to become, at times, a catalogue of Lincolniana rather like the human catalogues of Sandburg's poetry, which overwhelmed by their sheer mass and, as William Carlos Williams observed, were as "formless as a drift of desert sand."[26]

In Lincoln, Sandburg had a historical vehicle to convey the democratic message of his poetry. Lincoln was the common man writ large; every hunkie, hogbutcher, cornhusker, cop, and prostitute felt affirmed in him. The poet's feelings about the past were mixed. On the one hand, he could announce "the past is a bucket of ashes," and "there is nothing in the world / only an ocean of tomorrows." On the other hand, he acknowledged the danger of forgetfulness and the value of remembrance.

> When I, the people, learn to remember, when I, the People, use the lessons of yesterday and no longer forget who robbed me last year, who played me for a fool—then there will be no speaker in all the world say the name: "The People," with any fleck of a sneer in his voice or any far-off smile of derision.

There was no nostalgia in Sandburg. Remembrance of the past was a condition of freedom. *The Prairie Years* helped to fill the historical void of the poetry. So, too, did his interest in American folklore, which led to the publication of *The American Songbag* in 1927. One section was entitled "The Lincolns and the Hankses," which included songs they supposedly heard and sang.

Not surprisingly, the Lincoln Sandburg most cherished was the folklore Lincoln. *The Prairie Years* was, in fact, the summation of the long tradition of folk memory and reminiscence. Here was no godlike man. As Harry Hansen wrote, "Out of the pages of this book emerges no heroic figure, no epic character, no titan towering above puny men. This is the book of the railsplitter, of the country storekeeper, the young lawyer, the frontier advocate, the practical backwoods politician." And this was well, Hansen thought, for the people could identify with him—he was Everyman—though not with the mythic hero and plaster saint. In this respect, as in others, however, the book was retrogressive, epitomizing the past rather than opening new paths toward the understanding of Abraham Lincoln.[27]

In constructing his portrait, Sandburg used whatever came to hand without much regard to authenticity. He drew freely upon Herndon's writings and gave currency to newer, less familiar, sources, such as T. G. Onstot's *Pioneers of Menard and Mason Counties* (1902). He used Eleanor Atkinson's interview with the octogenarian Dennis Hanks in 1889, unpublished until twenty years later. He, like Stephenson, used Austin Gollaher's reminiscences; he repeated on John Hanks's authority Lincoln's anti-slavery vow, "hit it hard," and had Lincoln tell Mrs. Crawford, "Me? I'm going to be President of the United States." This oral testimony, with much like it, was "deemed authentic" by the author since nothing contradicted it. For the most part, the sources were unacknowledged and unknown to the reader. *The Prairie Years* was published without footnotes or bibliography. Where there was no evidence to speak of, as on Nancy Hanks, Sandburg gave free rein to his imagination.

> She believed in God, in the Bible, in mankind, in the past and the future, in babies, people, animals, flowers, fishes, in foundations and roofs, in time and the eternities outside of time; she was a believer, keeping in silence behind her gray eyes more beliefs than she spoke. She knew . . . so much of what she believed was yonder—always yonder.

As Nancy Hanks cooked and scrubbed and patched, she sang traditional ballads and hymns, among them "From Greenland's Icy Mountains," though, in fact, it was not written until after her death. Sandburg went halfway toward Warren's gentrified view of Thomas Lincoln, conceding he was a respectable citizen and taxpayer, yet thought him more a hindrance than a help to his son.[28]

Sandburg could not resist a good story; if it wasn't true, so much the worse for it. And so young Abe Lincoln, storekeeper, discovered Blackstone's *Commentaries* at the bottom of a junk-barrel, where it seemed to beg, "Take me and read me; you were made for a lawyer." In retelling the story of the Ann Rutledge romance, Sandburg said that the duo sang from *The Missouri Harmony Songbook,* indeed devoted four pages to the subject, yet admitted in passing that Lincoln himself never alluded to the book. Sandburg accepted the tradition of the "phantom wedding," January 1, 1841, as if Ida Tarbell had never written. He credited Henry C. Whitney's version of the Lost Speech and Joseph Medill's recollection of Lincoln's strategy in the Freeport debate.

On the perplexities of Lincoln's reading, the author reported Swett's statement about the fifty-mile range of his access to books in Indiana, and said he later "read as he chose" in Herndon's library. But what did he choose? According to Sandburg, he read carefully a second time Robert Chambers's *Vestiges of Creation,* a work of natural religion, yet Sandburg adduced no proof of either a first or a second reading. It was essential for Sandburg to believe that Lincoln read and responded to *Leaves of Grass.* Herndon had been strangely silent on the subject, but Sandburg cited Rankin's testimony that the senior partner read in the book often.[29]

Because the author's interest was in making "a certain portrait," he blurred the line between verifiable and imagined truth, and glided over important historical questions that depended upon analysis of cause and effect relationships. Thus no one would, or should, turn to *The Prairie Years* as an authoritative source on the historical Lincoln. In three or four chapters interspersed throughout the work, Sandburg chronicled in broad strokes the larger events and developments of Lincoln's lifetime. The purpose, clearly, was to relate that life to the growth of the country, thus to make it the country's epic, but these chapters stood apart from the biography, despite the contrivances employed to relate them, for instance sitting Lincoln in his "rattletrap buggy" on circuit as the historical canvas, circa 1850, unfolded before him. Sandburg's Lincoln was a profoundly indigenous character, one who grew from the common folk and the great national experience of pioneering. His genius was rooted in physical nature.

> He lived with trees, with the bush wet with shining raindrops, with the burning bush of autumn, with the lone wild duck riding a north wind and crying down on a line north to south, the faces of open sky and weather,

> the ax which is an individual one-man instrument, these he had for companions, books, friends, talkers, chums of his endless changing soliloquies. . . . He grew as hickory grows, the torso lengthening and toughening. The sap mounted, the branches spread, leaves came with wind clamor in them.

Except for the poet's touch, this was scarcely an original conception. What distinguished Sandburg's portrait, above all, was its massive and colorful detail. Sometimes, as Mark Van Doren said, Sandburg became "drunk with data." In one of his longest chapters (137), devoted to characterizing the mature Lincoln, he stitched all the reminiscences and anecdotes he had left over into a veritable crazy quilt. By 1854, in the author's view, Lincoln had become the representative man of the country. "He had arrived at a sense of history," having seized with "stubborn grandeur" the meaning of the Declaration of Independence for the burgeoning crisis over slavery. And like the representative American, Lincoln was tall, elusive, quizzical, sad, comic, and mysterious. "Everybody knew him and nobody knew him." He was, said Sandburg, the Strange Friend and the Friendly Stranger.[30]

Sandburg ended *The Prairie Years* with Lincoln's farewell to Springfield. He considered the work complete and had no intention of going beyond it. He actually wrote a special preface which surveyed the balance of Lincoln's life backward from his death to his inauguration as President. But author and publisher decided not to use this. And so overwhelming was the success of the biography that Sandburg soon found himself laboring on the sequel, *The War Years*. "Ain't it hell the way a book walks up to you and makes you write it," he wrote to Amy Lowell.[31]

Publication of the book was preceded by serialization of less than one-fifth of it in the *Pictorial Review*. For this Sandburg received $21,600, more money than he had ever seen before. Critical acclaim followed. The book sold forty-eight thousand copies at ten dollars the set during the first year. Reviewers lauded the work for its lyrical prose—"the best, the noblest, poem that Carl Sandburg has yet written"—for the extraordinary vividness of its Lincoln portrait, and as an American epic. Reading the advance sheets, Tarbell immediately recognized the unconventionality of the biography. "*I believe you've done a new kind of book,*" she wrote approvingly. Professional historians, generally, disapproved of the work. Whatever else it was, poetry or fiction or "literary grab bag," said Milo Quaife, editor of the *Mississippi Valley Historical Review*, it was not history as the term was understood among scholars. But, of course, the work was not written for historians or Lincoln specialists. To their criticisms H. L. Mencken brayed, "Are the facts all respected? Is the narrative satisfactory to the professors of Lincolnology? To hell with the professors of Lincolnology."[32]

ALBERT J. BEVERIDGE, another midwesterner, seems not to have crossed paths with his rival in Lincoln biography. Apparently Beveridge wrote to him in 1925, inviting correspondence, but Sandburg, having heard from various persons on the Lincoln grapevine that Beveridge had disparaged his work as inauthentic, wrote a stinging letter in reply, then thought better of it and made no response. Born in 1862, the son of a Union soldier, Beveridge grew up a fervid Republican. He worked his way through DePauw University, practiced law in Indianapolis, and acquired a reputation for flag-waving oratory. In 1898, the year Sandburg enlisted to fight the Spaniards, Beveridge spoke on Lincoln at the annual banquet of the National Republican Club in New York City; some months later he stampeded the Republican state convention with his oration, "The March of the Flag," and in 1899, at the age of thirty-six, bolted into the United States Senate. There he was an expansionist in foreign affairs and a champion of "progressive" legislation at home. He became a leader of the "insurgency" against the Republican Old Guard during the Taft administration. This led to his defeat for a third term in 1911. Two years later Beveridge embarked on a new career in American history and biography.[33]

He undertook a major biography of John Marshall, the great Chief Justice of the Supreme Court. Marshall became a vehicle for charting the rise of the American nation. Always Beveridge's heroes were ardent nationalists, Bismarck and Disraeli in Europe, and in the United States Alexander Hamilton above all others. But for Frederick Scott Oliver's study of Hamilton in 1906, he would surely have chosen to write an epic biography of him. Beveridge began as a rank amateur, just as Sandburg would nine years later, but unlike the poet-turned-biographer, the politician, Beveridge, admired professional historians and sought to emulate them. *The Life of John Marshall* was completed in four volumes in 1919. Marshall died in 1835, when the future of American nationalism was clouded; obviously the theme should be pursued through the Civil War, and Lincoln was the obvious vehicle.

Lincoln appealed to Beveridge not only as Savior of the Union but also as an immensely human subject, like Marshall, whose life was full of movement and color. Seeking to convey the personality of a man little known to history outside the pages of the Supreme Court Reports, Beveridge actually wrote, "We must imagine a person very much like Abraham Lincoln." And he sketched an extended parallel. Marshall and Lincoln were remarkably alike physically and intellectually; both were fun-loving and companionable; both were logically acute, lucid, and penetrating; both had a genius for managing men; both were men of the people. Beveridge expected that a biography of Lincoln would fall into place as felicitously as had the *Life of Marshall*. He conceived of it as "a companion piece" and planned it on the same scale. Politics reared its head again in 1922, as Beveridge sought to return to the Senate, but no sooner had he been defeated that he began work on Lincoln.[34]

The biographer moved heaven and earth to gain access to the papers under Robert Todd Lincoln's control, but was repeatedly rebuffed. Jesse Weik, a fellow Hoosier, was more obliging, placing in the author's hands the huge collection of Herndon manuscripts in his possession. Weik had again drawn on the papers for *The Real Lincoln*, which Beveridge had persuaded his own publisher, Houghton Mifflin Company, to bring out. The Herndon archive was growing stale. Every biographer used it. Sandburg drew many of his stories from Herndon. And now Beveridge relied implicitly upon these materials. William E. Barton, who would review the biography with faint praise in a historical journal, regretted the author had been led astray by Herndon and wished Weik had withheld the archive from him.

Beveridge made himself acquainted with most of the leaders of the Lincoln guild. He met Warren and Townsend during his travels in Kentucky; he knew Inglehart and others of the Hoosier inquiry; in Chicago he met Barrett and marveled at the wonders of his collection; in Springfield he made friends with Paul Angle. Dismayed by how little solid research had been done on Lincoln's Springfield years, he enlisted Angle's aid. The "dear boy," as Beveridge called him, was then working on Lincoln's law practice and so proved a godsend. At the University of Illinois, Beveridge got Theodore C. Pease, a specialist on antebellum Illinois, to organize systematic research in the newspapers. But of the guild generally, he formed a low opinion. Barton was a preacher, who hunted up evidence to support his preconceived theories. The worst was Ida Tarbell. Beveridge tagged her "Mid-Victorian" and said, "The dear girl's efforts to fumigate are pathetic." He turned her name into a verb, "to Tarbellize," meaning to elevate and sanitize. Most of the biographies only got in the way of the truth.

After eighteen months of hard work, Beveridge felt desperate. When he embarked on the project, he had supposed most of the facts had been dug out and he had only to assemble them according to his plan. But this was not the case. Unlike Marshall, Lincoln remained a mystery. Moreover, the Lincoln he was discovering did not conform to the conception he had formed of him. "Lord! Pease," he wrote in August 1925, "what a job it is. I almost despair; and I am sometimes so appalled at the magnitude of the task that I well-nigh feel like yielding to the pressure . . . to get into Indiana politics again just as a reason to give up this undertaking." He persevered, yet remained baffled about Lincoln and wondered if he would ever be able to explain his greatness.[35]

No biographer ever labored more diligently than Beveridge. He arose at 6:00 a.m. and was soon at his desk, where he worked for ten to twelve hours a day. The biography became his life. He declined invitations to speak or lecture and put aside all writing unrelated to the biography. An intense man, bursting with energy, Beveridge conveyed something of his personality in his animated literary style. To organize that energy for the slow and plodding work of historical biography could not have come easily, yet Beveridge accomplished it with a discipline rarely equaled. He had an absolute passion for accuracy. As he said to Angle, when asking him to read several chapters in draft, "I am determined that every statement shall be bomb proof." Ordinarily, he expected to revise every chapter nine or ten times, not just to make improvements of his own but to take account of criticisms and suggestions of readers in the historical fraternity. Nothing about his manner of working stood in sharper

Albert J. Beveridge. (Illinois State Historical Library)

contrast to Sandburg's than his deference to the judgment of professional historians. Beveridge was an active member of the American Historical Association and wished his work to pass muster with that guild regardless of the Lincolnians. In undertaking the *Life of Marshall*, he had gone to J. Franklin Jameson, one of the leaders of the profession, then with the Carnegie Institution in Washington, and asked, and received, his aid as a scholarly counselor. The world of academic scholarship thus opened to him, Beveridge formed the habit of mimeographing his chapters and circulating them for review by up to twenty readers, most of them university professors. He continued this practice with the Lincoln biography. The list of his correspondents reads like a who's who of the guild: Charles A. Beard, Edward Channing, Arthur C. Cole, Hamilton J. Eckenrode, Frank Owsley, among others, as well as such non-academic dignitaries as Justice Oliver Wendell Holmes and Ellery Sedgwick.

The "history sharps," as Beveridge called them, found much to praise in his work. Beard congratulated him on masterly analysis through abundant factual detail. His method was to arrange the facts into a narrative which spoke for itself, without overt interpretation or literary embellishment. When Sedgwick suggested he carried the method too far, Beveridge replied, "Of course the mere fact that anything whatever needs explanation, shows that it is defective." In keeping with his compulsion for fullness and accuracy, Beveridge overburdened his narrative with footnotes, as many as ten to a page. This bordered on pedantry. But it helped to secure his acceptance by the guild. As Benjamin Thomas later wrote, "He was the first biographer to apply the techniques of the trained historian to Lincoln research, though he was not a professionally trained historian."[36]

A good example of his methodical treatment, as compared to Sandburg's, were the chapters devoted to Lincoln's career in the Illinois legislature. Of course, a biographer with Beveridge's political experience might have been expected to take a special interest in this subject. But nothing had prepared him to think it of much importance. He told Barrett he had expected to dispose of Lincoln the legislator in a few pages. Digging into it, however, he discovered that everything to come was anticipated in that career. Whereas Sandburg had been content with a few splashes of color thrown around the experience of Lincoln and the "Long Nine," Beveridge created a vivid drama with a full cast of characters and plots and counterplots. He made extensive use of newspapers and was the first biographer to exploit the *House Journal*. In all his legislative maneuvering, cajoling, and logrolling, Lincoln's chief purpose was to secure removal of the capital from Vandalia to Springfield. It was achieved in 1837. When he moved to Springfield in that year, and became the partner of John Stuart, one of the ablest lawyers in the state, the twenty-eight-year-old Lincoln stood forth as the victorious leader of his party in the legislature. "Astounding progress!" Beveridge marveled. "But yesterday pottering about New Salem in contact only with little things and crude surroundings, heavily in debt and with crude prospects for advancement; today starting on the high road of advancement and achievement!" Beard called the chapter "a revolutionary piece of historical writing," one that "revealed the early Lincoln in a blinding light."[37]

Dwelling on politics, Beveridge built up a realistic portrait in contrast to Sandburg's poetic one. As he followed Lincoln to Congress in 1847, he did not hesitate to criticize him for opportunistically changing his opinion on the Mexican War. He had defended the war to get elected, but upon discovering the Whig opposition in Congress turned against it. His famous speech denouncing Polk and the war rested upon the false idea that the Mexican War was a war of conquest instigated by the Slave Power. Present-day historians, Beveridge said, had exploded this view as Whig and abolitionist propaganda. Lincoln's speech proved his "political undoing" in Illinois; a promising career in national politics ended abruptly. "Measured by any standard, his term in Congress had been a failure." An expansionist in his own time, Beveridge could not look complacently on Lincoln's opposition to Manifest Destiny. Sandburg, on the other hand, said "there were no zigzags" in Lincoln's political course; moreover, the Whigs had been right—Lincoln had been right—in denouncing the war; finally, the notorious speech had little to do with his decision not to seek reelection.[38]

Beveridge confounded his readers by his sympathetic treatment of the South in the sectional conflict of the 1850s and his partiality for Douglas in the political combat with Lincoln. The second volume opened with a glowing picture of plantation slavery, based upon travelers' accounts and the recent work of southern historians like Owsley and Ulrich B. Phillips. Perhaps the interpretation was all a matter of selection and arrangement of data, but Beveridge seemed to make the planters' opinions his own. Thus he said that the slaves were content in their bondage, while the free Negroes were "the most vicious and corrupting element in Southern life." Thus he argued that the southern chivalry would have straightened things out among themselves if only the abolitionists had left them alone. Professor Owsley of Vanderbilt was thrilled by these chapters, saying to the author, "You have caught the *psychology* of the South."

Justice Holmes, on the other hand, who was a summer neighbor at Beverly Farms north of Boston, thought it all humbug. He confessed that Lincoln had not been a personal hero, though in middle age he recognized the growth of the myth, and now accepted the popular judgment of his greatness.*[39] It was a drawback in a biographer, Holmes averred, not to like his subject, and he began to wonder if this was not the case with Beveridge. In reply, Beveridge said he was sensible to "the southern tone"—he had got an earful from Cole, Jameson, Beard, and others about that—but the facts justified, indeed required it. This had come as a shock to him, as it must to any son of a Union soldier and birthright Republican. "When one has in his mind, and even in his blood, every phase of the whole controversy as put out by sheer propaganda, and then runs up against the facts as they actually were, one is in rather a distressing state of mind." But realizing he had been deceived, he now listened to the scholars rather than the propagandists. Whether or not he liked or disliked Lincoln had nothing to do with it.[40]

The hero of the second volume was Stephen A. Douglas, remembered as the other man in the famous debate of 1858 or, perhaps, as the man who held Lincoln's hat at his inauguration. Beveridge had grown up thinking Douglas had secured repeal of the Missouri Compromise restriction of slavery at the bidding of the Slave Power. Now he discovered that, too, was abolitionist propaganda. Influenced by the seminal work of Professor Frank H. Hodder—"one of the biggest contributions to American history . . . ever made," Beveridge said—he now understood that railroad expansion, not slavery expansion, lay behind the Kansas-Nebraska Act. The teaching of his youth that Douglas had sold himself to the South to win the presidency was bunk. And while Lincoln was declaring the Union could not continue to exist half slave and half free, Douglas was winning the critical battle over "Bleeding Kansas." In the Senator's insurgency against the Pierce and Buchanan administrations, Beveridge could see something of himself in 1910: bright, ardent, aggressive, a crusader for the right, and, like Douglas, blessed with a second wife of great charm and beauty at his side. Compared to Douglas, Lincoln was pale and clumsy. It was a myth that Lincoln had discredited Douglas in the South by the famous interrogatory at Freeport; and Lincoln's charge that the Dred Scott decision resulted from a conspiracy

---

*Justice Holmes contributed his own personal reminiscence to the stock of Lincoln lore. When Fort Stevens, on the outskirts of Washington, came under enemy fire in 1864, the President rode out to view the action. So eager was he to observe everything that he climbed on a parapet where he became an easy target in his stovepipe hat. Captain Holmes, from nearby, yelled, "Get down, you damn fool, before you get shot!" only then to recognize that the civilian he had so impetuously commanded was the President. Some fifty years later Holmes told the story to Harold Laski, and in 1938 Alexander Woollcott wrote it up with embellishments in the *Atlantic Monthly*. The literal truth of this "choice bit of Americana," as it has been called, soon ceased to matter.

involving Douglas was pitifully weak and inconsequential, for Beveridge had discovered that the decision caused scarcely a ripple in the North. The ordinary reader, Jameson said of Beveridge's unfinished account of the debates, must view Douglas as the hero, which was surprising in a biography of Lincoln.[41]

Upon publication Beveridge's *Lincoln* was as lavishly praised as *The Prairie Years*, though for different reasons. Samuel Eliot Morison, a rising star at Harvard, pronounced it "incomparably the greatest story of Lincoln's life before his presidency." Claude Bowers, a Hoosier protégé who shared his idol's outrage over victimization by propaganda, hailed the biography as a fundamental work of demythicization. It discarded the image of the boy with the ax (the real boy hated work), the boy reading before the fire (he read very little), the crusader against slavery, the great lawyer, and so on. Reviewers applauded the exhaustive and painstaking research. One demurred that the text was too overloaded to sustain narrative interest; and for too much of it, said another—for instance the first 237 pages of the second volume—Lincoln disappeared from view. Everyone expressed regret that so masterly a work should remain but a fragment and lamented the author's loss to American scholarship.[42]

The biography was acclaimed in the South, especially by southerners who had never recognized any moral grandeur in Lincoln and wished to topple him from his pedestal. David R. Barbee, a North Carolina newspaperman who had undertaken to educate Beveridge about the Old South, mourned his death as "a terrible loss." Had he lived he would have gone on to reveal the war President for what he was, a coarse, godless, power-driven man. William Townsend, with whom Beveridge had toured the Kentucky Lincoln country, resented the imputation. When diverse southerners came to him saying Beveridge had shown "ugly spots" on Lincoln's character, Townsend cited chapter and verse to refute the charge, but was then told that the author had made "terrible admissions" in his correspondence. The Lexington lawyer wrote to Pease and others closer to the lamented author than he to ask if there was any truth in the allegations. Pease replied that, although Lincoln was not a demigod to Beveridge, neither did he entertain prejudiced southern opinions of him. For a time Townsend waged a personal crusade against "the notorious professional Confederate" Barbee, and Lincoln-baiters like Lyon G. Tyler.[43]

Beveridge obviously had difficulty envisioning the Lincoln of 1861 or 1865 from the life he traced into the forty-ninth year. Worthington C. Ford, his friend and counselor at the Massachusetts Historical Society, who guided the unfinished manuscript through publication, and for a time thought of carrying the biography to conclusion, doubted that Beveridge was the right person to portray Lincoln the President. Writing Barrett in 1939, the year of the publication of Sandburg's *Abraham Lincoln: The War Years*, Ford observed, "He was not a little discouraged in approaching the war period by the difficulty of making his Lincoln of the earlier years into the Lincoln of martyrdom." Besides, had Beveridge continued the work on the same scale, "his asides would have been endless—good but too embracing." Sandburg himself agreed. He thought Beveridge had chosen the wrong subject. "He should have done Douglas who was handsome, fluent and born under a star like Beveridge," Sandburg wrote with a trace of acid on his pen. "He could have gotten inside Douglas and given the breath of a man. He rattles the dry bones of Lincoln." Perhaps, and perhaps Beveridge would have failed in treating Lincoln the President. It seems more likely, however, that the nationalist compulsion that had led him to Lincoln in the first place would have led him to recognize the full measure of his greatness. When Jameson voiced alarm over the direction of the biography, Beveridge referred him to a little piece he had written for the *Saturday Evening Post*, wherein he characterized biography as a form of historical drama in which the hero forges to the front only at the end.[44]

THE UNPRECEDENTED ABUNDANCE of books about Lincoln in the years between 1926 and 1933 included not only a number of biographies but historical fiction, juvenalia, poetry, special studies such as Lloyd Lewis's *Myths After Lincoln*—the first serious study of that subject—and a spate of books on Mary Todd Lincoln.

It was a manic period for biographies of all kinds. Among the Lincoln authors were Don C. Seitz, a journalist and biographer with a good grasp of Lincoln the politician; Emil Ludwig, the German biographer already well known for his lives of Goethe and Napoleon; Emanuel Hertz, earlier noted; Dale Carnegie, who later gained fame as the author of self-help success manuals; and William E. Lilly, whose *Set My People Free* was the first substantial Lincoln biography by a Negro.

Of special interest was *Lincoln, A Psycho-biography*, the first of the kind, by Dr. L. Pierce Clark, a practicing psychiatrist in New York City. Two years earlier, in 1931, he had joined in public protest of a paper delivered by Dr. A. A. Brill, one of the pioneers of psychoanalysis in the United States, at the annual meeting of the American Psychiatric Association. Brill analyzed Lincoln as a schizoid personality who suffered from a manic-depressive psychosis. His emotional highs and lows stemmed from the conflicting natures got from his parents. This, according to Clark, implied that Lincoln could not cope with reality, which was false. As early as 1919 Clark had advanced the thesis that Lincoln suffered from a bad mother-fixation complex. Her loss was his "psychic wound." Feeling no love for his father, he internalized the Madonna image of his mother; henceforth, every loved object, Ann Rutledge for instance, was a mother substitute. Lincoln's depression, which was mildly neurotic rather than psychotic, arose from the need to be loved totally. Because of fear of his father, he also suffered from a weak ego. His humor was compensatory. As President, his personality was fulfilled, and it became whole with the religious faith found after Willie's death.

Clark's book impressed some readers, while others thought it hopelessly diffuse and opaque. A sounder book, published in the same year, was *Lincoln and the Doctors* by Milton H. Shutes, a physician. He examined Lincoln's medical history as far as it was known and gave him a clean bill of health. The only exception was the hypochondria he suffered after the broken engagement with Mary Todd. Of course, everyone who knew him said he was inclined to melancholia, but this was not inhibiting. Ophthalmologists had noticed, in photographs, a squint in Lincoln's left eye, and speculated that it might have caused minor discomfort or depression or incidents of double vision. Shutes thought any malady more apparent than real. In a later study of Lincoln's emotional life, he maintained that any notion of "abnormality" was untenable. "The evidence indicates a depressive type of psychoneurotic within the bounds of so-called normality."[45]

The most remarkable new biography after Beveridge's was *Lincoln the Man* by Edgar Lee Masters. Indeed, Masters accepted the former's factual narrative as the basis for a frontal assault on the myth Beveridge had attacked obliquely. Masters had grown up in Illinois. His grandfather, like Lincoln's and Herndon's, had been part of the exodus out of Virginia into the Ohio Valley, and had settled near Petersburg. He had known Lincoln, and liked him, though they differed in politics. Through his father, a lawyer like himself, young Masters came naturally by his dogmatic Jeffersonianism. In Chicago he met Sandburg and discovered a common interest in poetry. *Spoon River Anthology*, in 1915, made Masters a literary celebrity. The illinois literary triumvirate of Masters, Sandburg, and Lindsay all lived with memories of Lincoln that shaped the ways they viewed their own times.

Masters dedicated *Lincoln the Man* to his hero, Thomas Jefferson. He stood preeminent as the philosopher-statesman of the democratic republic Lincoln destroyed during the Civil War, preparing the way for the "imperial America" of the twentieth century. The salvation of the country required demolition of the Lincoln myth, Masters believed. Although reputedly a disciple of Jefferson, Lincoln had not a drop of the founder's blood in his veins. "He was a Hamiltonian always, though his awkwardness and poverty, and somewhat gregarious nature and democratic words seemed to mark him as a son of Jefferson." War and consolidation and executive tyranny were all subversive of the Jeffersonian state. Nothing in the man redeemed him from his political betrayal. Masters described him as lazy, selfish, and ignorant. He knew nothing of the history of liberty. Whom had he read? Not Aristotle or Locke or Montesquieu, or Hallam or Grote or Tocqueville or Macaulay. In fact, he was ignorant of most of Shakespeare. Hardly a figure for the emulation of American citizens! "Lincoln had done as much as any prominent figure . . . to instruct the youth of limited advantage and outlook to loaf and trust to God. . . . He was a lazy mind."[46]

Members of the Lincoln fraternity were livid over Masters's book. "What are you going to do about [it]?" Hertz implored Louis Warren. "It is simply unthinkable to allow this grotesque libel to stand." Few reviewers had a kind word for the book. Harry Hansen, in the *New York World*, called it "the most bitter attack on Lincoln . . . since the days of Secession." Legislation was introduced in Congress to bar the book from the mails. Lindsay, attempting to account for Masters's tirade, said his disposition was just "plain sick." Hounded by the early fame of *Spoon River*, he never wrote anything to equal it; unhappy in love, he was divorced and exchanged Chicago for the strangeness of New York; he hated Illinois, most of all Sandburg and his *Prairie Years;* he hated preachers, Negroes, and prohibitionists; and he was "as angry as Simon Legree" at the consecration of a lazy sneak without brains, Abe Lincoln. Andrew Lytle, the southern writer, among the last defenders of the Lost Cause, praised the book in the *Virginia Quarterly Review*. And Claude Bowers, who shared Masters's political sympathies, thought the work brilliant, despite the excess of rage, and justified by the process of demythicization Beveridge had begun.[47]

The biographical virus spread to Mary Todd Lincoln. Herndon had fixed her image as a modern Xanthippe—a shrew, a curse, a haughty fool. She had been treated kindly by Noah Brooks and by William O. Stoddard, a Lincoln White House secretary, in his memoir; and her little sister, Emilie Todd Helm, prompted by Ida Tarbell, had offered a happy portrait of the marriage in *McClure's*. But there were hundreds of brutal words for every kind one. Old Henry Rankin had liked Mrs. Lincoln, and thought that she had been essential to her husband's success. The chapter Rankin devoted to her in his *Recollections* impressed a leading Lincolnian as "the first gallant and chivalrous thing ever written about her." In 1925 Gamaliel Bradford, who never ventured beyond a few lines on Lincoln, wrote one of his "psychographs" on Lincoln's wife.

Like everyone else he was puzzled by the love between this oddly matched couple, "the long and the short of it," as Lincoln joked. Mary's inner life was virtually impenetrable, partly because of the absence of personal letters, which, if gossip could be credited, her eldest son had destroyed. Bradford thought her most attractive as a mother. He could not imagine much spiritual sympathy between husband and wife; still, Lincoln was not "a woman's man," and Bradford doubted he would have been happier or greater with another mate. Monographic studies like Barton's *The Women Lincoln Loved*—he counted fourteen—and the biographer's Kentucky friend Townsend's *Lincoln in His Wife's Home Town* helped to fill out the portrait of Mary Todd Lincoln.[48]

The first book-length biographies appeared in 1928. One was by Katherine Helm, a niece and the daughter of Emilie Helm. Given the family auspices, as well as the aim to counter the Herndon image, the book contributed little to understanding of the subject. On the question of who jilted whom on January 1, 1841, and why, Helm divided the blame by locating the problem in Mary's alleged flirtation with Douglas and Lincoln's jealous reaction to it. Honoré Morrow's *Mary Todd Lincoln* was a by-product of her fictionalized biography of the President, to be discussed below, and written in the same vein. It was avowedly a tribute to a woman wronged in history.

At last, in 1932, Mary Lincoln was the subject of two biographies without obvious bias or preconception. Carl Sandburg and Paul Angle collaborated on *Mary Lincoln, Wife and Widow*, the former writing the narrative, the latter editing a series of letters, newly discovered, between Mary Levering and Mary Todd and between the former and her fiancé during twelve months in 1840–41. The evidence of these letters overturned all the more dramatic explanations of the aborted wedding—that she had jilted him or he had left her at the altar—and replaced them with Lincoln's sense of his own unreadiness for marriage and Mary's sympathetic acquiescence. For the rest, the portrait was rather grim and unsurprising. William A. Evans, a physician, attempted to compile a documented psychological history of Mary Lincoln. On his evidence, she had been a spoiled brat, always self-centered, and by 1861 her mental health was so shattered she could no longer be held accountable for her behavior. Obviously, then, the First Lady had been deserving of sympathy rather than censure. Regardless of her peculiarities, Evans did not think she was a significant influence on the mind and personality of President Lincoln.[49]

Dale Carnegie's *Lincoln the Unknown* was actually the biography of a marriage, written, moreover, in complete accord with Herndon. "The great tragedy of Lincoln's life was not his assassination but his marriage," said Carnegie. The book cannot be understood apart from the author's life. It appeared a year after Carnegie terminated a marriage he described as a decade in hell. "The book is strictly autobiographical in every respect." His second wife wrote a self-help manual, *How to Help Your Husband Get Ahead in His Social and Business Life*, for which, one might think, Mary Lincoln offered exemplary guidance, just as her husband most assuredly did for Carnegie's million-copy best-seller, *How to Win Friends and Influence People*, in 1936. Lincoln was the model for the message. In the episode of the lampoon of James Shields—the "Rebecca" letters—which led to a duel averted only at the last minute, Lincoln learned a fundamental lesson: curb your anger and never ridicule anyone. And so, when he exploded at General Meade for not pursuing the enemy after Gettysburg, he put his anger into a letter which he then suppressed. From Lincoln's proverbs, for instance "A drop of honey catches more flies than a gallon of gall," Carnegie formulated most of his rules for success in business and life. Among these rules, got from the abuse Lincoln took from his wife, was "Don't, don't nag!!!"[50]

Among several portrayals of Lincoln in fictional biography or historical fiction, Honoré Morrow's was the most ambitious and most successful. The novels of Irving Bachellor and Katharine H. Brown continued the tradition of casting Lincoln as a sterling presence in the storied lives of other people, in both these instances fictitious families that migrated to Illinois from the East. Morrow was the first author to make Lincoln the protagonist of the story. Born and educated in the Middle West, she achieved literary success with a series of novels set in the Southwest. She had a hankering for history, however, and this, combined with fascination about Lincoln got from her father, a drummer boy at Shiloh who later observed him in all his magnificence—"Emperor of the commonalty"—in 1865, led her in 1920 to embark on the work that would become the trilogy *Great Captain*, on Lincoln the President.

The first volume, *Forever Free*, appeared in 1927. The plot revolves around a fictitious character, Miss Ford, loosely based on Rose Greenhow. A Virginia-born Washington socialite with Confederate sympathies, she becomes secretary to the First Lady. She charms the President, who employs her in a successful mission to Unionists in western Virginia, even as he suspects her of spying for the rebels. Like the author, Miss Ford is emotionally drawn to Lincoln; indeed she passionately declares her love for him. No biographer had ever suggested he possessed this power with women. When she opposes the proposed Emancipation Proclamation, Lincoln sees her love as a mask of treason and banishes her under escort to Richmond. She returns secretly to Washington on Jefferson Davis's orders, and in an act of madness tries to shoot Mrs.

Lincoln. She is imprisoned in the Old Capitol but escapes and plots to kidnap the President before he can issue the fateful proclamation on January 1, 1863, and loses her life in the attempt. Quite apart from the romantic character of Miss Ford, Morrow repeatedly sacrifices historical verisimilitude by the invention of episodes—Lincoln the target of a rebel assassin at the Soldiers' Home, Walt Whitman being introduced to Lincoln—out of whole cloth.[51]

Morrow discarded most of the melodrama in the second and third volumes. But if better as history, they are poorer as novels. *With Malice Toward None* leaps ahead to 1864. It treats such subjects as Lincoln's reelection and Reconstruction. Mrs. Lincoln is revealed as a skillful politician in her husband's interest. But the author's focus is on Lincoln's complicated relationship with Charles Sumner, the Massachusetts senator. Despite suffering humiliating defeat at Sumner's hands on the Louisiana Government Bill, the President will not give him up. Sumner, moved by his affection, agrees to resign from the Senate to become Secretary of State in the second administration, thereby removing himself as an obstacle to Lincoln's moderate course. "Mary," says the President, "Sumner has filled my cup to overflowing." *The Last Full Measure* takes the reader back to September 1864 and through the maze of events occupying Lincoln traces Booth's conspiracy to assassinate him. The last half of the book is a prolonged meditation on Lincoln's rendezvous with death. Morrow's Lincoln is neither the folk character nor the Titan. He is a commanding figure, but also a loving and lovable one. Readers who desired the lilt of romance to make history palatable and to vivify a character like Lincoln were not badly served by *Great Captain*.[52]

Stephen Vincent Benét's *John Brown's Body*, a long narrative poem on the Civil War, in 1928, contained approximately seven hundred lines about Lincoln. As the title makes clear, it is the blood-mad prophet, not the Great Emancipator, who stands at the center of the work, and it is his soul that goes marching on. Benét's Lincoln,

> The lank man, knotty and tough as a hickory rail,
> Whose hands were always too big for white-kid gloves,
> Whose wit was a coonskin sack of dry, tall tales,
> Whose weathered face was homely as a plowed field,

comes straight out of Sandburg's biography. Lincoln appears repeatedly in the disjointed narrative. About halfway through Benét composes a long soliloquy for him that is better than anything Sandburg wrote in the same vein.

What is God's will?
They come to me and talk about God's will
In righteous deputations and platoons,
Day after day, laymen and ministers.

They all knew God's will; Lincoln says he is the only one who does not know it. The words of the soliloquy are Lincoln's, mainly; the poet only arranged them in rhymed verse. As he went on, however, it occurred to Benét that Lincoln, who told stories to everyone else when worried, might tell a story to God. And so he had Lincoln tell God about a man near Pigeon Creek who had a dog he would never part with, "an old half-deaf foolish-looking hound" who wasn't much for speed but was "hell on a cold scent" and death to his prey.

I am that old, deaf hunting-dog, O Lord,
And the world's kennel holds ten thousand hounds
Smarter and faster with finer coats
To hunt your hidden purpose up the wind
And bell upon the trace you leave behind.
But, when even they fail and lose the scent,
I will keep on because I must keep on
Until You utterly reveal Yourself
And sink my teeth in justice soon or late.
There is no more to ask of earth or fire
And water only runs between my hands,
But in the air, I'll look, in the blue air,
The old dog, muzzle down to the cold scent,
Day after day, until the tired years
Crackle beneath his feet like broken sticks
And the last barren bush consumes with peace.

And so he would go on, forever faithful to the vision of a Union free of slavery.[53]

The Minor Affair

On June 27, 1928, Wilma Frances Minor, in San Diego, wrote to the editor of the *Atlantic Monthly*, in Boston, to inquire about the submission of a manuscript purporting to be the true story of the romance between Abraham Lincoln and Ann Rutledge for the biennial non-fiction prize awarded by the Atlantic Monthly Press. "Held sacredly in the family archives for many years were all the original love letters, diaries, and keepsakes of Abe and Anne, and the friends of the two in New Salem days." The treasure had been handed down to Miss Minor's mother, thence to her. Another publisher, Harper's, had expressed keen interest in the manuscript, she said, but the prize offered a strong inducement to publish with the *Atlantic Monthly.* Ellery Sedgwick, the stout fifty-six-year-old editor and publisher of the distinguished monthly, much doubted the manuscript could fulfill the claim made for it, and his skepticism was shared by the manager of the book division, Edward Weeks. But wouldn't it be amazing if this was, in fact, the true story of that fabled love? All his life he had followed the meteor of Lincoln's fame; recently he had read his friend Beveridge's chapters in manuscript, and regretted missing the chance to serialize that work. Within a week Sedgwick telegraphed Minor to send the manuscript, together with photostats of key documents. This material arrived in due course, and on first inspection appeared credible.

The author was invited to make a personal visit to Boston. She arrived in September with her mother, Mrs. De Boyer, and a twelve-year-old half-sister. Wilma Minor, described by Weeks as a lithesome, comely woman, some forty years of age, with large Hollywood eyes, blossomed under the attention. She, it turned out, had once been an actress but now worked as a journalist with literary aspirations. Sedgwick found her charming. Recalling his apprenticeship on *McClure's* magazine when Ida Tarbell's "Early Life of Lincoln" astonished everyone, Sedgwick saw an opportunity to repeat the performance. Minor was given a contract for the book, with an advance of $1,000, and the promise of serialization in the *Atlantic Monthly.* Sedgwick thus set himself up for one of the most humiliating fiascoes in the history of magazine publication.[54]

The Ann Rutledge legend was certainly an inviting subject for con artists and forgers. Launched on its public career by William H. Herndon in 1866, it was embellished, stretched, and inflated over the next sixty years until it became not only a classic tale of lost love but the emotional turning point in Lincoln's life. For Herndon the overwhelming significance of the affair lay in the cloud of melancholy that invaded Lincoln's soul upon Ann's death, that led him to William Knox's poem, that made his marriage a farce, and that haunted him to the hour of his death. It was, as Tarbell said, "a study in morbidity." There were other interpretations, however. In Denton Snider's blank-verse epic *Lincoln and Ann Rutledge* (1912), Lincoln is reborn in the loss of his beloved. Ann, on her deathbed, says to him:

> "Hear me, henceforth my love is not
> merely in me to be bounded,

But to the Love of all people will rise up
thy love of Ann Rutledge."

Abe understands. "This is my marriage eternal," he vows, founded on a love neither of self nor of Ann but of humanity.

> Every day he has to enact her life and her death too,
> Harmonizing the scission of soul whereof she has perished,
> Suffering fully her fate in his own for higher fulfillment,
> Living her tragedy over and feeling its throes in each heart-throb
> That he may rise above it the victor by loyal endurance.

Another author saw deep religious significance in this redirection of the self to the love of that which cannot die.

In still another conception, Ann becomes Lincoln's guardian angel. Typically, as in Bernie Babcock's fictional *The Soul of Abe Lincoln*, the President is moved to respond to some lover's or sister's or mother's appeal to his mercy by the pang of Ann Rutledge's memory. In Richard H. Little's popular story *Better Angels*, set in the borderland between folklore and fiction, "Mammy Jinny" is the cook of an army regiment stationed in Washington where the President likes to go to eat. From a colonel she learns of his "secret sorrow." So when the son of her old mistress is about to be shot, Mammy Jinny gets his fiancée to go to him, fall on her knees, and say the soldier's death will kill her too. She goes. Lincoln stares upon her golden hair in the moonlight, mutters "Ann, Ann, Ann," and grants the pardon. A drama, *Spirit of Ann Rutledge*, uses William Rutledge, a cousin, as an intermediary. In one act he visits Lincoln just before his inauguration and urges him to pray to Ann's memory; in another, while on the train to Gettysburg with the President, he is visited by Ann's spirit and says it has lighted his way constantly; finally, at Ford's Theater, he is rejoined with Ann, who bids him to become "a living spirit."[55]

The epiphany of the theme was Masters's verse epitaph for Ann Rutledge in *Spoon River Anthology*.

> Out of me unworthy and unknown
> The vibrations of deathless music;
> "With malice toward none, with charity for all."
> Out of me the forgiveness of millions toward millions,
> And the beneficent face of a nation
> Shining with justice and truth.
> I am Anne Rutledge who sleep beneath these weeds,

Beloved in life of Abraham Lincoln,
Wedded to him, not through union,
But through separation.
Bloom forever, O Republic,
From the dust of my bosom!

The sentiment became hateful to the author; he no longer believed it when he wrote *Lincoln the Man*. In 1890 Ann's remains were removed from their secluded grave in Concord to a new cemetery above the winding Sangamon in Petersburg with a view, it was said, to attracting clients; and some years after the publication of *Spoon River Anthology*, the lines of the poem were chiseled on a large granite monument over her grave. What sacrilege! thought Ross Lockridge, a young Indiana writer, after *Lincoln the Man* appeared. "I would be only too glad to join a pilgrimage of devoted souls, with mallets and chisels to remove that epitaph."

Quite apart from anger at Masters, there were a growing number of souls skeptical of the legend. None had gone so far in debunking it as Paul Angle. In 1926 he wrote a little critique entitled "Lincoln's First Love," but fearing his case was less conclusive than required to undermine the legend, he asked Beveridge what he thought of it. In reply, Beveridge agreed with Angle, the legend was "mostly rot," though he refrained from saying so in his impartial review of the evidence in the biography. Angle published the article in the *Bulletin of the Lincoln Centennial Association* in 1927. His main argument was that the romance rested on nothing more substantial than Herndon's report of what informants told him some thirty years after the fact, and Herndon, he said, was notoriously unreliable when not himself a witness. No documentary evidence had been found. Popular writers like Eleanor Atkinson assumed that the lovers had exchanged letters, and they had simply disappeared. Of course, they might yet be found. Wilma Minor and her mother, if they read Atkinson's story in the *Ladies' Home Journal*, might have carried the thought one step further: and if not found, the letters might be imagined and forged.[56]

During the fall of 1928, Sedgwick, Weeks, and staff were busy checking the facts in Minor's text and the authenticity of the supporting documents. The paper and ink passed laboratory tests with flying colors. Worthington Ford, at the Massachusetts Historical Society, was one of the first persons Sedgwick turned to. Having just finished seeing Beveridge's biography through the press, Ford knew the subject; moreover, as the former chief of the manuscripts Division of the Library of Congress, he qualified as an expert on handwriting. With the aid of a curator at the Huntington Library, he had exposed an important letter attributed to

Lincoln as the forgery of Henry Cleveland. Sooner or later Sedgwick consulted all the leading authorities in the Lincoln field: Angle, Barrett, Warren, Barton, Sandburg, Tarbell. The result was a split decision on the series the magazine had decided to call "Lincoln the Lover."

The first three filed strong dissents. Angle, already on record against the legend, spotted numerous historical errors, discrepancies, anachronisms, and contradictions in the manuscript, and publicly denounced it a hoax in an Associated Press dispatch. A special bulletin of the Lincoln Centennial Association declared, "Atlantic Monthly Letters Spurious." "It's the biggest thing that ever happened to me," Angle gloated. "One doesn't get a chance very often to put the magazine of the country in the frying pan and cook it brown." Ford also went public with his criticism, opposing the judgment of one old Boston literary institution against another. He was incredulous when Sedgwick proceeded posthaste with publication. "Have you gone insane, or have I?" he demanded near the end of November. "You are putting over one of the crudest forgeries I have known and must expect criticism." Barton was equivocal when he first saw the photostats, but after further reflection and a meeting with Wilma Minor in California, he denounced the series without reserve. The Lincoln letters and other documents, he said, "are forgeries and she [Minor] knows it." This was a serious accusation. Teresa Fitzpatrick, who edited the series, advised Minor to sue Barton for libel. The prospect was bleak, but Sedgwick took courage from the positive assessments of Sandburg and Tarbell. Both spent many hours with Sedgwick going over the documents and independently vouched for their authenticity, according to Sedgwick, though Tarbell later said in her defense she had only advised proceeding with the task of verification. Sandburg stated, "These new Lincoln letters seem entirely authentic—and preciously and wonderfully coordinate and chime with all else known of Lincoln." The editor committed Sandburg publicly to support of the series.[57]

Introducing the series to the magazine's readers in the December issue, Sedgwick said that Lincoln's was the one life about which Americans wished to know everything, and a fascinating question mark had long been the Rutledge romance. Now an amazing set of documents, including letters between the lovers, had turned up to provide historical confirmation of the legend. "What a collection! Here is the human Lincoln, before the sterility of deification." Sedgwick went on to explain the efforts taken to authenticate the documents. He recounted the lineage of the collection, which included treasured relics as well as letters and diaries and memoranda. It descended from Matilda Cameron and "Sally" Calhoun, who had been Ann's confidantes, through many hands to Mrs. De Boyer, finally to the author, Wilma Minor, who herself claimed descent from the Anderson neighbors of the young Lincoln on Anderson Creek, in Indiana, and from Major Anderson who fought in the Black Hawk War and commanded at Fort Sumter in 1861. The line of descent checked out, said Sedgwick. "It is a chain of actual flesh and blood."[58]

The first of three installments treated "The Setting—New Salem," the second "The Courtship," the third "The Tragedy." All the principal documents were published in facsimile. Among the first introduced were two letters from Lincoln to John Calhoun, the surveyor who had employed him in New Salem, when he was in Congress. A telling argument of the experts was that the handwriting of the letters ascribed to Lincoln appeared cramped, rakish, and uneven, whereas in fact his script was smooth and evenly spaced. There were peculiarities of diction as well. Even the choice of Calhoun as a correspondent was suspect, since he was a Democrat and no letters had been known to pass between them. Minor employed the device of the father, Calhoun, dictating his memories of Lincoln to his daughter Sarah, or "Sally," Ann's friend. But as far as Angle could discover, no daughter of that name belonged to the Calhoun family.

In the article Ann and Abe meet at the Rutledge Tavern in 1832. She is described as red-haired, although in the tradition from Herndon her hair was light, flaxen, or golden. Abe is smitten at once, and is dismayed to learn that Ann is engaged to the absent John McNamar. Nonetheless, romance blossoms. In the summer of the next year they are all but openly engaged. "Abe and Ann are awful in love he rites her letters," "Mat" Cameron jots in her diary. They study *Kirkham's Grammar* together and sing *Missouri Harmony* hymns at meeting, and after Lincoln finds Blackstone's *Commentaries* in the bottom of the barrel, he studies it sprawled out on the grass at Ann's feet. This was a picture straight from Atkinson. The same barrel, according to Mat, yielded a breastpin, a gift from Abe to Ann, which had been preserved in the collection. Abe also recited to Ann "O why shud the spiret of mortale be proud?" omitting the two explicitly religious stanzas. Of course, according to Herndon's biography, the poem took hold of Lincoln in the grief *after* her death. Ann tries to make a Christian of Abe. She lends him her mother's Bible. This, too, is part of the collection, and it shows that Abe encircled several verses of the Song of Songs, beginning "Behold, thou art fair, my love." The lovers plan to marry after Lincoln returns from Vandalia in the summer of 1835. But he has no money and Ann aspires to a college education. Besides, she is still pledged to McNamar. Fate intervenes; she dies, and Lincoln is plunged into lasting grief. "Abe is luny," Mat writes in her diary. But it is all for the best—a necessary step in his rise to greatness.[59]

The *Atlantic Monthly* was reluctant to admit it had been the dupe of a blatant fraud. The drumbeat of criticism steadily increased, however. After publication of the second installment in January, even Sandburg withdrew his support. Wilma Minor could not convincingly answer the critics. Sedgwick had no choice but to retreat. He let the series run its course and asked Angle, the sharpest of the critics, to write an article exposing the fraud for the April issue of the magazine. He traveled to California, talked to Minor and the detectives who had been hired to investigate her, and terminated any further publication. Upon his return to Boston, Sedgwick released a statement to this effect and caught the first ocean liner for Europe.

Angle's article, predictably, was devastating. He credited the forgers with a soupçon of cleverness but found the whole clumsy, even ludicrous. Presumably it was written to brighten the aura of romance around Lincoln; instead, said Angle, it portrayed a sorry character. "What he [Lincoln] wrote was full of inflated sentimentality, and . . . suggested a man no more than half literate." Barton, perhaps affronted by the choice of Angle to write the rebuttal article, berated Sedgwick for never offering an apology to his readers. All he had done, in the February number, was to plead "good faith" and acknowledge the challenge of the experts. He neither confessed error nor explained the mystery of the affair. "Why doesn't Ellery Sedgwick frankly, fully, and finally confess?" asked the director of the John Hay Library at Brown University. "The answer would seem to be that he is unwilling to say what he knows."[60]

No confession or apology or explanation ever appeared in the *Atlantic Monthly*. What it finally learned was too embarrassing for publication. In a confession given to Teresa Fitzpatrick in Los Angeles in July, Wilma Minor said:

> I went to see Scott Greene [son of Billy Greene, who knew Lincoln in New Salem] and got his story [for a newspaper] and went home to Mama and said to her, Mama at last our faith of a lifetime has led to something. It has been given us for a divine purpose. On another plane those people (Lincoln and Ann and those other people) must exist. We have talked to many others, our family and close friends, and I said to Mama, Don't you think I have earned the right to be the channel to tell that real story to the world? Mama said, I don't know, darling we can try. Mama had always been the medium through whom the spirits had spoken. . . . I then began with a series of questions. I would write out the questions. I would hand them to Mother then in trance; the spirit would come, whoever it might be, and fill out the answer. . . . Every word in Matilda Cameron's Diary is verbatim as given by the guide. Every word written through my mother as the Medium. . . . I would die on the gallows that the spirit of Ann and

> Abe were speaking through my mother and me, so that my gifts as a writer combined with her gifts as a medium could hand in something worthwhile to the world.

So the hoax was perpetrated by misguided spiritualists rather than con artists or professional forgers. Sedgwick's pride would not even permit him to mention the subject in his autobiography.[61]

What in the end was the significance of the Minor Affair for the Lincoln image? One thing it did not do was to dispel the Ann Rutledge legend. By definition a legend does not require authentication to be believed. Even the foremost of the disbelievers, Angle, backed away from any discussion of the truth or falsity of the romance in his critique of "Lincoln the Lover." Nevertheless, the pathetic weakness of this attempt to provide a documentary basis for the legend lessened public confidence in it. The scandal revealed the thinness and superficiality of knowledge about Lincoln even among educated men and women. Did the author, the editors, and the readers really think that he was such a surveyor of the public land as to refer to "Section 40," or that boats plying the Sangamon brought him books from Springfield, or that he said goodbye to a family going to "Kansas" in 1834? Fact and fiction had been so freely intermingled in writing about the young Lincoln that almost anything passed as creditable. Finally, the Minor Affair was a watershed between the traditional approach to Lincoln and the disciplined scholarly approach. Milo Quaife, speaking for the historical profession, raked the *Atlantic Monthly* over the coals and pointed a finger of shame at the popular biographers, Sandburg and Tarbell, for being fooled by a cheap fraud. For Sandburg, certainly, it proved a chastening experience. And because of it he became a more cautious and responsible biographer, though as his friend Angle confessed in retrospect, "He never had any critical sense."[62]

The Historians' Lincoln

In his begrudging review of the "noble fragment" that was Beveridge's *Abraham Lincoln*, William E. Barton concluded: "And now where is the man who shall do what Beveridge set out to do, and who with trained legal mind and painstaking analysis shall show us how the Constitution as John Marshall interpreted it, was maintained and consistently broadened and permanently established in the work of Abraham Lincoln? That task awaits the man who can perform it, and it is a great one." The man to perform it was at the time a forty-eight-year-old associate professor of history at the University of Illinois, James G. Randall, the same who five years later would sound a clarion call to historians to make Lincoln an object of inquiry. Randall was not alone, for Carl Sandburg had already embarked on the work that became *Abraham Lincoln: The War Years* in 1939. It was, of course, very different from anything Beveridge might have done or from the work of his scholarly friend Randall. The first two volumes of his four-volume *Lincoln the President* appeared in 1945. But Randall's influence extended beyond the books he wrote. The first historian to build an academic career around Lincoln, he was a stimulating teacher and a mentor to many students he initiated into Lincoln scholarship. Moreover, by setting exacting standards Randall elevated the level of that scholarship. Paul Angle, in a guide to the Lincoln literature, asked himself if there were any legitimate challengers to the claim of priority advanced for Randall. He acknowledged Nathaniel Wright Stephenson, who had held academic positions; but Stephenson's highest degree was a B.A., and having come out of the world of journalism, he qualified as a historian "only in the occupational sense," said Angle.[63]

Born in Indianapolis in 1881, Randall graduated from the local university, Butler, and went on to the University of Chicago for graduate work. He obtained his doctorate in history in 1911 with a dissertation on "The Confiscation of Property During the Civil War." It was a small topic, chiefly interesting for the light it shed on the problem of suppressing a rebellion as if it were a war; and he was well advised by Andrew Mclaughlin, the constitutional historian, to make it the point of departure for a full-scale study of the constitutional problems of the war. He found a teaching position at Roanoke College, in Virginia, and there met Ruth Painter, who became his second wife and, in time, his valued co-worker on Lincoln. After a stint with the United States government during the World War, Randall taught briefly in Richmond College before securing a firm rung on the academic ladder as an assistant professor at the University of Illinois at Urbana. His first book (omitting his own printing of a condensed version of the dissertation), *Constitutional Problems Under Lincoln*, was published in 1926, when he turned forty-five. If his progress had been slow, it had also been sound.[64]

The book immediately became the standard work on the subject, superseding John W. Burgess's *Civil War and the Constitution*. Burgess, a quarter-century earlier, had still been absorbed by the justification of national supremacy and a Constitution boldly empowered to meet every emergency. For Randall this was only the starting point for systematic analysis of such specific problems as suspension of the writ of habeas corpus, arbitrary arrests, martial law, confiscation, conscription, and so on. By wedding the Commander in Chief clause and the "take care" clause ("he shall take Care that the Laws be faithfully executed"), Lincoln invented the presidential war powers of the Constitution. The supreme example of its exercise was the Emancipation Proclamation, which Lincoln admitted had no constitutional basis apart from military necessity. In other hands, Randall argued, the exercise of such sweeping powers would almost certainly have been dangerous to the republic. But Lincoln proceeded with circumspection and "a wholesome regard for individual liberty." He wielded dictatorial power, yet he was not a dictator because of his sterling humanity, lawyerly caution, and absence of malice.

Having served in the war administration of President Wilson, Randall was influenced in his judgment of Lincoln by his firsthand observation of one who faced similar problems and who was the historian's second hero. History does not repeat itself, it has been said, but it rhymes. Randall wrote an article rhyming "Lincoln's Task and Wilson's." Both wartime presidents enormously expanded executive authority though in different ways, Lincoln independently of Congress, Wilson with legislative sanction. Tragically, neither witnessed the fruition of his policy. "The Vindictives [Radical Republicans] of 1865, with all their lip service to Lincoln's memory, were as savage in wrecking the fallen leader's generous plans for reconstructing the Southern States as were the opponents of Wilson in defeating his work in the field of international reconstruction. In each case the lofty resolves of war were added to the other casualties. Both men were misunderstood, not alone in their own day, but by posterity." There could be no better instance of the relevance of historical understanding to contemporary problems, Randall thought, and he returned to this admonitory theme in the midst of the Second World War.[65]

Randall did not set out to become a Lincoln scholar. His first book occupied the field of constitutional history, albeit Lincoln figured largely in it; he planned to continue in that field, and indeed spent his sabbatical year, 1928–29, in Washington doing research for a constitutional history of the United States. But there he met Allen Johnson, editor of the newly launched *Dictionary of American Biography*, and was invited to contribute the article on Lincoln. The discerning reader may find in his fifteen-thousand-word article the germ of ideas, such as the idea of Lincoln as "liberal statesman," later matured under Randall's pen. Still he was not fully committed to Lincoln as his primary subject. In 1930, at the urging of Allan Nevins, the Columbia University historian, he agreed to write a textbook history of the era of Civil War and Reconstruction for use in college courses. Published in 1937, *Civil War and Reconstruction* helped to shape the "revisionist" thinking of young scholars and, of course, superseded the work of historians like James Ford Rhodes.

James G. Randall. (Illinois State Historical Library)

Meanwhile, Randall had nailed his flag to the mast with his paper on the "Lincoln Theme" before the American Historical Association in 1934. At the same meeting he and Angle agreed to collaborate on a one-volume biography of Lincoln. Angle, who was to write the first part, went to work with enthusiasm, but soon realized that his professional responsibilities would not permit him to continue. At this juncture Randall entered into a contract with Dodd, Mead and Company for a one-volume biography of Lincoln the President. Sandburg was only two years away from completion of *The War Years*. It would surely smother any single volume, which was discouraging. In the end, Randall, too, wrote a four-volume series, though he did not live to finish the final volume. *Lincoln the President* complemented *The War Years*, yet in tone, method, and style was entirely different from it.[66]

Except for his rising prominence as a Lincoln specialist, Randall lived the conventional life of a professor in a major university. He taught large numbers of students, guided doctoral dissertations, reviewed books, and engaged in a variety of professional activities. Given his scholarly interests and his presence in Illinois, he was naturally drawn into the work of the Abraham Lincoln Association and the Illinois State Historical Society. Aside from his publications, he exerted influence through his students. A number of them—Harry Pratt, William E. Baringer, David Donald—became Lincoln scholars in their own right. Pratt succeeded Benjamin Thomas as executive secretary of the Abraham Lincoln Association and became a prolific author. Baringer, who would succeed Pratt, authored *Lincoln's Rise to Power* in 1937. Donald, Mississippi-born and -educated, became a favorite student as well as incomparable research assistant and, with the publication of the brilliant *Lincoln's Herndon*, in 1948, the leading protégé. But the response to Randall's challenge in 1934 was not confined to Illinois precincts. The profession was awash with Lincoln scholarship. In 1945 the editor of the *American Historical Review* declared "a moratorium on Lincoln articles."[67]

Randall was a leader of the so-called revisionist school on the historiography of the Civil War. With regard to the coming of the war, revisionists denied it had been "irrepressible" and that slavery had been its primary cause. With regard to Reconstruction, revisionists blamed its failure not on an obstinate South or a refractory President but on the Radical Republican leadership in Congress. Randall boldly set forth his views on the causes of the war in his presidential address, "A Blundering Generation," before the Mississippi Valley Historical Association in 1940. It was destined to be widely read and discussed. The essay was not about Lincoln, rather his generation. But by blaming the war chiefly on fanaticism, in both sections, and characterizing Lincoln, by contrast, as "moderate, temperate, and far-seeing," Randall seemed to exempt him from the irrational psychosis that gripped the nation. The war was "needless," he declared repeatedly, because it could have been averted by more moderate statesmanship and less angry agitation and propaganda.

In his Fleming Lectures at Louisiana State University on the theme "Lincoln and the South," Randall exaggerated the southerness of Lincoln's origins, character, rhetoric, and politics in order to portray the President as one who understood and sympathized with the slave states. Southerners as diverse as Henry Watterson and Thomas Dixon had always believed this. For the distinguished professor of history from the University of Illinois to confirm it was somewhat surprising. This view also shaped Randall's interpretation of Reconstruction.

In the two years 1929 and 1930, three revisionist studies of Reconstruction appeared: *The Tragic Era* by Claude Bowers, *The Critical Year* [1866] by Howard K. Beale, and *The Age of Hate: Andrew Johnson and the Radicals* by George Fort Milton. Each told its own story, but all three books bludgeoned the Radicals and vindicated Johnson and the South. Randall agreed and, recurring to the moot question of how Lincoln would have fared had he lived, held he would have suffered the same assault from the Radicals, in fact had suffered it in life, that President Johnson experienced. Reconstruction was a tragedy, not because it failed the Negro, but because it enthroned a party, no longer Lincoln's, bent on revenge, plunder, and domination.[68]

Although Randall wrote comparatively little on the pre-presidential Lincoln, his revisionist zeal led him to single out for attack William H. Herndon for his picturesque but false conception. (Randall actually preferred the term "restoration," rather than "revision," since the aim was to clear away "the historical debris, and . . . to restore events and essential situations of the past.") Herndon's stature was slipping among Lincolnians. In 1938 Hertz published a mishmash of letters, memoranda, and other materials drawn mainly from the Lamon Papers in the Huntington Library, which he called *The Hidden Lincoln*. Better it should have been entitled *The Hidden Herndon*, said Louis Warren, who had long since dismissed the sham Boswell as "an irresponsible gatherer of folklore and traditions." Bullard thought the book "scandalous," not alone for its hundreds of errors but also for the publication of Herndon's most flagitious opinions, for instance that Lincoln had contracted syphilis at an early age. Randall had a poor opinion of Hertz, according him "the distinction of being the world's most tiresome writer on Lincoln," and he wrote a devastating review of *The Hidden Lincoln* for the *American Historical Review*, to which Hertz took sharp exception. After the Library of Congress acquired the Herndon-Welk Collection, Randall went through it page by page. In the first volume of *Lincoln the President*, he deprecated Herndon's intuitive, capricious, and fantacizing approach, yet conceded the truth of some of the evidence and admitted "a photographic quality in his record." Randall adopted Herndon's opinion that the senior partner was a better appellate than trial lawyer. But he was unsparing in his criticism of Herndon's account of Lincoln's courtship of Mary Todd and of their marriage. The picture of the defaulting bridegroom was "made up," the "tigress" portrait of her malicious.[69]

The plan of Randall's book, cast against the background of Springfield, did not allow for treatment of the Ann Rutledge romance within the narrative, but it was so central to the matter of Herndon's credibility that critical evaluation of it ought not be avoided. So Randall wrote a twenty-page essay, "Sifting the Ann Rutledge Evidence," and consigned it to an appendix.*

The mischief of the legend continued. Only a year before, in 1944,

---

*Mrs. Randall, it should be pointed out, collaborated on both the courtship and marriage chapters and the appendix. She unquestionably influenced her husband's gracious interpretation of Mary Todd Lincoln, whose biography she would write.

"new light" had supposedly been shed on the romance. A Petersburg newspaper, it was discovered, had alluded to the story three years before Herndon got wind of it. The editor, who was a Democrat, wrote in a satirical vein of a sorry affair between a "love-sick swain," later revealed as the young Lincoln, and an "angelic young lady," unnamed, that ended with her death. Randall subjected the testimony of each of Herndon's principal informants to critical scrutiny. It was rife with gossip, falsehood, and wild conjecture. Of one statement on the alleged engagement, Randall wrote: "Here is one person reporting what another person had written him concerning what that person recollected he had inferred from something that Ann had casually said to him more than thirty-one years before!" It was impossible to disprove the legend, but Randall had no difficulty concluding it lacked historicity, and he threw it into the limbo of the unproved. Warren rejoiced. The appendix was "a revolutionary pronouncement in the field of Lincoln biography," and he predicted it would have more influence and do more good than the text of the book. From the perspective of a present-day student seeking to revive Herndon's reputation, Randall's essay was the "capstone contribution" to a campaign to discredit the Rutledge legend, and its success was virtually complete among historians.[70]

THAT THERE WAS more than one way to write Lincoln biography, the major works of Randall and Sandburg amply demonstrated. One was skillfully crafted, the other wallowed in the chaos that surrounded Lincoln; one proceeded in controlled and analytic fashion, the other adopted what Nevins called "the method of Niagara"; one aimed to clarify historical understanding of important questions and issues, the other aimed at pictorial vividness, creating an album of stories and images that belonged to the collective memory of the people; one was revisionist in its point of view toward the Civil War, the other was written in the nationalist tradition. Lincoln was a hero—not without flaws—to both authors, though for Randall the liberal statesman stood foremost while for Sandburg Lincoln personified the deepest thoughts and sentiments, dreams and passions, of the American people.[71]

Sandburg's *War Years* was a whale of a biography: four fat volumes, twenty-five hundred pages, one and a quarter million words, and hundreds of illustrations. He had put eleven years into it, compared with only three for *The Prairie Years*. "It took him to every part of the country," Benjamin Thomas observed, "through mountains of newspapers, floods of letters, diaries, pamphlets, posters, proclamations, handbills, pictures, cartoons. He floundered neck-deep through official records and Congressional debates. He bought books by the hundreds: ripped them apart to obtain the pages he would use and stacked the gutted remnants in his barn. Two copyists worked on a glassed-in portion of his home [in Harbert, Michigan] while his wife and three daughters helped file and organize."

By historiographic standards, *The War Years* was a better and stronger work than its predecessor. The materials were so rich and abundant, Sandburg had no need to fantasize; and because Lincoln stood at the center of events, his biography and the nation's epic were all one. A more self-conscious historian than before, Sandburg consulted scholarly monographs, asked the experts to check chapters for accuracy, and in his marshaling of evidence and incident from a thousand observers and memorialists even deigned, as he said, "to appraise the goddam witness." Still the author's indiscriminate use of sources, together with his failure to cite them, maddened serious readers. Randall himself, not only a serious reader but a friendly one, pointed out that the work was a virtual "catalogue of quotations" artfully strung together to make a narrative. Both biographers drew heavily upon one authentic new source, *The Diary of Orville H. Browning*, recently published by the Illinois State Historical Society. But Sandburg in his amorphous fashion often gave the same credibility to rumor and hearsay as to an eyewitness like Browning. If a story was told, somebody believed it, so it had that measure of truth. Thus Sandburg wrote, on the word of a leading Republican, that there was a secret move to impeach Lincoln late in 1862; and at the same time, in order to squelch rumors of Mrs. Lincoln's disloyalty, the President himself appeared before the Senate Committee on the Conduct of the War. The larger truth—that Lincoln was under fire and his wife was suspected—redeemed any falsity of detail.

As before, Sandburg was sparing in his judgments on persistent issues. Was Lincoln too lenient? Probably, or so one would assume from the roster of examples. But it may not have mattered, for Sandburg quoted General Sherman's alleged answer to the question of how he evaded the President's proclivity to pardon slackers and deserters: "I shot them first!" Was Lincoln a military genius? Probably not, though genius of another kind counted for more. Sandburg was impressed by what Douglas Southall Freeman wrote in his biography of Robert E. Lee, sometimes compared to Sandburg's *Lincoln*, and quoted him at length. Picturing the Confederate general balancing the ponderables in the military scales in 1864, Freeman astutely observed:

> He could not realize, and few even in Washington could see, that an imponderable was tipping the beam. The imponderable was the influence of President Lincoln. . . . The Confederate people had mocked him, had

> despised him, and had hated him. Lee himself, though he had avoided unworthy personal animosities and doubtless had included Mr. Lincoln in his prayers for all his enemies, had made the most of the President's military blunders and fears. He was much more interested in the Federal field commanders than in the commander in chief. After the late winter of 1863–64, had he known all the facts, he would have given as much care to the study of the mind of the Federal President as to the analysis of the strategical methods of his immediate adversaries. For that remarkable man, who had never wavered in his purpose to preserve the Union, had now mustered all his resources of patience and determination. Those who had sought commonly to lead him, slowly found that he was leading them. His unconquerable spirit, in some mysterious manner, was being infused into the North as spring approached.

This was penetrating, and Sandburg, who could not have written it, concurred entirely in its truth.[72]

While less poet than chronicler in *The War Years*, Sandburg could not separate his heroic conception of Lincoln from his Whitmanesque faith in American democracy. As he worked at it, he also wrote the long rhapsodic series *The People, Yes*, published in 1936. A gathering of the lore of Everyman and a great shout for democracy, the book hinted at the analogy between Lincoln's fight to eradicate slavery and the present-day struggle of the masses of people against capitalist exploitation. Section 57 is devoted to Lincoln:

> Lincoln?
> He was a mystery in smoke and flags
> saying yes to the smoke, yes to the flags,
> yes to the paradoxes of democracy,
> yes to the hopes of government
> of the people by the people for the people.

Sandburg found his text in fragments of Lincoln's utterance:

> "I hold,
> if the Almighty had ever made a set of men
> that should do all the eating
> and none of the work,
> he would have made them
> with mouths only, and no hands;
> and if he had ever made another class,
> that he had intended should do all the work
> and none of the eating,

he would have made them
without mouths and all hands."

In one passage the poet spoke of the war:

Death was in the air.
So was birth
What was dying few could say,
What was being born none could know.

He repeated the same litany in *The War Years*. At the close of a mammoth chapter, "The Man in the White house," Sandburg wrote of the emergent "All American" President, who blended the Yankee and the pioneer and who was coming to epitomize the democratic ideal.

> Also around Lincoln gathered some of the hope that a democracy can choose a man, set him up high with power and honor, and the very act does something to the man himself, raises up new gifts, modulations, controls, outlooks, wisdoms, inside the man, so that he is something else again than he was before they sifted him out and anointed him to take an oath and solemnly sign himself for the hard and terrible, eye-filling and center-staged, role of Head of the Nation.

At the time of his death, victory won, this "mystery in smoke and flags" had become "the Father of his People." "It had no voice but Lincoln's," Sandburg declared. Surely the Radical Republicans, embarked on a fruitless quest for abstract justice for the Negro at the risk of turning the South into "a vast graveyard of slaughtered whites," would crumble before it. The author did not elaborate, but brought his biography to completion with a majestic account of the Martyr President's death and apotheosis.[73]

The publication of *Lincoln: The War Years* at the onset of the Second World War was a national event. Sandburg's portrait occupied the cover of *Time* magazine. The four-volume blockbuster stunned reviewers. It was "a mountain range of a book," said Stephen Vincent Benét, and Allan Nevins pronounced it "one of the greatest of American biographies." It was certainly that. Apart from the initial sale, Harcourt, Brace & Company would sell forty thousand copies of the complete six-volume biography. In retrospect, it has sometimes seemed the literary counterpart of the Lincoln Memorial. Alfred Kazin, in his path-breaking *On Native Grounds*, viewed Sandburg's *Lincoln* as an authentic sign of the nation's new historical consciousness and acclaimed it, not as a work of art, but as the capstone of that "greatest of all American works of art, the people's memory of Lincoln." On the other hand, Edmund Wilson, another man of letters, confessed the opinion that "Carl Sandburg is the worst thing that has happened to Lincoln since Booth shot him."[74]

Professor Randall reviewed Sandburg's book in the *American Historical Review*. "I really deal too leniently with him," he remarked to Bullard, who had himself gathered a cask of errors weighty enough to sink most books. "There seems to be a tendency, almost a conspiracy, among reviewers to do this." Randall appended his list of errata and ventured to suggest that the work would have been improved if cut in half, which would have left it about the size of his own *Lincoln the President*. The first two volumes of that work, in 1945, were subtitled *Springfield to Gettysburg*. Randall, too, rated a magazine cover, in this instance the *Saturday Review of Literature*. There was only one Sandburg, of course, and the professor felt no rivalry with him. Of the many ways in which his work differed from Sandburg's, one was that it was fundamentally an internal political history of Lincoln's presidency, while another was that it undertook the difficult task of critically reassessing a man encircled by a halo and draped in myth.

The revisionist mode of the book was set in Randall's account of Lincoln's conflict with Douglas in the 1850s. An attentive young scholar read it thus: "that Lincoln was essentially a Douglas Democrat." Although a liberal in his values and outlook, said Randall, Lincoln was a conservative on matters of slavery and sectional adjustment. The biographer had some difficulty explaining the "House Divided" speech on these terms. Viewing the broad problem of slavery and race relations as a social rather than a moral issue, Randall said that Lincoln and Douglas "did not fundamentally differ," which could be true only if one found no difference between a careless indifference to whether slavery was voted up or down and a commitment to fulfillment of the promise of freedom and equality for all people, black and white. Impressed by Lincoln's moderation in the secession crisis of 1860–61, Randall characterized the Inaugural Address as pacific and conciliatory and the provisioning of Fort Sumter as a non-aggressive act.[75]

Randall perceived in Lincoln's presidency "a fateful dualism" between moderate and radical Republicanism and offered this as "a veritable *leitmotif*" of his analysis. Keeping to the moderate course, Lincoln saved Kentucky for the Union. On the other hand, in a mistaken effort to assuage Radical hostility, Lincoln meddled harmfully with General McClellan and contributed to the failure of the Peninsular Campaign. The General's just fame had been blackened by Nicolay and Hay and other historians and biographers, including Sandburg, who supposed that loyalty to Lincoln demanded hostility to McClellan. Randall tried to set the record straight. Lincoln's final removal of McClellan after Antietam, he said, was "the act of a buffeted President"—buffeted by the Radicals—and it raised serious doubts about his competency in military matters.

Randall's treatment of Lincoln and emancipation was a primary case of how, in the President's own words, events had controlled him. He offered his "blueprint for freedom" in the special message to Congress of March 6, 1862. A "pathetically earnest" plea, said Randall, for gradual and compensated emancipation linked to colonization of the freed Negroes, beginning with action by the border states, it found no immediate support and was overtaken by events. The Emancipation Proclamation itself did not supersede the blueprint, at least not in the President's mind, for it was strictly an act of wartime "military necessity." Focusing on the limits rather than the scope of the proclamation, Randall denigrated its significance. It offered at most "a limping freedom." The popular notion that it broke the shackles of the slaves, "the stock image" of Lincoln the Great Emancipator, was myth. A thoughtful Negro reviewer, Carter G. Woodson, found in Randall's work "unconcealed antipathy" to the tradition of emancipation.[76]

In the third volume, *Midstream*, Randall departed from his chronological narrative to treat various topics, some concerning Lincoln's personal life and character, others historical events, such as the Vallandingham case. The volume resembled the series of essays *Lincoln the Liberal Statesman*, earlier published. Viewing the President's leadership in shifting contexts, Lincoln characterized it not in terms of "mastery" (he took no notice of Alonzo Rothschild's work) but in terms of grace and tact. When a delegation of Republican senators called on him in December 1862 and demanded radical changes in his cabinet, beginning with the removal of Secretary of State Seward, the President parried the thrust and kept the game in his own hands. "Such was Lincoln's adroit planning and exquisite tact in this crisis," Randall wrote, "that the complaining senators . . . found after they had shot their bolt that the President was still in command, with the cabinet unchanged."[77]

Randall died in 1953 leaving the final volume, *Last Full Measure*, but half finished. A young Lincoln scholar, Richard N. Current, completed the work. More than the previous volumes, this one relied upon new scholarly monographs streaming from the presses, for instance Jay Monaghan's *Diplomat in Carpet Slippers*, which magnified Lincoln's role in foreign affairs, and *Lincoln and the Patronage*, by Harry Carmen and Reinhard Luthin, which further attested to his political skills. The President's renomination and reelection in the face of Radical discontent, on one side, and the Democratic challenge of General McClellan, on the other, was treated at length. The leitmotif of the work culminated in the conflict over Reconstruction. The President's plan, in Randall's view, sought to rally the latent Unionism of white southerners, beginning with only a tithe of the loyal. It paid little heed to the civil rights of the freedmen. Except for emancipation, which the Thirteenth Amendment made permanent and complete, Lincoln had more in common with McClellan Democrats than with Radical Republicans, the authors maintained, and had the General been elected the outcome of the Union cause would not have been different. As to Andrew Johnson, "His restoration policy was essentially the same as Lincoln's and . . . the same type of postwar radical opposition and bitter congressional obstruction would have confronted Lincoln if he had lived and adhered to the policies as stated during his presidency and down to the time of his last speech on April 11, 1865."[78]

7

Zenith

THE CONCENTRATION OF writing about Lincoln in the several years before and after 1930 was paralleled, to a degree, by the artists' interest in him as a subject of heroic sculpture. New effigies of Lincoln burst upon the landscape. Franklin B. Meade, in his pioneering study *Heroic Statues in Bronze of Abraham Lincoln*, in 1932, began with the latest, Paul Manship's *Hoosier Youth*, and classified the rest under four headings: the Lincoln of Illinois, Lincoln the Emancipator, Lincoln the Orator, and Lincoln the President. The conception of the Emancipator, after once dominating the field, had become moribund; not for many years had any sculptor been inspired by it. The young Lincoln proved a challenging subject, especially in Illinois, which offered the most fertile ground for monuments. The Chicago sculptor Charles J. Mulligan had led the way with his lean and muscular youth, ax in hand, *The Railsplitter*, in 1911. Twenty-six years later Leonard Crunelle, a student of Lorado Taft at the Chicago Art Institute, created for the city of Dixon a portrait of Lincoln as *The Captain*, dubiously accoutered with scabbard and sword, in the Black Hawk War. Taft himself carved for Urbana, one of the county seats on the old Eighth Circuit, a strikingly handsome Lincoln in a speaking pose. And there were bronzes of *Lincoln Entering Illinois*, of *Lincoln at Twenty-One*, and of *The Debater*.

The outstanding bronze of the period, however, was the work of James Earle Fraser, distinguished for historical portraiture as well as for his poetic masterpiece, *The End of the Trail*. Commissioned by the Jersey City Lincoln Association, the oldest such body in the country, the statue was dedicated on its sixty-fifth anniversary in the city's Lincoln Park. Sometimes called "The Statesmen," sometimes "The Mystic," the seated Lincoln, nobly pedestaled, is shown solemn and meditative on the eve of his presidential career. Of quite a different order, of course, was Gutzon Borglum's Mount Rushmore Memorial in the Black Hills of South Dakota. It was as much an authentic product of these years as Sandburg's biography and qualified as a heroic achievement in its own right. The gigantic head of Lincoln in the group that upon completion would include four Presidents was dedicated in 1937.[1]

Lincoln's fame rose to zenith in the three decades that culminated in the sesquicentennial of his birth in 1959. This continuing, even enhanced, fascination with Lincoln's life and meaning stemmed, in part, from the felt need for symbols of democracy and leadership when the country faced grave problems of depression and discord at home and the threat of totalitarianism abroad. Alfred Kreymborg, in his "Ballad of the Lincoln Penny," captured the faith of the people in the hero's simple greatness.

> Now, when you hold a penny, look at Lincoln's face!
> See how up and down the land Lincoln saved the race!
> Look at that small penny, hold it close to you—
> And if you ever lose your way, Abe will lead you through—
> Abe will lead you through.

Instead of "In God We Trust," the message was "In Abe Lincoln We Trust." The fascination stemmed as well from the ongoing discovery of America's cultural past, in which Lincoln figured as a true hero. Sandburg fixed him in the imagination, and throughout these years the prairie poet was the dominant voice for a Lincoln who embodied indigenous American wit and wisdom. One of the passions of this movement was historic preservation. Naturally, Lincoln became an object of preservation. The Lincoln Home in Springfield was researched, restored, and refurnished with legitimate or period relics. In 1955, for the first time in sixty-eight years, the entire house was opened to the public; some years later it came under the jurisdiction of the National Park Service as the Lincoln Home National Historic Site.[2]

Although the study of Lincoln fell more and more to professional scholars, popular interest in him did not abate. The sheer bulk of writing about him, far greater than for any other American, continued to astonish. The newest *Lincoln Bibliography*, compiled by Jay Monaghan at the Illinois State Historical Library and published in 1945, cointained 3,958 titles, of which 212 were in foreign languages. Later surveying the har-

TOP LEFT: *The Railsplitter*, by Charles J. Mulligan, Chicago, 1911. (The Lincoln Museum, Fort Wayne, a part of Lincoln National Corporation)

TOP RIGHT: *Lincoln the Lawyer*, by Lorado Taft, Urbana, Illinois, 1927. (The Lincoln Museum)

BOTTOM: *Lincoln the Mystic*, by James E. Fraser, Jersey City, New Jersey, 1930. (The Lincoln Museum, a part of Lincoln National Corporation)

vest of a single year, 1950, Monaghan counted thirty book titles. Eight contained a hundred or more pages, while one, Allan Nevins's *Emergence of Lincoln*, was a blockbuster. Two were reprints, two were compilations, one was for children, and five titles, all published by the State Department—a new trend in Lincolniana—were in foreign languages. Ralph Newman's Abraham Lincoln Book Shop, in Chicago, was the foremost vendor of Lincolniana. Some five hundred persons, it was estimated, made their living from Lincoln. In addition to book dealers, the industry included custodians of monuments, curators, librarians, souvenir sellers, and others. While the great collectors were few, perhaps as many as ten thousand people collected Lincolniana as a hobby. Lincoln was, indeed, "everybody's business."[3]

Men and women still lived who had shaken the hand of Lincoln. In 1963 Governor Nelson Rockefeller of New York paid a birthday visit to a 114-year-old man, Henry Hendron, who remembered Lincoln stopping at his family's farm during the war. "I can't get over the fact that I've actually shaken the hand that shook hands with Abraham Lincoln," the Governor remarked. "I can't get over it." Lonely survivors like Hendron were reminders of how brief was the country's history. Increasingly, however, Lincoln's place in American memory depended less upon reminiscence and tradition than upon critical understanding of the past. And since historical scholarship proceeded in a dialectical fashion, images of Lincoln became more complex, fractured, and blurred. In theory, the scientific historian sought to understand the past on its own terms, regardless of its service to anything else living or dead. But Lincoln was too deeply embedded in the moral consciousness of the nation, too pervasive a metaphor of what the nation should be, to permit that as to his own history. People wanted to know what he had to teach the present and how his legacy might be renewed and revalidated. During these years the name of Lincoln, with all it symbolized, was more bandied about the political forum than at any previous time in his posthumous career.[4]

"What Would Lincoln Do?"

Presumably it was a rhetorical question. Yet some politicians made bold to answer it, almost always, of course, putting Lincoln's signature to their own agendas. Senator George W. Norris, once asked in the depths of the Great Depression what Lincoln would do, paused, then blurted out, "Lincoln would be just like me. He wouldn't know what the hell to do." Especially notable was the response, perhaps apocryphal, of Gerald Ford, the thirty-eighth President of the United States: "If Lincoln were alive today, he would roll over in his grave." The persistent question expressed the wish for inspiration and guidance in a time of trial not unlike that of the Civil War. Only poets could fathom it.

> O Father Abraham;
> O Liberator,
> What message for us now?

The oracle replies that the greatness was not his but the people's, and the living people must answer. Other poets wished for something more palpable than the invocation of democracy.

> We need a leader—one who knows not fear
> of Man or Gold, Lincoln, you should be here!

Another pleaded:

> Lincoln, come back, rebuild our broken world.
> Shattered and torn, the nations face the night;
> Our leaders halt, bereft of wisdom's light
> Humanity's fair banner now is furled;
> The peoples of the earth are leaderless;
> They ask a Friend, and there is none to bless.
> Lincoln, come back, rebuild our broken world.

The reverence of these verse prayers is most manifest in a sonnet of unknown authorship that recalled Whitman's "O Captain! My Captain!" and appeared in the *New York Herald Tribune* in 1937.

> Lincoln, thou shouldst be living at this hour,
> Son of the soil, brother of poverty,
> Those hard sharers of great destiny;
> Exemplar of humility and power,
> Walking alone to meet thy waiting fate
> Whose shadow was reflected on thy brow,
> Lincoln, thy people invoke thy spirit now—
> Preserve, protect, defend our sovereign state!
> Lover of justice and the common good,
> Despiser of lies, from thy yonder solitude
> Consider the land of thine and freedom's birth—
> Cry out: It shall not perish from the earth!
> Engrave upon our hearts that holy vow.
> Spirit of Lincoln, thy country needs thee now.

This was an invocation of the templed god of the Lincoln Memorial.[5]

As the Great Depression began, the Republican party still claimed to be the political house of Lincoln. His birthday was the Republican feast day throughout the country. On February 12, 1931, President Herbert Hoover delivered a nation-wide radio address from the White House. He spoke of Lincoln's invisible presence in the corridors of power. Hoover had claimed for his study the spacious room on the southeast quarter of the second floor, where Lincoln's office had been, and hung above the mantel a fine engraving of *The First Reading of the Emancipation Proclamation*. It was from this room, the setting of that portrait, that Hoover addressed the nation. He lauded Lincoln as a builder and liberator and warned of the dangers of centralization of power and invasions of individual liberty in the current economic crisis. This was the President's constant theme. In June he spoke at the dedication of the remodeled tomb beneath the Lincoln Monument in Springfield. The occasion invited rededication to Lincoln's ideals. On Lincoln Day 1932, Hoover again addressed the nation from the Lincoln Room of the White House. He implored the people to show the same commitment, courage, and resourcefulness as they had under Lincoln's leadership. In the presidential election that fall, Hoover's Republican followers drew the parallel to 1864; but in the eyes of the electorate the analogy ill became him. On Lincoln Day 1933, the defeated President delivered his farewell before the National Republican Club in New York City. Will Rogers quipped, "If it hadn't been for Lincoln the Republicans . . . would be short of a cause for celebrating."[6]

President Franklin D. Roosevelt aspired to be "the new Jefferson" of a reborn Democratic party. The Roosevelt New Deal suppressed the "least government" clauses of the Jeffersonian creed in pursuit of a government strong enough to secure the welfare of all the citizens in a mature and heartless industrial society. And to dramatize the conviction that the New Deal was an affirmation rather than a negation of Jeffersonian ideals, Roosevelt and his friends built a great memorial to the party's patron saint on the Tidal Basin of the Potomac askant from the Lincoln Memorial. The stage seemed to be set for a confrontation of political symbols; it did not occur, however, for at this juncture in the history of the parties, both confusingly claimed descent from both of the sainted founders. As Michigan's Senator Arthur Vandenburg observed, "It is not easy to pursue political genealogy in this puzzling, volatile age."

Skirmishes over symbols between the parties were often more amusing than enlightening. In 1934 Ohio Republicans erected a bronze tablet at the state capitol on the spot where Lincoln had made a famous speech seventy-five years before. A woman who as a five-year-old girl had bounced on Lincoln's knee presented the tablet in the name of the Young Men's Republican Club of Columbus, successor to the body that had invited Lincoln. No sooner had the tablet been dedicated than the state's Director of Public Works, a Democrat, objected to the presence of the word "Republican" in the inscription. The tablet was taken down and recast, purged of any reference to the Republican party, and returned to the east portal of the capitol. "Here Stood Lincoln," the memorial declared, but nobody would know that he spoke as a leader of the Republican party. In 1938 several Republican congressmen protested the demotion of Lincoln on the nation's postage. He had yielded the place of honor, on the most common first-class stamp, to George Washington in the first President's bicentennial year, 1932. Roosevelt, himself a philatelist, called for a general revision of the postage designs. The new Presidential Series of 1938 correlated the sequence of the Presidents with the postage denominations, which put Lincoln's portrait on a new sixteen-cent stamp colored black. New York Representative Bruce Barton denounced the series as a scheme to erase Lincoln, with other Republican presidents, from American memory. "Abraham Lincoln committed the great political offence of opposing the Democratic party, wherefore he is punished by being stricken off the three cent stamp. Thomas Jefferson takes his place." In 1954, when the Republicans returned to power in the Capitol, they took the opportunity presented by an increase in the first-class rate to four cents to ensure that Lincoln was again the most visible effigy on the nation's postage.[7]

The New Deal under the Roosevelt leadership threw the Republican party into disarray. In June 1935, GOP leaders from ten midwestern states met in Springfield and, amid Lincoln shrines and Lincoln banners, resolved to rally "grass roots" opposition to the New Deal under the appellation of the Constitution party. Justus Johnson, the Republican state chairman of Illinois who organized this "grass roots roundup," said he was inspired by the words Lincoln uttered in reply to William E. Dodge, the New York merchant prince, in February 1861. A delegate to the Peace Congress, Dodge had an interview with the President-elect soon after his arrival in Washington during which he urged conciliation of the South lest grass grow in the streets of northern commercial cities. To this plea Lincoln replied, according to Johnson's report, that the Constitution would prevail, "let the grass grow where it may." And this was the battle cry of the Springfield conference. Its only effect was to stiffen the party's opposition to the New Deal on constitutional grounds.[8]

Lincoln Day, as an annual rite of the Republican Party, was never more widely observed than during these years. The week of February 12 scattered Republican senators and representatives to hundreds of dinners across the country. By a gentleman's agreement the Democratic leadership of Congress agreed not to call up important legislation for vote while so many of the opposition party were absent. On occasion not enough Republicans remained in Washington to render homage on the floor of Congress; and the Polish patriot Thaddeus Kosciuszko, whose natal day also fell on February 12, was better remembered than the putative father of the Republican party. Only in 1941 did the Republicans institute an annual Lincoln Day dinner in the capital. In the election year 1936, the party launched a political blitz on February 12. There were some three thousand meetings nationwide, and the traditional banquet of the National Republican Club in New York, with one thousand guests, was the largest in its history. The party, it was repeatedly said, must stand by the principles and the virtues exemplified by Lincoln. Had he not brought the nation through the ordeal of the Civil War without rending its constitutional fabric? But terrible defeat at the polls led liberal Republicans, like Governor George D. Aiken of Vermont, to question whether the party had got the right message from Lincoln. "The greatest praise I can give to Lincoln . . . ," he told the National Republican Club in 1938, "is to say that he would be ashamed of his party's leadership today."

Generally, the leadership from Hoover through Thomas E. Dewey and Robert A. Taft espoused the conservative line. Hardy perennials of the Lincoln Day circuit included such figures as H. Styles Bridges, the New Hampshire senator, and Joseph W. Martin, minority leader in the House and sometime national chairman of the party. "Republicans must be devoted to their traditions if our great party is to survive," Martin declared in 1941 after another crushing defeat in the presidential election. This became a kind of mantra. Among younger Republican leaders, none was more devoted to the party's founder than Illinois Congressman Everett McKinley Dirksen. Well versed in Lincoln lore, Dirksen quoted him at every opportunity, often frequented the Lincoln Memorial, and next to God, according to his wife, "worshiped Lincoln most." He began one Lincoln Day speech by recalling the admonition of the Lord to Moses, "And this day shall be unto you for a memorial," then proceeded to assail the Roosevelt administration in the sacred name.

Bruce Barton, the New Yorker, was among the most ingratiating Republican orators. Lincoln had been a saint in his father's house, and so Barton spoke eloquently of his moral and spiritual grandeur. His Lincoln Day dinner address, "The Faith of Abraham Lincoln," at Buffalo in 1940, broadcast nationally, was a tour de force. He began by mimicking the language President Roosevelt had used to describe the plight of the nation's poor: "We are met here to honor the memory of an American who was ill-fed, ill-clothed, ill-housed—and did not know it." Instead of surrendering to failure or calling on the government for help, young Lincoln said "root, hog, or die" and triumphed over the little accidents of birth and upbringing. Indeed he made those supposed deprivations pillars of strength. Barton conceded that the struggle was more difficult today, but Lincoln's example of self-reliance was still good. Strength of character, not material well-being, was what counted in America. "Inwardly warmed by spiritual fires, [Lincoln] was unconscious of being ill-fed . . . , ill-clothed . . . , or ill-housed."[9]

Everyone understood, of course, that Lincoln was not and could not be the possession of a single party. Parties all across the political spectrum claimed to speak for him. Like Karl Marx before him, Earl Browder, chief of the Communist party, recognized in Lincoln "the single-minded son of the working class," a genuine revolutionary. In the mouths of Republicans and Democrats his words were hollow. "Today it is left to the Communist Party to revive the words of Lincoln," said Browder. Democrats began seriously to contest the Republican title to the Lincoln tradition after Roosevelt's reelection in 1936. The following year the Grover Cleveland Democratic Club of New York City honored the Republican patron saint on his anniversary, the first time such an association had chosen to do so, according to Senator Robert F. Wagner. Roosevelt's second term began with his plan to "pack" the Supreme Court. Again, as before the Civil War by the Dred Scott decision, the court threw itself in the path of reform. The shrewd and witty mayor of New York City, Fiorello La Guardia, drew the parallel between the court's holding in Dred Scott that a Negro did not gain citizenship by going to a free state and the recent Schechter decision voiding the National Recovery Act on the grounds "that a chicken born in Iowa became a citizen of New York the moment it entered the State." Lincoln would surely have seen the fitness of the analogy. Nothing seemed more appropriate to the issue than the sixteenth President's statement in his Inaugural Address: "The candid citizen must confess that if the policy of the government on vital questions, affecting the whole people, is to be irrevocably fixed by decisions of the Supreme Court, the instant they are made, in ordinary litigation between parties, in personal actions, the people will have ceased to be their own rulers, having to that extent practically resigned their government into the hands of that eminent tribunal." In 1939 the editor of the *Forum* addressed the question "Was Lincoln a Democrat?" The next year James A. Farley, Postmaster General, Democratic party chairman, and prospective presidential nominee of his party, occupied the Lincoln Day platform with Alfred M. Landon in Springfield, Illinois.[10]

The Dutchess County Roosevelts, unlike the Oyster Bay branch of the family, had been Democrats from way back, and so the President came by his Jeffersonian graces quite honestly. Lincoln, on the other hand, was an acquired taste which he indulged for calculated political effect. With most Americans he had been taught to revere Lincoln. On each February 12 of his presidency, except for the rare occasion when he was away from the capital, Roosevelt visited the Lincoln Memorial and, as aides placed a wreath before the great statue, stood at the steps below, uncovered, in silent tribute. Like his predecessors, he visited the shrines in Springfield and Hodgenville. In his last speech of the 1936 campaign, Roosevelt went beyond ceremonial nods to Lincoln and seized upon his words to fire a shot across the Republicans' bow. At Wilmington, Delaware, home of the Du Pont Corporation, which had earned the President's enmity as a major contributor to the right-wing American Liberty League, he recalled Lincoln's musings on the word "liberty" in a little address in 1864. All of us are for liberty, he had observed, but we mean different things by the name, and he told a fable worthy of Aesop.

> The shepherd drives the wolf from the sheep's throat, for which the sheep thanks the shepherd as a *liberator*, while the wolf denounces the same act as the destroyer of liberty, especially as the sheep was a black one. Plainly the sheep and the wolf are not agreed upon a definition of the word liberty; and precisely the same difference prevails to-day among us human creatures, even in the North, and all professing to love liberty. Hence we behold the processes by which thousands are daily passing from under the yoke of bondage, hailed by some as the advance of liberty, and bewailed by others as the destruction of all liberty.

Roosevelt drew the moral. "My friends, today, in 1936, the people have again been doing something to define liberty. And the wolf's definition has again been repudiated." (Curiously, he repeated the speech at each of the two successive times he closed his campaign at Wilmington.) Thereafter Roosevelt intermingled the Lincoln and the Jefferson symbols in his public discourse. At the laying of the cornerstone of the Jefferson Memorial, he held that the two great democratic leaders were, after all, much alike. "Lincoln, too, was a many-sided man. Pioneer of the wilderness, counselor for the under-privileged, soldier in an Indian war, master of the English tongue, rallying point for a torn nation, emancipator—not of slaves alone, but of those of heavy heart everywhere—foe of malice and teacher of good-will."[11]

From the Fourth of July platform at Gettysburg in 1938, Roosevelt addressed the tiresome question "What Would Lincoln Do?" in compelling fashion. "It seldom helps to wonder how a statesman of one generation would surmount the crisis of another," he said. "A statesman deals with concrete difficulties—with things which must be done from day to day. Not often can he frame conscious patterns for the far-off future." But Lincoln's stature in the great conflict of his time broke the bounds of history and invited Americans ever after to turn to him for help. "For the issue which he restated on this spot seventy-five years ago," and which Roosevelt then restated in his own terms, "will be the continuing issue before this nation so long as we cling to the purposes for which it was founded—to preserve under the changing conditions of each generation a people's government for a people's good." Although the task takes different shapes at different times, the challenge is always the same. And today, the President continued, "it is another conflict, as fundamental as Lincoln's, fought not with the glint of steel but with appeals to reason and justice on a thousand fronts—seeking to save for our common country opportunity and security for citizens in a free society. We are near to winning this battle. In this winning and through the years may we live by the wisdom and humanity of the heart of Abraham Lincoln."[12]

As the Second World War came on, Roosevelt saw the challenge in global terms and grew increasingly conscious of treading in Lincoln's footsteps. The timely appearance of Sandburg's *War Years* contributed to this. The President and the poet-biographer liked each other. In 1935 Sandburg told Roosevelt he was the best light of democracy in the White House since Lincoln. In his column in the *Chicago Daily News* and in his speeches he drew the parallel between Lincoln's "irrepressible conflict" and Roosevelt's.

The New York playwright Robert E. Sherwood was making the same connection. His drama *Abe Lincoln in Illinois* opened on Broadway October 15, 1938. It was a big hit, in part because, like John Drinkwater's play twenty years before, it struck a political nerve. From Lincoln's life and Lincoln's words came the renewal of democratic faith. Sherwood had been inspired by Sandburg's *Prairie Years*, which he read until he knew it by heart. He felt little need for dramatic invention, for the life Lincoln lived was a "work of art . . . a veritable allegory of the growth of the democratic spirit." The play itself was somewhat allegorical, for the audience was invited to construct the great and good Lincoln from twelve emblematic scenes placed between 1831 and 1861. The climax of the play occurs in the ninth scene devoted to the Lincoln and Douglas debates, when the challenger addresses a ten-minute-long speech directly to the audience. He assails slavery in terms of the ongoing conflict between human rights and property rights and swears allegiance to fundamental American values in a context more allusive to the imminent threat of totalitarianism than to the southern slavocracy. Opening in the week of the Munich crisis, the play was like a tonic to democratic despair. Raymond Massey, for whom Sherwood had written it and whose portrayal of Lincoln contributed much to its success, remarked in an interview, "If you substitute the word dictatorship for the word slavery throughout Sherwood's script, it becomes electric for our time." And so it did. In the last scene, at the Springfield railroad station, Lincoln's Farewell reaches beyond the historic challenge that awaits him in Washington and encompasses the question of whether the democracy Americans had won was fit to survive in a dangerous world.

The play altered Sherwood's life. It turned him into a public man. He abjured the pacifist views he had formerly held and became a militant interventionist abroad as well as a spokesman for liberal democracy at home. In 1940, after the motion picture version of the drama appeared, Sherwood joined the White House staff as a speech-writer, and he remained on active service throughout the Second World War.[13]

It was a time when many thoughtful Americans saluted Lincoln. Bernard DeVoto, who occupied "The Easy Chair" in *Harper's* magazine, wondering about the phenomenon, said that the nation in crisis turns to the man who represents the highest reach of American character and personifies the strength of its democracy. "What America finds in Lincoln is confirmation of the best it has dared to believe of itself. . . . His life, his understanding, and his triumph compose a symbol which stands for the justification of American democracy." Ralph H. Gabriel, a professor of history at Yale University, backed this judgment in his path-breaking book *The Course of American Democratic Thought*. Lincoln stood at the center of "the new American symbolism," said Gabriel. "He personifies that faith upon which the Republic rests."

The leftist political commentator Max Lerner, who, with others, had thought that Lincoln's directive force had been spent in modern America, observed the emergence of a new image, "one almost providentially made for a national crisis, Lincoln the democratic idealist who had himself been a great national leader in time of peril." In such an image, reformed and revitalized by Sandburg and Sherwood, Lerner suggested, Americans might find "power to steel our will and give it a quiet strength." He speculated, too, on the comparison of Lincoln and President Roosevelt. "How much of Lincoln does Roosevelt have in him? More, I am convinced, than any President since Lincoln or before." Although the two men were poles apart in external characteristics, each was obliged by events "to complete an interior crisis out of which his real greatness emerged."

On the eve of the election of 1940, Sandburg was accorded the honor of the last word on a two-hour national radio broadcast by Roosevelt's advocates. He recalled the critical election of 1864, drew appropriate quotations from his bottomless bag of Lincoln lore, and measured the President against Lincoln for strength and wisdom. Roosevelt was pleased.[14]

The shadow of Lincoln the war President was never far from Roosevelt's mind. In August 1941, at his first press conference after the announcement of the Atlantic Charter in which he and British Prime Minister Winston Churchill formulated Anglo-American post-war aims, the President turned to the first volume of *War Years* and quoted Lincoln's response to a delegation of Sanitary Commission women who begged words of encouragement from him at a dark time. "I have no word of encouragement to give!" he snapped. After a painful silence, he continued, "The fact is the people have not made up their minds that we are at war with the South." And he elaborated on the desperate unpreparedness of the people. Mr. President, a reporter piped, if you were going to write a lead . . . ? "I'd say," Roosevelt interrupted, 'President Quotes Lincoln'—(Laughter)—'and Draws Parallel.' " Even before war was declared, Roosevelt found solace and support in Lincoln's invasion of "the dark continent" of executive war powers. Accepting the nomination of his party for the presidency in 1944, he closed his address with the peroration of the Second Inaugural.

But death took from him, as it had from Lincoln, the opportunity to "achieve and cherish a just and lasting peace, among ourselves, and with all nations." Sandburg wrote a poem eulogizing the beloved commander who died in another Easter season. Paul Buck, professor of history at Harvard University, meeting his class the day after Roosevelt died, quoted lines from Walt Whitman and remarked, "As one studies history, the stature of a man is judged by what he does to build or destroy the faith by which men live. . . . Mr. Roosevelt was great because he, like Lincoln, restored men's faith."[15]

Roosevelt's successors ruled in *his* shadow, as William E. Leuchtenburg has shown, yet all of them had their moments with Lincoln. For Harry S. Truman the moment came in April 1951, when he dismissed General Douglas B. MacArthur from command of American and United Nations forces in Korea after he threatened to expand the war to Communist China. MacArthur was a military hero, especially popular with Republicans, and they rained brickbats upon President Truman. As a Civil War buff he knew that Lincoln had fired a succession of generals. Of them General McClellan was most analogous to MacArthur, for he had, at least in Truman's opinion, allowed his politics to interfere with his duty in the field. The analogy was somewhat imperfect, however, for Lincoln had complained of McClellan's timidity, while Truman sought to restrain MacArthur.[16]

Truman was a White House buff, too, and took pride in the extensive renovation of the historic mansion during his administration. Among the showiest of White House antiques was the monumental "Lincoln Bed" acquired by Mrs. Lincoln in 1861. No one knew whether or not Lincoln had ever slept in it, but tradition said he had. When President Truman told his aged mother, an unreconstructed Confederate, that she would sleep in Lincoln's bed when visiting him in the capital, she told him in no uncertain terms she would sleep on the floor instead. In the renewed White House, the southeast room that had been the Lincoln study became "Lincoln's Bedroom," with his bed and other furnishings of his time, though it had never been his bedroom. Mrs. John F. Kennedy would remember it as her favorite room to which she often retired to meditate.* Other First Ladies, and Presidents, have testified to the pungent sense of Lincoln's presence in the mansion. In some of the stories it is an eerie presence. "Sometimes," Mrs. Roosevelt said, "when I worked at my desk late at night I'd get a feeling that someone was standing behind me. I'd have to turn around and look." Maids, butlers, and guests swore they had seen Lincoln's ghost. When Maureen Reagan, after a night in the Lincoln Bedroom, said as much to her father, the President, he asked her the next time he appeared to send him down the hall, because "I've got a few questions I'd like to ask him."[17]

Lincoln was a storybook hero for Dwight D. Eisenhower. A copy of the famous Alexander Hesler photograph of 1860 hung from the wall of his office, and Eisenhower painted a portrait after an Alexander Gardner likeness of the President that hung in the Cabinet Room. He kept Lincoln's *Collected Works* in his office; he worshiped in Lincoln's pew at the New York Avenue Presbyterian Church; and he quoted him repeatedly. Most often quoted was the so-called Lincoln dictum, from a fragment on government usually dated 1854: "The legitimate object of government, is to do for a community of people, whatever they need to have done, but can not do, *at all*, or can not, *so well do*, for themselves—in their separate, and individual capacities. In all that the people can individually do as well for themselves, government ought not to interfere." This was "Father Ike's" definition of modern Republicanism and "the formula," as a speech-writer named it, for deciding questions

---

*"It was the one room in the White House with a link to the past. It gave me great comfort. I love the Lincoln Room the most, even though it isn't really Lincoln's bedroom. But it has his things in it. When you see that great bed, it looks like a cathedral. To touch something I knew he had touched was a real link with him. The kind of peace I felt in that room was what you feel when going into a church. I used to feel his strength. I'd sort of be talking with him." Quoted in Hugh Sidey, *John F. Kennedy, President*, new ed. (New York, 1964), 231.

of government power. Of course, the difficulty came with the application of the principle; and Roosevelt, who had also invoked Lincoln's rule, applied it differently than Eisenhower.

The President liked to talk about Lincoln. Having himself commanded a great army, he sympathized with Lincoln as Commander in Chief and often quoted his forbearing remark about McClellan: "I would hold the General's horse if he would just win a victory." Emmet John Hughes, who worked in the Eisenhower White House, once asked the President to describe Lincoln's greatest qualities of leadership, and he replied, "Dedicated, selfless, so modest and humble." No mention was made of intelligence, vision, toughness, and tenacity. "The Eisenhower appreciation of Lincoln, in short," said Hughes, "reflected one sovereign attitude: all esteemed qualities . . . were personal and individual, and not one was political or historical." On this test the presidency should be an exercise not of political will and power but of personal virtue, and this, Hughes concluded, described President Eisenhower.[18]

Governor Adlai E. Stevenson of Illinois, Eisenhower's Democratic opponent in two elections, probably had more feeling for Lincoln than any man who occupied the office of President. His great-grandfather Jesse Fell had been among the first to propose Lincoln's candidacy in 1860, though Stevenson's more immediate forebears were Democrats. In regard to Lincoln, however, a certain ecumenism prevailed in Illinois politics, as exemplified in Stevenson's gubernatorial predecessor, Henry Horner. Stevenson knew all the Lincoln luminaries and spoke from all the Lincoln platforms. At Gettysburg, on the eighty-eighth anniversary of Lincoln's address, he linked the issue of the Civil War to the hopes of all peoples for freedom, which the blustery editor of the *Chicago Tribune*, Colonel Robert R. McCormick, labeled "preposterous." Speaking on "Lincoln as a Political Leader" before the annual meeting of the Abraham Lincoln Association in 1952, Stevenson began with the observation "A man in public office can find no surer guide than Lincoln," then rooted his excellence in the ability to lead the people toward a true understanding of their own will, hence to shape their own destiny. Although not a deep student of Lincoln, Stevenson felt a special affinity for him. Returning from Chicago after his nomination for the presidency in 1952, he slipped out of the governor's mansion around midnight in order to commune with Lincoln in the home on Eighth and Jackson. "For an hour he sat alone in Lincoln's rocking chair," the caretaker said. The affinity came to be widely recognized. Years later, when he was Ambassador to the United Nations, Stevenson was chosen to narrate Aaron Copland's *Lincoln Portrait* at the gala opening of the Lincoln Center for the Performing Arts in New York City.

For all their differences, reporters dwelt upon certain resemblances between Lincoln and Stevenson, literary flair, for instance, and above all humor. In the car taking him to the headquarters hotel in Springfield to make his concession speech on election night in 1952, the Governor remembered something Lincoln had said after losing the senatorial election to Douglas in 1858. With him in the car was Benjamin P. Thomas, the Lincoln biographer, who verified the candidate's response to questions about how he felt. And so Stevenson described his own feelings in Lincoln's words. "He said he felt like a little boy who stubbed his toe in the dark. He said that he was too old to cry, but it hurt too much to laugh."* When Stevenson sometimes made light of weighty issues, he cited Lincoln on humor as a vent and deplored "the Republican law of gravity." Upon his death in 1965, the State of Illinois placed his coffin on the catafalque that had borne the Martyr President from Washington, and he lay in state in the capitol prior to burial in his hometown, Bloomington.[19]

Neither John F. Kennedy nor Lyndon B. Johnson nor Richard M. Nixon had any special affinity with Lincoln. Kennedy owed his narrow victory over Nixon in 1960, in part, to the public verdict for him in the celebrated television debates that recalled the Lincoln-Douglas debates but, everyone agreed, scarcely compared with them. Kennedy admired Lincoln as an orator and repeatedly instructed his speech-writer, Theodore Sorensen, to study his speeches as models for his own, for instance the Gettysburg Address in connection with Kennedy's inaugural address. The Lincoln moment for each of these Presidents, though never sharply defined, occurred in relation to the civil rights revolution. The assassination of President Kennedy on November 22, 1963, was a Lincoln moment of another kind. The American people had not been so devastated since Lincoln's assassination, and many parallels would be drawn between these events. Cardinal Cushing, in Boston, eulogized Kennedy as "a youthful Lincoln" who gave the world hope. Bill Mauldin's cartoon of a weeping Lincoln published the morning after Kennedy's death, and widely reproduced, summed up the nation's sorrow. Appropriately, President Johnson chose to close the month of national mourning for Kennedy in a candlelight service at the Lincoln Memorial.[20]

*This was, in fact, the recollection of Lincoln's response by Charles S. Zane in 1912. Zane had practiced law in the same courts with Lincoln. Thomas, in the biography he published in 1954, preferred to give the response John Hay recorded in his diary six years after the event. As he walked home through the wet, gloomy streets on the night of Douglas's election, Lincoln said, "My foot slipped from under me, knocking the other out of the way; but I recovered and said to myself, 'It's a slip and not a fall.' " A more sanguine Stevenson might have remembered that.

Weeping Lincoln. Cartoon by Bill Mauldin, 1963. (*Chicago Sun-Times* and Illinois State Historical Library)

Historians' Encounters

On July 26, 1947, at one minute past midnight—twenty-one years to the day after the death of Robert Todd Lincoln—the great collection of his father's papers was unveiled to the public with appropriate ceremony at the Library of Congress in Washington. Luther H. Evans, the Librarian, had invited some thirty Lincoln scholars to a gala dinner in the Whittall Pavilion of the Library the evening before the opening. Carl Sandburg entertained the party with song and story, and William H. Townsend paid tribute to the donor. As the midnight hour approached, the special guests trooped across the street to the Annex and proceeded to the Manuscripts Division on the third floor where about two hundred people—staff, reporters, onlookers—had gathered. Ruth Randall remembered the event as the most exciting moment of her life. Evans intoned the deed of gift, and as newsreel cameras whirred and flashbulbs popped, the doors of the vaults housing the treasure swung open, disclosing row upon row of tall volumes bound in blue-shaded black buckram with red labels. Instantly, the scholars rushed to the card catalogue index and opened the volumes of manuscripts, sampling and searching until daylight. CBS recorded the unveiling for broadcast on that day. Its correspondent, John Daly, interviewed four of the scholars, Sandburg, James G. Randall, Paul Angle, and Jay Monaghan, for first impressions of the collection. Randall was rhapsodic, quoting lines from Keats's sonnet "On First Looking into Chapman's Homer":

> Then felt I like some watcher of the skies
> When a new planet swims into his ken.[21]

The Manuscripts Division had been hard at work organizing, mounting, and cataloguing the collection since its return to the Library of Congress from safekeeping at the University of Virginia during the Second World War. An index of 38,403 entries guided researchers to the documents contained in the 194 folio volumes. Only about nine hundred of the documents were Lincoln holographs; very few antedated 1860; and the bulk of the collection consisted of incoming correspondence. In all this the eager eyes of the scholars who first viewed the documents found nothing startling: no skeleton in the closet, no new light on the Ann Rutledge romance, no evidence of cabinet treason against the President, and nothing implicating Secretary of War Edwin M. Stanton in the plot to assassinate him. As a result, some were disappointed.

Early reports in the press suggested that the unveiling failed to live up to its billing. "Among the remarkable things concerning the Robert Todd Lincoln Collection . . . ," Randall wrote to Randolph Adams, a distinguished bibliophile, "is the manufacture of another Lincoln myth before our very eyes, *i.e.* the myth that the whole affair fell flat, that there is nothing worth while in the papers, *et cetera*." Stefan Lorant, a newcomer to the Lincoln field, asked in *Life* magazine, "Where Are the Lincoln Papers?" He accused the Library of perpetrating an empty "publicity stunt," and credited Nicholas Murray Butler's lurid story to account for what was missing. Robert Lincoln must have burned the most important papers. (The notion persisted even in high places. President Truman, for instance, went to his death believing the son burned "half to two-thirds" of his father's papers.) In his own syndicated report, Sandburg conceded that nothing spectacular or mystifying had turned up, yet he found many things "that go to root the Lincoln legend deeper."[22]

Randall was in Washington that summer working on the third volume of *Lincoln the President*. Ecstatic over the collection, he plunged into it for two months, and after returning to the University of Illinois that fall continued to work from one of the microfilm copies—two miles of microfilm—that the Library supplied to meet a large demand. For a scholar like Randall this was the very stuff of history, fresh and awaiting his inquiring mind and shaping hand. He had the sense of "living with Lincoln, handling the very papers he handled, sharing his deep concern over events and issues, noting his patience when complaints poured in, hearing a Lincolnian laugh." The age had come to life. What could be more exhilarating! Nicolay and Hay, it now appeared, had used very little of the collection. Among its riches, Randall thought, were the manuscript of the Farewell to Springfield as written down on the moving train partly in Lincoln's hand and partly in Nicolay's; the worksheets and rough drafts of the First Inaugural and of various messages to Congress, which invited the kind of textual investigation interesting to literary scholars as well as to historians; and in the mass of incoming material, poignant letters that revealed as never before Lincoln's relations to the people.

Among the newly discovered documents in Lincoln's hand, Randall was particularly taken with the President's "exquisite reprimand" of an overzealous young captain, J. Madison Cutts, a brother-in-law of the late Senator Douglas. Cutts, in 1863, had been court-martialed and sentenced to dismissal from the service for an altercation with a fellow officer. However, the sentence was commuted, and he was ordered to rejoin his regiment after an oral reprimand by the President. In *Midstream*, Randall introduced the document, which took the form of a memorandum of the President's interview with Cutts, as Exhibit A of Lincoln's mastery of the art of human relations.* It was, said Randall, "a classic comment on quarreling." To all the great things Lincoln did and said must be added this "priceless element of human thoughtfulness and personal understanding." Thus it was that the Lincoln Papers, without any sensational disclosures, opened a new phase in the appreciation of the man. "Little of what was basically known has had to be discarded," Randall concluded, "but in vividness and glowing detail we have much more of his period than before."[23]

Even before this event the Abraham Lincoln Association, in Springfield, had undertaken to realize its dream of a complete and authoritative edition of Lincoln's speeches and writings. Roy P. Basler, who suc-

* "The advice of a father to his son 'Beware of entrance to a quarrel, but being in it, bear it that the opposed may beware of thee' is good, and yet not the best. Quarrel not at all. No man resolved to make the most of himself can spare time for personal contention. Still less can he afford to take all the consequences, including the vitiating of his temper, and the loss of self-control.

"Yield larger things to which you can show no more than equal right, and yield lesser ones, though clearly your own. Better give your path to the dog than be bitten by him in contesting for the right. Even killing the dog would not cure the bite."

ceeded to the office of executive secretary of the association in 1947, became editor as well of what would be named *The Collected Works of Abraham Lincoln*. A forty-three-year-old professor of English at George Peabody College, in Tennessee, Basler had earned his Ph.D. at Duke University with a dissertation on Lincoln in literature, subsequently published in revised form as *The Lincoln Legend*. Slowly, as the Illinois luminaries took notice of it, Basler gained entry into their circle. The publication in 1946 of *Abraham Lincoln: His Speeches and Writings*, with an Introduction by Carl Sandburg, proved Basler's worth both as a scholar and an editor. For years, under four previous secretaries, the association had been compiling copies of Lincoln manuscripts and newspaper records of speeches and writings. It had diverted precious funds to this project. It had searched far and wide for Lincoln documents and sought permission to publish them in the planned edition. The Library of Congress, which had contemplated an edition that would incorporate the Lincoln Papers, agreed to cooperate fully in the ALA project. Rutgers University Press, whose editor, Earl Schenk Miers, was a Lincoln enthusiast, entered into a contract to publish the work. Basler thus became the chief of what has been described as "a collective effort" of the ALA for several years.

Apparently no serious thought was given to following the new path pioneered by Julian P. Boyd in the instance of Thomas Jefferson of editing the complete papers, including incoming correspondence, of major historical figures. This is surprising, especially in view of the richness of the materials in the newly opened collection. But the association had long had its mind set on a new edition of Lincoln's writings, and the editor, Basler, was more literary scholar than historian. Financial considerations, too, may have barred the more ambitious approach. The $100,000 expended to edit the work was modest, if not downright parsimonious. The association met over half of this cost by Rockefeller Foundation grants. Unfortunately, after a dispute with the press, the association was also compelled to assume a large part of the costs of publication. Miers had departed, and his successor at Rutgers took a dim view of Lincoln's salability. Basler searched for an alternative publisher but "struck out," as he put it, time after time.[24]

The *Collected Works* appeared in nine volumes on February 12, 1953. Because of Lincoln's importance to the nation, the publication was both a civic and a literary event. In a formal presentation at the White House, President Eisenhower vowed to keep the volume close at hand in his office. Scholars were pleased with the *Collected Works*. Its 6,870 items more than tripled the number in the fullest edition of the old *Complete Works*, and added 3,312 items never before collected. Basler made no claim for the definitiveness of the work. A man who stuffed papers in his hat, in carpetbags, and in pigeonholes mocked claims to completeness. One vast uncharted domain, Lincoln's law cases and the documents related thereto, had been excluded with a view to separate publication at a later date. Alas, it would be a long way off. *The Collected Works* depleted the association's treasury, forcing it to discontinue *The Abraham Lincoln Quarterly*, to close its office, and to transfer its files to the Illinois State Historical Library. Rutgers University Press, on the other hand, made out handsomely on the *Collected Works*. Thanks to an assist from the History Book Club, over forty thousand copies were sold during a period of twenty years. Not a single Lincoln penny of royalty went to the broken ALA.[25]

Published almost simultaneously with the *Collected Works* was the extraordinary new biography of Lincoln by Benjamin P. Thomas. The two works came out of the same shop. In his earlier post as executive secretary of the ALA, Thomas had helped to assemble Lincoln's papers, and he served as an editorial adviser to Basler and his staff. He was the first biographer of Lincoln since Nicolay and Hay to use the Robert Todd Lincoln Collection. Blessed with expert knowledge and a facile pen, Thomas sought to meet the need for a compact biography in one volume that would supersede Lord Charnwood's thirty-five-year-old *Abraham Lincoln*, still widely read in a cheap paperback edition. He succeeded admirably, in part by sacrificing scholarly punctilio, for in-

Benjamin P. Thomas. (Illinois State Historical Library)

stance dispensing with footnotes, to readability. Thomas's aim, clearly, was not to offer new views but to provide a synthesis of modern scholarship that was accurate, balanced, lucid, and uncomplicated. Critics were all but unanimous in their praise. Writing in the *New York Herald Tribune*, Mark Van Doren predicted that Thomas's *Lincoln* would be "the standard, indispensable one-volume biography" for the next generation, though he would keep Charnwood on the shelf for its literary genius.[26]

Thomas had no rivals, for the course of scholarship had turned away from biography to monographic study of discrete themes and facets of Lincoln's life. Every other title took the form of "Lincoln and . . .". Thus there was *Lincoln and the Radicals* (1941) and *Lincoln and His Generals* (1952), both by T. Harry Williams; *Lincoln and the Patronage* (1943) by Harry Carmen and Reinhard H. Luthin; William B. Hesseltine's *Lincoln and the War Governors* (1948), Robert S. Harper's *Lincoln and the Press* (1941), David M. Potter's *Lincoln and His Party in the Secession Crisis* (1942), Robert Bruce's *Lincoln and the Tools of War* (1956), and many others. Interest centered on the presidential years, but there were important monographs on Lincoln's career before 1861 as well. With all this scholarly production, Richard Current observed, it ought to be possible to assemble, at last, the definitive Lincoln image. "The trouble is, the scholars disagree." Paradoxically, the more they wrote about Lincoln the more blurred, confused, and problematic the image became. It almost seemed as if the modus operandi was to disagree, thereby to keep the enterprise going. All the work devoted to illuminating and elucidating particular subjects tended to obscure the forest for the trees.

Two of the foremost Lincoln scholars of the younger generation, Current and David Donald, adopted the essay form in wide-ranging reassessments of basic issues. Donald, the biographer of Herndon, collected a number of these in *Lincoln Reconsidered* in 1956. Writing on Lincoln the politician, he questioned the evidence of his popular appeal—the common notion recently restated by Thomas that it was "the people" who sustained him—and instead found the secret of his success in the fact that "he was an astute and dexterous operator of the political machine." The studies by Carmen and Luthin and by Hesseltine offered persuasive support for this view. A parallel essay by Donald so much emphasized Lincoln's political flexibility, opportunism, and pragmatism that he was left with no ideological undergirding at all. Current wrote *The Lincoln Nobody Knows* not to unravel the mystery of the man but to confirm and even exalt it. Eleven deftly etched chapters ranged over controverted areas of Lincoln's life in order to explain to a general audience why he remained an enigma to scholars.[27]

In the opinion of American historians, Lincoln was the greatest of the country's Presidents. Harvard Professor Arthur M. Schlesinger, Sr., asked members of the guild to rate the Presidents in 1948 and published the results of the survey in *Life*. The top five were Lincoln, Washington, Franklin D. Roosevelt, Woodrow Wilson, and Thomas Jefferson in that order. Schlesinger took another poll twelve years later with the same result, at least in the category of "great" Presidents. Other and later surveys, after the civil rights struggle and the traumas of Vietnam and Watergate, ratified the order of the first three but showed small shifts in the positions of Jefferson, Wilson, and Theodore Roosevelt. How far the general public may have agreed with the historians' verdict is impossible to say.[28]

Political scientists as well as historians favored the model of a strong presidency. In 1948 one of the former, Clinton Rossiter, published an important book, *Constitutional Dictatorship: Crisis Government in the Modern Democracies*, in which he posed the compelling question Lincoln raised in 1861: "Is there in all republics this inherent and fatal weakness? Must a government of necessity be too *strong* for the liberties of the people, or too *weak* to maintain its own existence?" Lincoln found an effective answer, said Rossiter, in the executive "war powers." Randall, of course, had made much the same argument. But Rossiter felt no compunction about calling Lincoln a "dictator," moreover a self-appointed one prior to the convening of Congress on July 4, 1861. After his death the Supreme Court invalidated military trials under the President's authority, but "such fustian," as Edward S. Corwin labeled it, would have no effect in future emergencies because of Lincoln's towering reputation. Rossiter agreed without sharing Corwin's regrets. "Lincoln's actions form history's most illustrious precedent for constitutional dictatorship." Unfortunately, it was a precedent for bad men as well as good men, so Lincoln could not be absolved of responsibility for bad results. Corwin thought Roosevelt, by his breaches of the Constitution in the Second World War, was a case in point. Another political scientist, Gottfried Dietze, blamed the decline of government by law in the United States, as he perceived it, on Lincoln's violation of his own principles of liberal constitutionalism. His dictatorial actions revealed "a democratic Machiavellian whose latent desire to achieve immortality broke forth at the first opportunity offered by . . . the Civil War." Quite aside from the question of Lincoln's desire for immortality, this was an extreme position. Yet it appealed to spokesmen of the neo-conservative movement.[29] But for Lincoln's "repressive dictatorship," and the wounds he inflicted on the Constitution, wrote Frank S. Meyer in the *National Review*, "it would have been infinitely more difficult for Franklin Roosevelt to carry through his revolution, [and] for a coercive welfare state to come into being." The truth was, of course, that Lincoln always claimed reasoned constitutional warrant for his actions and believed that by saving the Union he was saving the Constitution. Charles A. Beard, though he shared Corwin's distrust of Roosevelt, thought it Lincoln's imperishable glory that he did *not* seize dictatorial power. Nor did he interfere with the democratic political process. His own bitterly contested reelection in 1864 was one of the best fruits of the war. "It has demonstrated" as Lincoln himself said, "that a people's government can sustain a national election in the midst of a great civil war. Until now, it had not been known to the world that this was a possibility." The dictatorship question would again come up for review during the controversy over "the imperial presidency" in the era of Vietnam and Watergate.[30]

The World War, with its strong moral imperatives, challenged the Civil War revisionism of historians like Randall. Still, the challenge was slow to materialize. One of the most influential statements of revisionism, Howard K. Beale's "What Historians Have Said About the Causes of the Civil War," appeared in 1946. In that same year Bernard DeVoto, himself a historian of the American West, raised a solitary voice of dissent in *Harper's*. True, the fundamental question of slavery was faced only on the periphery, in the territories, but that, DeVoto insisted, was precisely the point to be explained, not dismissed with denunciations of trouble-making agitators, as the revisionists had done.[31]

The brilliant young historian Arthur M. Schlesinger, Jr. continued the attack in an address before the Lincoln Group of Boston, later published in abbreviated form in *Partisan Review*. The entire "needless war" thesis was a piece of rationalistic sentimentality, in Schlesinger's opinion. He associated its popularity with that of the romantic novel *Gone with the Wind*, which was a case of "the victors paying for victory by pretending literary defeat." But history should not be vulnerable to such strategems. There was no escape from its insistent demand for moral decision. Lincoln understood that and assumed the burden. To say with the revisionists that the status of slavery in the territories was an "unreal" issue, Schlesinger observed, was equivalent to saying that the World War was fought over the "unreal" issue of the invasion of Poland. It was the issue upon which the battle was joined. It made the crucial moral choice inescapable. Thomas Pressly, in his thoughtful study *Americans Interpret Their Civil War*, did not disguise his distrust of revisionism. He hailed the early volumes of Allan Nevins's monumental history of the Civil

War era, *Ordeal of the Union*, for initiating a "new nationalist tradition." Other readers were less sure this was the thrust of the work. For Nevins, too, like the revisionists, seemed to think that with cooler heads and better leadership war could have been averted.[32]

In 1959 a young student of classical political theory, Harry V. Jaffa, not only assailed revisionism but restored the central importance of the natural rights doctrine of the Declaration of Independence—"this grand pertinacity," as Charles Sumner had called it—in Lincoln's politics. His book *Crisis of the House Divided* focused on the issues in the Lincoln-Douglas debates. Beveridge and Randall, it may be recalled, had dismissed the debates as little more than curious folklore and narrowed the differences between the candidates to the vanishing point. Jaffa pronounced this treatment "shocking." The issue between free soil and popular sovereignty in Kansas was crucial because the free states could not abandon their position "without losing the root of the conviction which was the foundation of their freedom." That root was the Declaration of Independence, which Lincoln transformed from a charter of individual liberty into something "organic and sacramental—a kind of political religion." It was prophetic and progressive, looking to the realization of freedom and equality for all. Jaffa came to this interpretation not through American history but through the study of Plato's *Republic* under the natural law theorist and scholar Leo Strauss, at the University of Chicago. "The beauty of Lincoln's argument," he would later write, "was that he made the case for popular government depend upon a standard of right and wrong, independent of mere opinion, and which was not and could not be justified by the counting of heads." With this outlook Jaffa could be a charter member of the conservative intellectual movement and a contributor to the *National Review*, although his belief that equality was a right to be made good in the unfolding promise of American life was branded heretical.[33]

The case of those who held Lincoln materially responsible for the war began with the "House Divided" speech and ended with the decision to provision Fort Sumter on April 6, 1861. Throughout Lincoln maintained that he never threatened the southern states or the institution of slavery in those states and sought always a peaceable resolution of the crisis. Almost a century later, however, some southerners continued to believe he deliberately provoked the war. By refusing conciliation and by reiterating "the idiotic sentence that all men are created equal," wrote one die-hard in 1942, he made the South secede, and by the Sumter expedition he made the South fight. Several years later another southerner, Professor Charles G. Tansill, speaking at a wreath-laying before Jefferson Davis's statue in the Capitol, declared that the responsibility for the Civil War rested squarely on the shoulders of one man, President Lincoln. This provoked outrage, as Tansill intended; both the sponsors, the Sons of Confederate Veterans and the United Daughters of the Confederacy, disavowed the speaker's sentiments.

In the historiography of the subject special significance attaches to Charles W. Ramsdell's article "Lincoln and Fort Sumter," published in the *Journal of Southern History* in 1937. To be sure, the President only provisioned the beleaguered fort and so informed Governor Pickens of South Carolina in advance; but the act, said Ramsdell, carried the threat of force if resisted, and therefore was a direct challenge. The essence of the stratagem was to give the state no choice between backing down and firing the first shot. As Nicolay and Hay had said, Lincoln was "master of the situation," for he would win either way. John S. Tilley, an Alabama lawyer, made the case at greater length and with less detachment in his book *Lincoln Takes Command*, in 1940. He accused the President of diabolical trickery. Unable to find any evidence in the record of the need to provision the fort, Tilley said that Lincoln created the ruse of "the starving garrison" as an excuse for the expedition. But when the Lincoln Papers were opened several years later, there was the pleading letter from Major Anderson to the President.[34]

Randall made the most effective response to Ramsdell. "To say that Lincoln meant that the first shot would be fired by the other side *if a shot was fired*, is by no means the equivalent of saying that he deliberately maneuvered to have the shot fired. This distinction is fundamental." David M. Potter elaborated on the distinction in his 1942 monograph devoted to the Sumter episode. This young historian portrayed Lincoln as weak, confused, and bungling at the start of his presidency, which did not set well with devotees like Louis Warren. Hoping for a peaceful resolution of the crisis, the President sought time to cultivate the pacific sentiments of northern Republicans and the closet Unionism of many in the South. The Sumter crisis broke unexpectedly; he had no choice but to deal with it, and the outcome was actually a terrible defeat for his policy. Kenneth Stampp, another young historian of the Civil War, reviewing Potter's book, said admirers of Lincoln were now left with "an ugly choice": either he was a crafty schemer or a pitiful bungler. Stampp thought he was neither. In *And the War Came*, in 1950, he argued that Lincoln was no pacifist and that he employed a "strategy of defense" to save the Union, avoiding the appearance of aggression but from the first taking the "calculated risk" of hostilities. In the end, Lincoln, like the Confederate leaders, preferred war to submission. It came logically, neither by blunder nor by trickery. With at least three interpretations of Lincoln's conduct in the Sumter crisis, it is not to be wondered that his image became blurred and confused in the public mind.[35]

The story of Lincoln as Commander in Chief received fresh attention from historians. Kenneth P. Williams, in his voluminous *Lincoln Finds a General*, berated McClellan and crowned Grant the hero of the war. In *Lincoln and His Generals*, T. Harry Williams acclaimed the President the greatest strategist of the war, out-praising Ballard, Maurice, Palmer and others from fifteen or twenty years before. With regard to McClellan, the President's only mistake was in yielding his judgment to the General's plan for the Peninsular Campaign and in generally over-indulging him. What Williams called a "modern command system" was put in place upon Grant's appointment as lieutenant general. But the notion that Lincoln withdrew from strategic decision-making upon Grant's appointment was false. That notion had been put forward by Grant himself when there was no conception or tradition of Lincoln as strategist. To the end, said Williams, the President remained "a better strategist than any of the generals." Robert Bruce's prizewinning book *Lincoln and the Tools of War* acclaimed the President's keen interest in the development of improved guns and ordnance as a major contribution to winning the war.[36]

An old theme that took a new turn about 1960 was that of Lincoln and the Radicals. Williams, in his book of this title in 1941, set forth a revisionist view of the subject. In the factional division within the Republican party, which loomed larger than the division between Republicans and Democrats, Lincoln was aligned with the conservatives or moderates. The Radicals constantly plotted against him. At times he yielded to them in order to survive. They, of course, triumphed upon his death. In *Lincoln Reconsidered*, Donald challenged this view. Opposition to Lincoln ought not to be equated with Radicalism, nor should Lincoln's politics be equated with opposition to the Radicals. Donald offered Charles Sumner, whose biography he was writing, as a paramount example of cooperation between Lincoln and the Radicals. Williams rejected "this revision of revisionism." Nevertheless, it marked the turn toward a more radical interpretation of Lincoln.

In their monograph *Politics, Principle, and Prejudice, 1865–1866*, La Wanda and J. H. Cox argued persuasively that had Lincoln lived he would have prevented the disastrous split between the factions of the Republican party. Lincoln, the Coxes maintained, "would never have driven [the Radicals] into a position from which there was no exit except by open and unrelenting conflict with the Executive." Several years later Hans Trefousse, in *The Radical Republicans: Lincoln's Vanguard for*

Racial Justice, turned Lincoln into a Radical. From as far back as 1857, in Illinois, Lincoln sided with the more far-reaching Republicans. Fundamentally, he and they had the same goal, "a free democracy untainted by slavery." This was a sharp departure from the views advanced by Bowers, Milton, Randall, Williams, and others. Here was a Lincoln for the Second Reconstruction no less than the first.[37]

Lincoln's assassination and the persons and events connected with it had long been a cult subject unto itself. In 1937 Otto Eisenschiml, a Chicago Lincolnite who was an industrial chemist by profession, found a large audience for his book *Why Was Lincoln Murdered?* A fascinating romp through a maze of circumstantial evidence leading nowhere but implying the complicity of Stanton in the murder, the book promised more than it delivered. The silence of the Lincoln Papers, the entire absence of creditable evidence of conspiracy in high places, did not dispel the myth. The assassination of President Kennedy, which also gave rise to lurid conspiracy theories, helped to revive it. George S. Bryan named it, in the title of his impartial survey, *The Great American Myth*. Others implicated from time to time in Lincoln's assassination were the Catholics and the Jews, the former mainly on Father Chiniquy's wicked testimony, revived in the twentieth century by the Ku Klux Klan, the latter by Father Charles Coughlin, the fascist radio-priest, who in 1939 declared that the Rothschilds murdered Lincoln because he opposed the power of international bankers in American affairs.

Among the curiosities of the mythology were Booth "survival tales." Long after his reported death, whether by Sergeant Boston Corbett's gun or by his own hand, Booth or his double was sighted in China, Ceylon, or some other place. In 1907 a sensational book, *The Escape and Suicide of John Wilkes Booth*, purportedly written by a Tennessee lawyer, Finis L. Bates, maintained that one John St. Helen, who committed suicide in Oklahoma four years earlier, had in fact been John Wilkes Booth. St. Helen's mortal remains, embalmed in arsenic, became the feature attraction of Jay Gould's Million Dollar Spectacle, a carnival touring the South and Midwest. Bates later acquired the mummy. It changed hands several times before a ghostlike full-page photograph appeared in *Life* in 1938.[38]

Lincoln's early career did not escape the attention of scholars during these years when special studies poured from the press in a steady stream. Much had been written about Lincoln the lawyer, but most of it was in the nature of reminiscence, and it depicted him as an amiable rustic or, as Frederick T. Hill said, "a sort of end-man with an itinerant minstrel show." When John J. Duff, a New York lawyer long interested in Lincoln, undertook the research that led to his *A. Lincoln, Prairie Lawyer*, in 1960, he realized that for all its faults William C. Whitney's old *Life on the Circuit* was still "incomparably the best" account. Duff corrected a number of errors and misconceptions, several of them stemming from William H. Herndon.

Herndon's reputation as Lincoln's Boswell continued to sink. Duff reflected the not uncommon opinion among scholars of his generation when he dismissed him as an ingrate and a windbag. Herndon unfairly denigrated Lincoln's skills as a technical lawyer, said Duff. The superabundance of "Lincoln legals"—writs, petitions, pleadings, and other drafts in his own hand—proved that he was as sound an office lawyer as he was on the circuit. It was also a myth that Lincoln shunned the practice of criminal law. And Duff blasted the notion that Honest Abe would not take an "unjust" cause or, if he chanced to, would throw it away rather than let injustice be done.

The Matson case of 1843 was usually cited as the primary evidence. Here Lincoln found himself unhappily, so it was said, on the side of a Kentucky slaveholder, Robert Matson, against a slave family he had employed on Illinois land and who, on that ground, claimed their freedom. Jesse Weik, the first to treat the case, had said that Lincoln, by a pitiably weak argument, "*gave his case away*," and Beveridge, among others, had agreed. This was "pure hogwash," according to Duff. Lincoln lost the case—the Matson slaves were freed—but not because he threw it away. Duff affirmed the judgment of Leonard Swett, Isaac Arnold, and a host of contemporary practitioners that Lincoln was at his best as a jury lawyer, though he concurred with Herndon that he exerted his greatest influence before the Illinois Supreme Court. "No Illinois lawyer of his day could do so many things so well," the author concluded. Following closely upon this study came another by John P. Frank, which was just as positive in its assessment and credited Lincoln's political talent for "speaking squarely to the ultimate jury of the American people," in some part, to years of practice on the prairie circuit.[39]

A young Illinois Democrat, Paul Simon, elected to the state house of representatives in 1955, was astonished to learn that no book had been written on the legislative career of the most famous man ever to serve in that body. So he supplied the deficiency with a sound and insightful study, *Lincoln's Preparation for Greatness*, in 1965. Simon was well aware of Beveridge's discussion, but he did not rely on it and repeatedly came to different conclusions. He denied that Lincoln unscrupulously traded votes to logroll the capital to Springfield, and that he became "the hero of the hour" in that infant metropolis. Lincoln learned much during eight years as a legislator, but the best grade Simon could give him was "above average."

A university professor, Donald W. Riddle, wrote a scholarly study *Congressman Abraham Lincoln* in 1957; and Don E. Fehrenbacher, in *Prelude to Greatness*, picked up the thread of his political career in the 1850s. Fehrenbacher, a Stanford University professor, also had a quarrel with Beveridge, for Beveridge had been unimpressed with the Lincoln of this decade. On the contrary, said Fehrenbacher, Lincoln demonstrated a capacity for growth and met the challenge of greatness in the years before he became President. It was in the political struggle of the 1850s, above all in the debates with Douglas, that his "moral stamina, humane judgment, and . . . profound sense of history" were called forth. In a keen analysis of the "House Divided" speech of 1858, Fehrenbacher established the linkage in the Illinoisan's mind between the founding principles of 1776, the exclusion of slavery from the territories, and its ultimate extinction. In doing so he added weight to Harry Jaffa's thesis. Fehrenbacher, who had no respect for Sandburg's work, dumped the railsplitter traditional together, and moved the conception of the heroic statesman back to the fifties. "Never in the presidency did he surpass the political skill with which he shaped the Republican party of Illinois, held it together, and made himself a leader. . . . The Lincoln of the 1860's was much the same man under greater challenge."[40]

ALL KINDS OF books were written and published about Abraham Lincoln, among them a 1,443-page tome, *Abraham Lincoln's Philosophy of Common Sense: An Analytical Biography of a Great Mind*, by an octogenarian psychiatrist, Edward J. Kempf. There were books that collected Lincoln's wit and humor, his favorite poems, and the poems written about him. The interest in Lincoln's early reading culminated in H. Jack Lang's *Lincoln's Fireside Reading*, which served as an introduction to five boxed volumes composing *Lincoln's Log Cabin Library: Aesop's Fables, Pilgrim's Progress, Robinson Crusoe*, Weems's *Washington*, and the Bible. There were children's books and juvenilia. There were such banal curiosities as *Abraham Lincoln: A Biographic Trilogy in Sonnet Sequence*, the work of an English teacher, Della Crowder Miller.

A *Lincoln Encyclopedia*, the brainchild of an Ohio newspaperman, Archer H. Shaw, made its appearance in 1950. Here, conveniently arranged from A to Y—from "A.B.C. Schools, attended by Lincoln" to "Young Men, attitude toward"—were the great man's spoken and written words for ready reference. "Mr. Lincoln is the most quotable notable in history," David Mearns opined in the Introduction. He might have added "one of the most fraudulently quoted" as well. Regrettably, some of these errors crept into the *Encyclopedia*. Here, for instance, was the oft-heard warning against "corporations enthroned" by the war, the letter to Colonel Taylor on the origin of Greenbacks, and an alleged plea to an Illinois jury in 1839 in defense of fifteen women on trial for saloon smashing. Protecting the Lincoln canon from spurious intruders was an ongoing problem.* [41]

There were also spurious photographs. In 1941 the *Saturday Evening Post* published a photograph purporting to be of the dead President in his coffin. Named the "Nelson Lincoln,"it was promptly discredited. Students had long known about a coffin photo surreptitiously taken by Jeremiah Gurney, Jr., in New York, which Stanton had ordered destroyed—plate, negatives, and prints—at the request of Mrs. Lincoln. But Stanton had saved one print, and his son gave it to John G. Nicolay, in whose papers it was rediscovered in 1952. This proved timely for Stefan Lorant, who included the photo in his popular *Lincoln, a Picture Story of His Life*, published in that year. Lorant, a refugee from Hitler's Germany, who had discovered Lincoln while imprisoned on the charge of high treason, had burst on the American scene in 1941 with *Lincoln, His Life in Photographs*, hailed as the first modern pictorial biography, hence the model for virtually a new industry. The sleuthing of Lorant and others over many years yielded important additions to the photographic canon. Frederick Hill Meserve, who had established it, gave his final accounting in 1955. With various supplements, he had identified 132 photographic images, though he recognized that some of these were near-duplicates. Lorant pared Meserve's list to under one hundred. When Charles Hamilton and Lloyd Ostendorf published their authoritative *Lincoln in Photographs* in 1963, they numbered 119 images. The Ostendorf numbers superseded Meserve's. Previously unknown prints or negatives, even authentic rarities, occasionally turned up. Photographs, it was generally believed, opened a window into Lincoln's soul. So did some works of sculpture. But the same interest did not pertain to painted and engraved life portraits, which Rufus R. Wilson, an indefatigable Lincoln researcher, catalogued in 1935.[42]

---

*In 1949 Ohio Congresswoman Frances P. Bolton read into the *Congressional Record* the set of ten self-help maxims attributed to Lincoln. The "Ten Cannots" began with "You cannot help men permanently by doing for them what they could or should do for themselves." *Look*, the weekly news pictorial, later published them full-page under a bust portrait of Lincoln. They had already been branded spurious by the Lincoln guardians at both Springfield and Fort Wayne. *Time* responded with a column, "Dishonest Abe," and another denial appeared in *Harper's*. The maxims were actually authored by John Henry Boetcker, who named them "The New Decalog," at the time of the First World War; and they were taken up and attributed to Lincoln by the Committee on Constitutional Government, in New York, in 1942. Despite the repeated exposure of their illegitimacy, the "cannots" continued to be put forward as Lincoln's, most egregiously by Ronald Reagan in his address before the Republican National Convention in 1992.

As historiography retreated from hagiolotry, the mythologized and idolatrized Lincoln depended increasingly on more plastic and dramatic modes of interpretation. In theory, as Arthur M. Schlesinger, Jr., has said, the literary imagination, being more closely allied to myth than to history, "may fertilize the historical mind and serve to stretch and enrich historical understanding." In the instance of Lincoln, poetry had done that, though he was ceasing to appeal to serious poets. Two dramas, those of Drinkwater and Sherwood, had vivified the Lincoln portrait. From the 1920s forward, radio plays, motion pictures, and television interpreted Lincoln to wider and wider audiences. Historical fiction, which might have been expected to sharpen and deepen understanding, rarely succeeded in doing so.

Paul Horgan, a fine writer, was obsessed with Lincoln but never got beyond literary vignettes. His most ambitious historical novel, *A Distant Trumpet*, in 1960, begins with such a vignette. Lincoln is an ideal, a beckoning, an inspiration, not a fully embodied character in the historical drama. In *Love is Eternal* the popular author Irving Stone, cued by Ruth P. Randall, turned the marriage of Abraham Lincoln and Mary Todd into the great American love story. Mary, of course, was the principal character. As one reviewer said, "It is not Lincoln who lifts himself to eminence by his bootstraps, but Mary who raises him with apron strings."[43]

Lincoln is an ominous presence, never part of the action, in Ben Ames Williams's huge *House-Divided*, presumably written for the same audience that devoured *Gone with the Wind*. The novel begins in 1783 on the Virginia frontier. Lucy Hanks gives birth to a daughter, the child of Anthony Currain, son of a patrician family in Williamsburg. When the Civil War comes, old Mrs. Currain, Tony's widow, discovers in the secret compartment of a desk a packet of faded letters in which Lucy had kept Tony abreast of her adventures in Kentucky, including their spurious daughter Nancy's marriage to Thomas Lincoln and the birth of Abraham in 1809. The gorilla just elected President, it appears, is the half-brother of Travis Currain, the present head of the family. "Great God Almighty!" one of them exclaims. They are responsible for the man who would destroy the South. On a mission of mercy to Washington, one of the Currains meets Lincoln and likes him. Travis actually grows proud of the blood he shares with him. "I'd like to tell him, some day," he says. Another Currain resolves to purge the family name of Lincoln's blood, and associates himself with John Wilkes Booth. Alas, the novel offers no resolution of this difficulty, though one leaves it expecting Tony Currain's wild seed to become the seed of national reconciliation.[44]

Radio aired dramatizations of Lincoln's life as early as 1925. A favorite subject was the Ann Rutledge romance. In 1929–30 Chicago's WLS, the *Prairie Farmer* station, offered a complete radio biography of Lincoln in half-hour broadcasts over sixty-one consecutive weeks. The best of the radio plays was Millard Lampell's "The Lonesome Train," a folk cantata with music composed by Earl Robinson, also recorded by Decca. It opens with the sounds of the Martyr President's funeral train:

> A lonesome train on a lonesome track,
> Seven coaches, painted black.

A ballad singer tells the story. Lincoln, it seems, was not on the train; he was in a Negro church in Alabama. The preacher says:

> He was a-lyin there,
> And the sky turned dark
> And seven angels leaped over the battlements of glory,
> And came down to get him;
> And just when they came near him, he rose,
> Yes, Lord, he rose up and walked down among us,
> Praise God,
> He walked back down among his people!

A woman cried, "Lord, he's living now!" And there he was sitting in the rear of the church. But no, he was also in other places with other people.

> When the funeral train pulled into New York,
> Lincoln was down in a Kansas town,
> Swinging his lady round and round!

When the train stopped in Cleveland, Lincoln was not on board but in a hospital nursing wounded soldiers. And at Springfield, he mixed with friends in the crowd, cracking jokes, laughing, draped in a shawl rather than a shroud.

> They were his people, he was their man,
> You couldn't quite tell where the people left off,
> And where Abraham Lincoln began.

In the same year as Sherwood's prizewinning drama, the Federal Theater play *Prologue to Glory*, by E. P. Conkle, also opened on Broadway. The "prologue," of course, was the Rutledge romance. Although it had a good run, and toured nineteen cities after Broadway, the play had only folk sentimentality to recommend it.[45]

The earliest motion picture life of Lincoln was produced by the Rockett Brothers in 1924. Deservedly remembered as the first attempt to film an American life, it is not memorable for its portrayal of Lincoln. D. W. Griffith's *Abraham Lincoln*, in 1930, was the first major historical film of the sound era. When the brilliant director failed to entice Sandburg to serve as consultant (the poet turned up his nose at $30,000), he hired Stephen Vincent Benét, who had just won the Pulitzer Prize for *John Brown's Body*, to write the screenplay. Walter Huston was cast in the lead. Actor and director tried to capture the sad and poetic young Lincoln, but the screen portrait is wooden, sentimental, and stale. It is Sandburg without grace and humor. A new twist is added to the legend of "the fatal first of January," when Lincoln, in agonizing uncertainty over the impending marriage, takes a photograph of Ann Rutledge from the drawer of his desk, stares at it, and marches off leaving Mary Todd at the altar. A master of set design, Griffith re-created the Telegraph Office, where Lincoln spent so many hours during the war, with cinematic artistry that made it unforgettable. There are battle scenes and cabinet scenes. The President drearily repeats, again and again, "The Union must be preserved!" For the second time Griffith faithfully re-created the assassination at Ford's Theater. The film closes with stills of the Birthplace Memorial and the Lincoln Memorial accompanied by the rising strains of "The Battle Hymn of the Republic." Critics who liked the film called it poetic.[46]

A decade later two motion pictures dramatizing Lincoln's life in Illinois were released within months of each other. John Ford's *Young Mr. Lincoln*, with a screenplay by Lamar Trotti, opens to the music of "The Battle Cry of Freedom" while the camera rolls over Rosemary Benét's plaintive lines:

> If Nancy Hanks
> Came back as a ghost,
> Seeking news
> Of what she loved most,
> She'd ask first
> "Where's my son?
> What's happened to Abe?
> What's he done?"

In the title role Henry Fonda is beguiling but unbelievable. Lincoln makes his maiden political speech in New Salem; he keeps store; he discovers Blackstone's *Commentaries*, not at the bottom of the barrel but right on top; he falls in love with Ann and over her grave resolves to pursue the destiny she had foreseen for him. Ford is at his cinematic best in dramatizing a rousing Independence Day in Springfield. Lincoln is shown courting Mary Todd, judging an apple pie contest, winning a railsplitting contest, and cheating, successfully, to prevail at tug-of-war. The comedy ends grimly in a murder. From this point the film centers on the murder trial, which bears a faint resemblance to the famous Duff Armstrong case. Without any pretensions to historical accuracy, Ford plays Lincoln for laughs. He carries his office in his hat; he strums "Dixie" on the jew's-harp; he is Sandburg's "Strange Friend and Friendly Stranger." At a party in the Edwards home, he is so backward Mary has to ask him for a dance. He accepts, saying he wants to dance "in the worst way." After a brief whirl Mary breaks off, agreeing that he did, indeed, dance "in the worst way." After all this folderol, the film ends with Lincoln trudging up a hill at dusk, lightning flashing, thunder cracking, "The Battle Hymn of the Republic" playing, as the image of French's great statue fills the screen.

Abe Lincoln in Illinois, on the other hand, takes itself seriously and is a better historical drama. Sherwood adapted his stage play for the screen; John Cromwell directed; Raymond Massey repeated his performance in the title role, while Ruth Gordon was a magical Mary Todd Lincoln. The motion picture, more than the play, carries out the romantic idea that Lincoln owed his greatness to two women. Mary is the driving force. Foreseeing greatness in the uncouth Springfield lawyer, she spurns Douglas and marries him partly to realize her own ambition but mainly "to fulfill his destiny." But Lincoln is full of self-doubts and walks out on the marriage. He is next seen wandering in the abandoned village of New Salem, where he has a vision of Ann. Although she died with McNamar's locket clutched in her hand, it now seems that Lincoln was her true love. Her specter commands him: "He that doeth the will of God abideth forever." And he returns to wed Mary. From this point the motion picture closely follows the play, including the "credo" speech that was the hit of the play.[47]

In some motion pictures it was the shadow of Lincoln rather than the substance, the templed god rather than the historic figure, the spiritual legacy of the words rather than the life, that mattered. Frank Capra's *Mr. Smith Goes to Washington*, in 1939, tells the story of a country bumpkin named Jefferson Smith, played by James Stewart, who is cynically appointed by the political bosses of his state to serve out a term harmlessly in the United States Senate. He is awed by Washington, filled to the brim with patriotism; but when he seeks passage of a bill to set aside certain lands for a boys' camp, he runs up against a corrupt scheme of the bosses, who move quickly to destroy him. Shattered, Mr. Smith makes a nocturnal visit to his favorite shrine, the Lincoln Memorial. He stands before the statue, he reads the words engraved on marble, and he is inspired to persevere. "That Lincoln Memorial—" he says to his hard-boiled aide Saunders, played by Jean Arthur, "gee whiz! Mr. Lincoln, there he is looking right straight at you as you come up the steps. Just sitting there, like he was waiting for somebody to come alone." And Saunders replies, "Mr. Lincoln . . . *was* waiting for somebody. He was waiting for you, Jeff." Lincoln, in sum, was a metaphor for the good and the honest in American politics.

The dramatic high point of Leo McCarey's *Ruggles of Red Gap* (1935), the hilarious farce about an English valet, Marmaduke Ruggles (Charles Laughton), dropped into the Wild West, comes when Ruggles, who has been searching for his new American identity, recites the Gettysburg Address in a crowded saloon. It is mesmerizing. The silence is finally broken when one of the cowboys says, "I'll buy a drink." Here the words of Lincoln alone, without the icon, become the action.

Television was not long in discovering Lincoln. In 1952 the producer of the Ford Foundation's *Omnibus* asked the writer James Agee, basking in the acclaim for the motion picture *The African Queen*, to write a five-part series on Lincoln's early life. The first of the thirty-minute episodes of Agee's *Abraham Lincoln—The Early Years* was aired the following February. The critic Meyer Levin hailed it as the most important dramatic work yet created for television. Agee's young Lincoln is a modern saint. By dramatizing characters and events embedded in the collective memory of Americans, Agee achieved what his biographer has called "hagiography of the highest order." The series ends with the death of Lincoln's beloved, Ann Rutledge. He blames himself—his cold ambition—for it. The message is the old one that from her sacrifice came his greatness. Allan Nevins, speaking for the history profession, protested the drama's presentation of myth as fact. The *Omnibus* producer, Robert Saudek, arranged a debate between Agee and Nevins. It was a debate across a void. Nevins insisted there was only one truth about the past—the record of what actually happened—while Agee pled the poet's order of truth in his defense.

In 1956 Lincoln's birthday fell on Sunday, with the result that American living rooms were bombarded with endless hours of Lincoln programs. "Oh, not Lincoln again!" children were heard to cry. There would be more as Lincoln's 150th anniversary approached. That year, 1959, saw the Project 20 production of *Meet Mr. Lincoln*, an enthralling documentary written by Richard Hanser. There was a new Broadway play as well, Norman Cousin's *The Rivalry*, starring Richard Boone, which turned on the Lincoln-Douglas debates. No drama for screen or television offered a substantial portrait of Lincoln the President.

Fascination with the final days had led Paul Horgan to write a three-act play, *Death, Mr. President*, which opened on Broadway in 1942 and starred Vincent Price, but closed after only two performances. Mark Van Doren's *The Last Days of Lincoln*, in 1959, was the fruition of what was conceived as a long narrative poem. After prolonged study of Lincoln, Van Doren, an Illinois native, concluded that he revealed himself best in the way he died. The middle scenes of the play, in which Lincoln speaks in prose, the other characters in blank verse, show the President trying to reason with the Radicals, conferring with General Grant at City Point, visiting Richmond, and looking forward to reconstruction of the Union. "I am free at last to be President of the United States. Think of it, Joshua! The United States." (Van Doren exercised his poet's license to place Joshua Speed in Washington.) These episodes are bracketed by scenes of the President on his deathbed. The play lacked dramatic movement. First produced at Florida State University in 1961, *The Last Days of Lincoln* has been rarely performed.[48]

Although Lincoln was among the most unmusical of statesmen—he could not read music, play an instrument, or carry a tune—books were written about his tastes and associations with the music of his time. Moreover, twentieth-century composers, recognizing Lincoln's importance as a symbol, wrote many works inspired by his life and his words. Included are symphonies (by Daniel Gregory Mason and Roy Harris), requiems (by Rubin Goldmark, Paul Hindemith, and Roger Sessions), and vocal and choral compositions (by Charles Ives and Aaron Copland). Copland's "A Lincoln Portrait" is unquestionably the most important and most performed work on the Lincoln theme. In this thirteen-minute musical portrait for orchestra and speaker, composed as a wartime morale booster in 1942, Copland sought something simple yet arresting and somehow equal to the greatness of the subject. He succeeded magnificently. An orchestral prelude combines stately chords with spirited folk melodies to evoke the dual aspects of Lincoln's character. A narrator's voice booms above the music: "Fellow-citizens, *we* cannot escape history." The powerful thought is condensed in several sentences from the annual message to Congress in 1862. Four successive passages from Lincoln's writings, each prefaced by the narrator's didactic "This is what he said," fill out the portrait. The last lines of the Gettysburg Address make for a soaring conclusion. The words are beautifully articulated with Copland's music. Of course, the Gettysburg Address has made a strong appeal to composers, and several have provided musical settings for its text.[49]

Civil Rights and Civil Religion

The image of Lincoln as the Great Emancipator endured among American Negroes into the middle of the twentieth century, then became somewhat attenuated as the modern civil rights movement matured in the era of the Second Reconstruction. Frederick Douglass had once said that Lincoln's famous prophecy for the United States, that it could not endure half slave and half free, was also true for the Negroes. "They cannot remain half slave and half free." Negro leaders could take pride in the progress the race had made since emancipation. It was most dramatic in education. According to the census of 1930, illiteracy had fallen to 10 percent. But the leaders measured progress less by how far the race had come than by how far it had yet to go to realize the promise of the Emancipation Proclamation. In that work Lincoln remained a moral hero to American Negroes and a potent symbol politically because of his stature among whites. He continued to be extolled on Emancipation Day, which some likened to Passover among the Jews, because the proclamation of January 1, 1863, for all its limitations, "heralded the sunrise of a new day." In an effort to strengthen this historical link, Carter G. Woodson, in 1926, inaugurated Negro History Week.

To many Negroes Lincoln was more than a cool or distant hero, more than a symbol. He was beloved. "I do love Abraham Lincoln," said Charles H. Hunter, a North Carolina leader of his race. "I love him," declared W. E. B. Du Bois, "not because he was perfect but because he was not and yet triumphed." He had inner reserves after the niceties of convention were stripped away. "There was something left, so that at the crisis he was big enough to be inconsistent—cruel, merciful, peace-loving, a fighter, despising Negroes and letting them fight and vote, protecting slavery, and freeing slaves. He was a man—a big, inconsistent brave man."

This love of the white Father Abraham, the white Moses, lessening with the passage of years, would be transferred to the black Moses, Martin Luther King, Jr., upon his martyrdom in 1968. Still, hallelujahs for Lincoln and jubilee could occasionally be heard on the somber plain. In 1986 a 115-year-old colored woman exulted before her death: "Oh, honey, I'm gwine to glory, an' when I get dar, I'm gwine to see Mass Lincoln. Yes, honey, jes as soon as I've had a chance to say howdy to dee good Lord, I' gwine to hunt up Mass Lincoln en' shake hands with him."[50]

Benjamin Quarles, a leading Negro historian, reaffirmed the tradition of the Great Emancipator in his book *Lincoln and the Negro*, published

Lincoln University Memorial. Mural by Thomas Hart Benton, 1955. (Lincoln University)

one hundred years after the preliminary proclamation of emancipation.* "To say . . . ," wrote the author, "that Lincoln lives in history is to say that he met head-on the greatest challenge to his country as the land of the free—the challenge of the Negro." He understood that the black man was "the touchstone of American democracy." He knew that slavery was the cause of the war and freedom and equality must be its outcome. So, of course, the Negroes were the first to make him a hero. "In

*The tradition was reaffirmed in other ways as well. In 1955 the predominantly Negro institution Lincoln University, in Missouri, dedicated a remarkable mural painted by Thomas Hart Benton in which Thomas Ball's iconic image of the Emancipator is placed amidst allusions to education, industry, and progress.

fine, Lincoln became Lincoln because of the Negro." Quarles was a careful historian, yet he could not escape the spell of the emancipation image. He accepted as true the apocryphal story of Lincoln's vow, after observing slavery in New Orleans, "If I ever get a chance to hit that thing I'll hit it hard," and he put a gloss on Lincoln's ambivalent feelings toward the Negro. While conceding he was no advocate of racial equality in 1858, Quarles denied he was "anti-Negro" and maintained that his grasp of the central doctrine of the American experiment, "that all men are created equal," and his decision to act upon it at the high moment of American destiny, gave to his life its abiding greatness.

Quarles called attention to the President's little-known actions against slavery, for instance the effective suppression of the foreign slave trade after a half-century of futility. This was perhaps his earliest and most complete victory. The historian was not uncritical of Lincoln's efforts at colonization of the Negroes, but pointed out that some of the race's own leaders supported these efforts. And paradoxically, "the man most responsible for the failure of Negro emigration was Lincoln himself," for emancipation doomed colonization. Negroes seized the heart of the Emancipation Proclamation and neglected the details. The "new birth of freedom" to which Lincoln dedicated the nation at Gettysburg showed that he knew what the war was about. Even before the martyrdom, he was installed as "the father image" of the Negroes. "We all knew Lincoln by heart," as Douglass remarked. Most Americans, devoted to his memory, came to share that affection; nevertheless, said Quarles, "it is a matter of historical record that they [the Negroes] loved him first and have loved him longest."[51]

By 1962, when Quarles wrote, the revisionist virus had already begun to sap this conception. An early sign was the chapter on Lincoln in Richard Hofstadter's influential book *The American Political Tradition*, in 1948. The author was not of the revisionist school and wrote from a liberal point of view, so it was surprising that he interpreted Lincoln as little more than a slick politician. In all his behavior, Hofstadter saw "a professional politician looking for votes." He was a Janus in the Lincoln-Douglas debates. His success in bridging the gap between northern hostility to slavery and antipathy to the Negro in 1860, Hofstadter wrote, "entitles him to a place among the world's great political propagandists." He became in 1862 a reluctant emancipationist, "a follower and not a leader of public opinion." His famous proclamation added nothing to what Congress had already done by the Confiscation Act. It was politically astute but morally callous, having "all the moral grandeur of a bill of lading." Jaffa, stung by this sneering estimate, said it was Hofstadter, not Lincoln, who was morally callous.

As the issue played itself out in historiography, at least four parts of Lincoln's civil rights record were brought into question. First, in the Lincoln-Douglas debates he had denied belief in or advocacy of social and political equality of the two races and maintained the superiority of the whites over the blacks. The principal text, cited again and again, not only by racist demagogues but also by sober historians, was Lincoln's speech in the fourth debate at Charleston. Second, Lincoln's preferred solution to the race problem was the voluntary separation and colonization of the Negroes in other lands. In this he occupied basically the same position as Thomas Jefferson and Henry Clay before him. Nor did he truly abandon it after the Emancipation Proclamation. Indeed, on the testimony of Benjamin F. Butler, credited by some scholars, he adhered to it probably to the hour of his death. Third, even Lincoln's greatest act, the Emancipation Proclamation, since it was grounded in military necessity, offered only the most guarded and reserved basis for the pursuit of Negro freedom and equality. Fourth, Lincoln failed to develop a plan of Reconstruction that would secure the equal rights of the freedmen. With regard to the suffrage, he made the suggestion of extending it to "very intelligent" Negroes in his letter to Governor Hahn in 1864, but nothing came of it. A stronger statement appeared in the published excerpt of a letter apparently written to General James S. Wadsworth, wherein the President proposed "universal suffrage, or, at least, suffrage on the basis of intelligence and military service," in return for universal amnesty. But the authenticity of the letter was questioned, and so itself became an issue in the larger debate.[52]

The modern civil rights movement of American Negroes took definite form in the 1950s, but before *Brown v. Board of Education* mandated desegregation in the schools, before the Montgomery bus boycott, before Martin Luther King, there were the significant gains, even more the rising expectations, of the Roosevelt era. In 1936, for the first time, a majority of Negro voters supported the Democratic party in a national election. By becoming part of the Roosevelt Coalition committed to liberal social change, the Negroes might work to move beyond the real economic gains they were making toward the eradication of racial injustice.

Unfortunately, the power of southern Democrats in Congress blocked legislative progress on this front. White supremacist demagogues like Senator Theodore G. Bilbo of Mississippi advocated in Lincoln's sacred name the voluntary "repatriation" of the Negroes to Africa. In 1939 Bilbo introduced the "Greater Liberia Act," which, he claimed, would fulfill at last Lincoln's "noblest aspiration." Liberal Democrats repeatedly proposed legislation to make lynching a federal crime and to outlaw disenfranchising poll taxes in the southern states. Both, it was said, profaned the shrine of Lincoln. But even if bills passed the House they were filibustered to death in the Senate. The Supreme Court, after it became the "Roosevelt Court," often upheld the rights of Negroes in civil rights cases. In 1938 it required the Law School of the University of Missouri to admit a qualified Negro applicant, thereby setting in motion the series of decisions that would terminate sixteen years later in the holding that segregated education at every level is inherently unequal and therefore in violation of the "equal protection" clause of the United States Constitution.

Although issues of civil rights were fairly low on President Roosevelt's agenda, in 1941, as the nation moved toward war, he issued the momentous Executive Order 8802 outlawing racial discrimination in employment on defense contracts. This achievement assumed symbolic significance for the civil rights movement, for the President signed the order under the threat of a march on Washington by fifty thousand Negroes led by A. Philip Randolph, head of the Brotherhood of Sleeping Car Porters—the original scenario for the celebrated march of August 28, 1963.[53]

Another event that loomed larger in retrospect was the concert by Marian Anderson at the Lincoln Memorial on Easter Sunday, April 9, 1939. The internationally acclaimed contralto with a voice "heard only once in a hundred years," according to Arturo Toscanini, was then at the height of her fame. Yet she had not performed in the nation's capital. Her agent, Sol Hurok, sought to book her for a concert in Constitution Hall, the capital's premier stage, owned by the Daughters of the American Revolution. The organization's president, Mrs. Henry M. Roberts, Jr., bluntly refused the request because of the artist's race. Charges of bigotry swirled around the DAR. In the ensuing controversy Eleanor Roosevelt resigned her membership. Meanwhile, Walter White, director of the National Association for the Advancement of Colored People, went to the Interior Department and asked if Marian Anderson might perform in Lafayette Park across from the White House. The park was deemed unsuitable, but White's heart skipped when the Lincoln Memorial was mentioned. "Oh, my God, if we could have her sing at the feet of Lincoln!" The recommendation was warmly approved by the Secretary of the Interior, Harold Ickes, thence by the President.

Some seventy-five thousand people, most of them black—one of the largest crowds ever seen in the capital—filled the space between the Memorial and Monument Hill half a mile away on this blustery but bright Easter afternoon. Marian Anderson had intended to give a concert in Washington; fortuitously, the concert had become something more: a public tribute to Lincoln by a great black singer rebuked and scorned on account of her race. Whether she wanted it or not, she became on this emotion-filled occasion a symbol of her people. Ickes underscored the meaning in introducing Miss Anderson. Lincoln had given his life to free the slaves; and today, at his memorial, "glorious tribute is rendered to his memory by a daughter of the race from which he struck the chains of slavery." Miss Anderson opened the concert, which was broadcast by NBC, with the National Anthem, followed by "America the Beautiful," and closed with the spiritual "Nobody Knows the Trouble I've Seen." A better stage than the Lincoln Memorial could scarcely be imagined. Civil rights leaders would use it again and again in their struggle.[54]

None of the movement's leaders had strong attachments intellectually or emotionally to Lincoln. The Reverend Dr. King, who emerged as the premier leader, had his roots in the tradition of the southern Negro evangelical churches. He brought the fervor of revivalism to his crusade and spoke the language of a biblical prophet. So, sometimes, had Lincoln, though King seems to have had no sense of this. Of course, he had learned to revere Lincoln as the Great Emancipator and he valued the sanction of his name. Rising to prominence during the Eisenhower years, King was disappointed in the Republican President whose response to pleas for civil rights legislation was "You can't legislate morality." King's response to that was to organize a Prayer Pilgrimage to Washington. The date fixed, May 17, 1957, was the third anniversary of the Brown decision. Clarence Mitchell, who directed the Washington bureau of the NAACP, requested use of the Lincoln Memorial, but was turned down by the National Park Service. Incensed and insulted, Mitchell marched into the director's office protesting, "All they wanted was to meet and pray on the steps of the monument of the Great Emancipator." Permission was granted. The plan of the pilgrimage followed that of Randolph's labor march in 1941. Some thirty thousand congregated before the Memorial. King and other leaders spoke. Mahalia Jackson sang "I Been 'Buked and I been Scorned," which was so true, wrote Langston Hughes, "even Abe Lincoln's statue nodded his head."

President Eisenhower took no notice of the event. But the administration soon proposed a civil rights bill to Congress, and it passed into law in debilitated form—the first civil rights legislation in eighty-two years—in August. The following month, when violence thwarted desegregation of Little Rock's Central High School, the President showed he was not all velvet by sending troops to keep the peace and enforce the law. A protracted conflict between state and national authorities ensued. Finally, in September 1958, the Supreme Court unanimously affirmed the supremacy of federal law, including the Court's desegregation decree. In a concurring opinion, Justice Felix Frankfurter invoked Lincoln's memory. "Lincoln's appeal to 'the better angels of our nature' failed to avert fratricidal war," he observed. "But the compassionate wisdom of Lincoln's First and Second Inaugurals bequeathed to the Union, cemented with blood, a moral heritage which, when drawn upon in times of stress and strife, is sure to find specific ways and means to surmount difficulties that may appear to be insurmountable."[55]

King hoped for bolder leadership from President John F. Kennedy. He had been elected by a paper-thin margin, so civil rights leaders could claim he owed his victory to Negro voters. By the same token, however, he was beholden to southern Democrats. Despite his coolness toward the civil rights agitation, key advisers saw to it that he listened to leaders like King. In the fall of 1961 he was the luncheon guest of President and Mrs. Kennedy. They conducted him on a tour of the White House. In the Lincoln Room, King noticed a framed engraving of the Emancipation Proclamation above the mantel and seized the opportunity to say, "Mr. President, I'd like to see you stand in this room and sign a Second Emancipation Proclamation outlawing segregation, one hundred years after Lincoln's. You could base it on the Fourteenth Amendment." Kennedy was non-committal in reply but suggested King might prepare a draft of such a proclamation. The country was in the midst of the Civil War Centennial; within a year or fifteen months it would be commemorating the one hundredth anniversary of the Negroes' charter of liberty. What could be more appropriate or more salutary than a presidential declaration of re-commitment to the promise of emancipation! Two months later King followed up the casual exchange in the White House with a telegram beseeching the President to issue "a Second Emancipation Proclamation to free all Negroes from second-class citizenship." He received no reply. Nevertheless, on the ensuing May 15 he delivered to the White House his draft of what he called a Second Emancipation Proclamation elegantly bound in leather.[56]

The document was actually a long brief, with appendix, for an executive order prohibiting segregation nationally and mobilizing the administration to enforce the ban. "The time has come, Mr. President," King said in this appeal, "to let those dawn-light rays of freedom, first glimpsed in 1863, fill the heavens with the noonday sunlight of complete human dignity." Thus far desegregation of schools, for all the Supreme Court's mandating of "all deliberate speed," had proceeded at a snail's pace. Only by vigorous executive leadership could it be achieved. The President must break through his political timidity. "We believe you, like Abraham Lincoln before you, stand at a historic crossroads in the life and conscience of our nation." As Taylor Branch has written, "The document carried ambitions far greater than a ceremonial tribute to the memory of Lincoln." King called upon Kennedy to slay segregation as Lincoln had slain slavery by an electrifying act that would shake the foundations of American society and reverberate around the world. King had even anticipated the dramatic requirements for the occasion, having requested the Secretary of the Interior to reserve the Lincoln Memorial for midnight ceremonies at the break of the new year one century after Lincoln's proclamation was signed.

But President Kennedy was as silent as the tomb. Although he dispatched troops to Oxford, Mississippi, to secure the desegregation of the state university, he was impassive toward white violence elsewhere in the South. The Southern Regional Council dubbed him "the Reluctant Emancipator." In this, perhaps, he was a little like Lincoln during the first year of the Civil War. While King continued to press his Second Emancipation Proclamation, the President sought to avoid even the appearance of association with that idea. Lincoln's preliminary proclamation was impressively celebrated at the Memorial on September 22. To it Kennedy sent a routine videotaped message. And he declined to issue a proclamation on January 1. Instead, he embraced the recommendation of his staff to substitute for the proclamation a big White House reception for civil rights leaders on Lincoln's next birthday. King boycotted the reception.[57]

The March on Washington for Jobs and Freedom, August 28, 1963, followed the strategy and the plan rehearsed twice before. On that sunny Wednesday some two hundred thousand people, black and white, converged on the capital and filled the vast space before the Lincoln Memorial, on both sides of the reflecting pool, up the hill to the Washington Monument. The marchers came to make a statement on behalf of the civil rights revolution and more particularly to lobby for passage of the bill Kennedy had, at last, sent to Congress to secure desegregation in interstate transportation, to protect Negro voting rights, and to hasten school desegregation. They listened to two and one-half hours of prayers, songs, and speeches. Some began to drift away before the last speaker, King, stopped them in their tracks with his sonorous voice. He held the audience rapt; indeed, said Eugene Patterson of the *Atlanta Constitution*, the throng's fervent response to him was the most notable event of the proceedings.

For the other speakers, the Lincoln Memorial seemed nothing more than a stage setting. But King began by acknowledging the hero in the pantheon behind him. "Fivescore years ago, a great American, in whose symbolic shadow we stand today, signed the Emancipation Proclamation. This momentous decree came as a great beacon of hope to millions

Lincoln the Icon Gazes upon the Marchers, 1963. (UPI/BETTMANN)

of Negro slaves who had been scarred in the flame of withering injustice. It came as a joyous daybreak to end the long night of their captivity." Sadly, one hundred years later the Negro still was not free. The oration became King's own Second Emancipation Proclamation. It rose to the lilting crescendo of "I have a Dream.": "I have a Dream that one day this nation will rise up and live out the true meaning of its creed: We hold these truths to be self-evident that all men are created equal." Thus did King, like Lincoln at Gettysburg, dedicate the country to a new birth of freedom in pursuit of the old dream. "This speech, more than any other single event," it has been said, "legitimized the ongoing black revolution in the eyes of most Americans and came to symbolize a historic national turning point, lifting King into the pantheon of great American heroes."[58]

King's life became an exhausting series of marches, capped by personal triumphs like the Voting Rights Act of 1965, for which he shared the honors with President Johnson; and it ended in the tragedy of his assassination in 1968. When the black Moses entered the pantheon, the myth of the white deliverer was lifted from the race. This development coincided with the black consciousness movement inspired by the civil rights revolution. From Douglass to Du Bois black leaders had felt a certain ambivalence toward Lincoln, but this had not compromised their gratitude and affection. Now the feeling grew among blacks that the image of the Great Emancipator diminished their own struggle. This was not confined to radicals like Malcolm X, who disapproved of King's non-violent methods, but was felt by King himself. "Exactly one hundred years after Abraham Lincoln wrote the Emancipation Proclamation *for them*, Negroes wrote their own document of freedom in their *own* way," he declared. Charles H. Wesley, in the *Negro History Bulletin*, described this kind of thinking as a healthy reaction against "the Great Man Theory of Emancipation." It was not that Lincoln was undeserving of the race's gratitude, but that emancipation was not his gift alone.[59]

Other blacks seemed anxious to smash the old idol altogether. They expressed anger for Lincoln. "How come it took him two whole years to free the slaves?" one asked. "His pen was sitting on his desk the whole time. All he had to do was get up one morning and say, 'Doggonit! I think I'm gon' free the slaves today.' " It was that simple, but Lincoln put white politics ahead of black freedom. The old myth was turned inside out. In place of Lincoln the emancipator was Lincoln the honkie, scarcely distinguishable from the Negrophobic image of Lincoln held by southern white supremacists from Thomas Dixon to Theodore Bilbo.

Lerone Bennett, Jr., a black historian, stated the case in a popular article, "Was Lincoln a White Supremacist?" in *Ebony* in 1968. He began by repeating a story Lincoln had supposedly told after his election in 1860. When someone said he would now have to attend to "the vexatious slavery matter," he was reminded of a Kentucky judge whose first case on the bench was a prosecution for abuse of slaves. "I will be damned," he remarked, "if I don't feel almost sorry for being elected when the niggers is the first thing I have to attend to." This was a shocking story, said Bennett, yet it was perfectly in character. Lincoln shared the racial prejudices of the mass of whites; he supported the oppressive Illinois black laws, as well as the Fugitive Slave Law; with regard to slavery he was a political opportunist. True, he grew in the presidency, but not by much. Battlefield reverses forced the Emancipation Proclamation upon him; and he never abandoned his belief that the blacks were "unassimilable aliens," therefore proper subjects for deportation. "Incredibly," by his plan for reconstruction of the rebel states upon 10 percent of the white vote, "the commander-in-chief . . . abandoned his black soldiers to the passions of Confederate veterans." If Lincoln was the hero of the American tradition, Bennett concluded, it was only because the tradition was racist through and through. A fellow Afro-American historian, Vincent Harding, expressed similar views but emphasized the privations of the myth. "The heart of the matter was this: while the concrete historical realities of the time testified to the costly, daring, courageous activities of hundreds of thousands of black people breaking loose from slavery and setting themselves free, the myth gave the credit for freedom to a white Republican president."[60]

Some blacks protested this counter-myth. "It needlessly gives the great Abraham Lincoln over to the enemy of mankind." In truth, the southern Citizens' Councils organized in opposition to desegregation, along with certified Yankee racists like Carleton Putnam, invoked the same Lincoln. The Citizens' Councils, in 1964, ran large newspaper advertisements with the heading "Lincoln's Hopes for the Negro—In His Own Words," in which he was exhibited as a white supremacist and black colonizationist. Putnam, too, heir of a distinguished New England family, ran similar advertisements and also wrote books portraying Lincoln as an opponent of racial equality. Lincoln's opinions, according to Putnam, ran directly counter to those of present-day liberals.

Curiously, this did not seem to make him more acceptable to the new conservatives. The Bible of their movement, Russell Kirk's *The Conservative Mind*, in 1954, found no place for Lincoln in the conservative tradition. He was more often censured than praised in the pages of the *National Review*. The sticking point for conservatives was Lincoln's commitment to equality. Thus Yale political scientist Willmore Kendall maintained that Lincoln "derailed" the American political tradition by according "constitutional status" to the principles of the Declaration of Independence. From this Lincolnian heresy of equality flowed all the evils of centralization, Caesarism, and state socialism.[61]

THE SEARCH FOR the soul of Abraham Lincoln was never ending. In the middle of the twentieth century, however, the question of Lincoln's own religion was superseded by the question of Lincoln's significance for American "civil religion." Not Lincoln's soul but the soul of the nation became the object of inquiry. The change reflected the moral and ideological stresses of the Cold War. The overt religiosity of the Eisenhower presidency together with the resurgent revivalism of Billy Graham called attention to the paradox that, for all the secularism and materialism of American civilization, despite the law and the tradition of separation of church and state, the United States was, as G. K. Chesterton had said, "a nation with the soul of a church," and its Christianity was concerned not alone with saving souls but with the perpetuation of the civil order. Speaking on the "Back to God" radio program of the American Legion in 1954, Eisenhower invoked the Pilgrim Fathers, General Washington at Valley Forge, Lincoln in the White House, and the four chaplains who went down on the USS *Dorchester* in 1943 as a continual tableau of prayerful devotion to national salvation.

Inevitably, Lincoln was conscripted in the Cold War crusade against Communism. The metaphor of a house divided was readily adapted to a world divided by an "iron curtain." The national commander of the American Legion, speaking at Lincoln's tomb in 1962, said that, could Lincoln but rise from the dead, he would prefer war to the prisoner's peace, which the Kremlin offered the world. Such righteous invocations of Lincoln's spirit assumed that God was on America's side, which, as Barbara Ward pointed out, belied Lincoln's vision of "all sinful humanity standing under Divine judgment."[62]

Among the new interpretations of Lincoln's religion, or Lincoln and religion—a somewhat different subject—the most admired was *The Almost Chosen People* by William J. Wolf in 1959. (The book was twice reprinted under two different titles: *The Religion of Abraham Lincoln* and *Lincoln's Religion*.) Wolf, professor of theology at the Episcopal Divinity School in Cambridge, Massachusetts, took his theme from Lincoln's brief address to the New Jersey senate on February 21, 1861. Recalling the state's part in the Revolutionary War, he spoke of the greater purpose beyond the independence of that war, and said, "I am exceedingly anxious that this Union, the Constitution, and the liberties of the people shall be perpetuated in accordance with the original idea for which the struggle was made, and I shall be happy indeed if I shall be a humble instrument in the hands of the Almighty, and of this, *his almost chosen people*, for perpetuating the object of that great struggle."

Like previous clerical interpreters, Wolf set out to wrap Lincoln in the folds of Christianity. His developmental view of Lincoln's spiritual pilgrimage differed little from William E. Barton's. Since Barton, however, one or two new discoveries had shed new light on Lincoln's early religious beliefs. In 1942 Harry Pratt, executive secretary of the Abraham Lincoln Association, turned up a handbill Lincoln had published just after the congressional election of 1846 in reply to opposition charges that he was an infidel. Heretofore it had been assumed that Lincoln never replied to those charges. But the handbill, Wolf and others now argued, mitigated or even refuted the notion of Lincoln's infidelity. Actually, Lincoln was intent in the handbill to say he had never been an open scoffer at Christianity nor one who denied the truth of Scripture. He also said he no longer espoused the "Doctrine of Necessity," yet did not even hint at what he believed. The supposed clarification, as Edmund Wilson observed, "does not commit Lincoln to anything." Wolf's book was more important in its civil dimension, for he maintained that

Lincoln's moral position against slavery was fundamentally theological. To be sure, he used the secular language of the Declaration of Independence, but unlike its author, Jefferson, he grounded the principles not in reason and nature but in the will of God. The national covenant owed more to the Pilgrim Fathers than to the Founding Fathers. Lincoln's biblical faith matured in the presidency. The "new birth" of the Gettysburg Address was essentially a Christian rebirth, and the nation's destiny was fittingly placed "under God." This, said Wolf, was a prophetic Christian interpretation of the nation's history.[63]

It became increasingly difficult to separate Lincoln's religion from his politics. Because he believed, as he often said during the war, that God rules in the affairs of men and that nations must answer for what they do, or fail to do, Lincoln's every decision had a spiritual aspect. "He fought repentantly. . . . He waged the war in the solemn consciousness that the heavy hand of God's condemnation rested upon himself and the whole nation, north as well as south." Any question of personal sin, personal piety, and personal salvation became irrelevant as Lincoln struggled to know and to do the will of God for "his almost chosen people," thereby to lead them out of the wilderness into the promised land of the restored and perfected Union. His Proclamation of Thanksgiving in 1863, which marked the commencement of the annual national observance of Thanksgiving, was cogent testimony to Lincoln's belief in the controlling power of Almighty God.

The danger of theological politics was the danger of self-righteousness. No one preached more perceptively on this theme than Reinhold Niebuhr, the foremost American Protestant theologian. A religious sense of ultimate judgment upon history was necessary to dissolve the pretensions, ironies, and paradoxes of civil society. No statesman had succeeded so well in attaining this, Niebuhr said in *The Irony of American History* (1952), as Abraham Lincoln. The "Lincoln model," as he described it, combined "moral resoluteness about the immediate issues with a religious awareness of another dimension of meaning and judgment." The model guarded against national pride and hypocrisy and ethnocentrism.

Some years later Niebuhr elaborated on Lincoln's example and its enduring value. "Above any statesman of the ancient and modern periods, Lincoln had a sense of historical meaning so high as to cast doubts on the intentions of both sides, to put the enemy into the same category of ambiguity as the nation to which his life was committed." The chief text for this sweeping statement was the Second Inaugural Address. Here Lincoln embraced the paradox of moral certainty and religious doubt.

His opposition to slavery was unequivocal. "If there is moral ambiguity in his position, it is an ambiguity which he shared with the founding fathers, indeed with the author of the Declaration of Independence, and, for that matter, with all responsible statesmen, who pursue their ideals within the frame of the harmony and survival of their community. In short," said Niebuhr, "he exhibited not his own ambiguity but the ambiguity of the political order itself." Since he did not pretend to know the will of God, the President was religiously superior to the abolitionists, who believed they were God's anointed, and free of their vindictiveness. Present-day civil rights activists, without relaxing their commitment, Niebuhr thought, might well be guided by Lincoln's recognition of moral ambiguity in the field of tactics, as also by his example of charity and magnanimity.[64]

The concept of American civil religion was formulated by Robert Bellah, a sociologist, in 1967. He maintained that attention to organized religions in America had obscured an unorganized but pervasive civil religion, though the existence of something on that order had not gone unnoticed. The origins of this civil religion traced back to the nation's founding, then took definite form during the Civil War in the life and death of President Lincoln, which conformed to the Christian archetype. The Gettysburg Address was the prophetic text. Other scholars, working with Bellah's idea, gave more specific content to Lincolnian civil religion. It sharpened and enhanced the American sense of mission, the original Puritan sense of being a "city upon a hill," which Lincoln caught up in his guarded "*almost* chosen people." It linked this mission to democratic government and the dignity of the common man. Finally, it rooted his democracy in the will of God, substituting a biblical language for the rational and deistic language of the core American faith.[65]

As the concept unfolded, it became apparent that everyone's favorite example was Abraham Lincoln. Indeed, he seemed to be the only American civil theologian worthy of respect, and a historian like John F. Wilson was entitled to wonder if the concept had not been invented to explain Lincoln rather than America. His case argued the exception, and his example, said Wilson, "stands alone as a remarkable monument to a tortured soul." In the history of the nation the evidence of civil religion was so thin, partial, and episodic as to cast doubt upon the validity of the idea. It became, moreover, a hydra of many heads: first, a folk religion, which featured the Lincoln of myth and legend; second, a prophetic religion of the republic, which was closest to what Bellah had in mind; third, a priestly religious nationalism, which presumably was what Alexander H. Stephens had ascribed to Lincoln in saying he invested the

Union with "the sublimity of religious mysticism"; fourth, a religion of democracy; and fifth, the traditional civic piety of American Protestantism.

All might agree that there was a religion *about* Lincoln in America. The sociologist W. Lloyd Warner, in his study of Memorial Day in the typical "Yankee City," said that "Lincoln loomed over the memorial rituals like some great demigod over the rites of classical antiquity." But making him the demigod of an American civil religion was problematic. Edmund Wilson, in *Patriotic Gore,* associated him with the apocalyptic vision of Julia Ward Howe's "Battle Hymn of the Republic." In it God speaks to Union soldiers:

> "As ye deal with my contemners, so with
> you my grace shall deal;
> Let the Hero, born of woman, crush the serpent
> with his heel,
> Since God is marching on."

But it was precisely this identification of God's will with the will of the nation that Lincoln rejected, if Wolf, Niebuhr, and most other students were to be believed. His religion had no place for a vengeful God, nor his civics any place for a state religion.[66]

Lincoln at 150

In the Eighty-fifth Congress several representatives and senators, some of them members of the Lincoln Group of Washington, D.C., introduced bills to establish a commission to commemorate the 150th anniversary of Lincoln's birth. Public Law 85-226 was enacted with President Eisenhower's signature, September 2, 1957. It created the Lincoln Sesquicentennial Commission composed of twenty-five members (with three ex-officio members) and charged it to develop a comprehensive program of civic commemoration. The George Washington Bicentennial of 1932, and to a lesser extent the wartime Thomas Jefferson Bicentennial of 1943, provided a model for events of this kind. The fact that Lincoln's 150th preceded the Civil War Centennial, already in planning, by only two years meant that it served as a prologue to that larger observance over which Lincoln cast his long shadow.

At its second meeting early in the new year, the commission elected Senator John Sherman Cooper of Kentucky chairman. The preliminary report to Congress sketched a broad program aimed at increasing American and world understanding of Lincoln's contributions to democracy and pointing up the application of his principles and ideals to the present. A theme, "Lincoln: Symbol of the Free Man," was adopted and inscribed on an official seal. With a modest appropriation from Congress, the commission proceeded to assemble a staff and establish an office, finally, in a renovated storage room of the National Archives. In June, William E. Baringer, formerly executive secretary of the Abraham Lincoln Association, now professor of history at the University of Florida, reported for duty as executive director of the commission. At the end of the year, President Eisenhower proclaimed 1959 "the Abraham Lincoln Sesquicentennial year" and called upon the people "to do honor to Lincoln's memory by appropriate activities and ceremonies, by a restudy of his life and written words, and by personal rededication to the practice of citizenship and the philosophy of government for which he gave 'the last full measure of devotion.' "[67]

It sometimes seemed that Lincoln would be swallowed up by the scholars who professed to study the past on its own terms and the antiquarians who treasured its relics, thereby obscuring the elements of myth and ideology and moral inspiration that had kept his memory green for the people. But this had not happened, not yet anyway. Lincoln's birthday, although a legal holiday in thirty or more states, was still not observed nationally. Periodic efforts in Congress to change that ended, finally, in the passage in 1968 of the "Monday Holiday Law," the third Monday of February, commonly called President's Day, whereby Lincoln's birthday was in a manner piggy-backed on George Washington's. The Lincoln name and effigy were more ubiquitous than ever. Lincoln was a transcontinental highway, an automobile, a brigade, a performing arts center. The state of Illinois, beginning in 1953, advertised itself as "Land of Lincoln" on its license plates. Over two million people a year—rising to 2,900,000 in the sesquicentennial year—visited the Lincoln Memorial, twice the number at any other attraction in the nation's capital.[68]

Why was there still such an aura around Lincoln? Because, the *New York Herald Tribune* editorialized, he was "our country's consummate democrat." The people believed in him as one of them, "the embodiment of the people's government." But that was not the whole of it. The people saw in him the vindication not only of American democracy but also of American character. In him every national quality or trait was perfected. "He is the most truly American American," declared Iowa Congressman Fred Schwengel, one of the leading celebrants in the anniversary season. This evoked Sandburg's Lincoln. So, too, did the dramatic advertisement of the John Hancock Mutual Life Insurance Company earlier in the decade. Skip all you've read about Lincoln, skip the Memorial, forget the Emancipation Proclamation and the Gettysburg

Two Insurance Companies Feature Lincoln in Advertisements.

LEFT: Lincoln National in 1939. *(The Lincoln Museum, Fort Wayne, a part of Lincoln National Corporation)*

BOTTOM: John Hancock in 1952. *(John Hancock Financial Services)*

Address. "Why do we love this man, dead long before our time, yet dear to us as a father?" the advertisement asked. And answered: "Abe Lincoln always did what most people would have done, said what most people wanted said, thought what most people thought when they stopped to think about it. He was everybody, grown a little taller—the warm and living proof of our American faith that greatness comes out of everywhere when it is free to come." Of course, as the anniversary testified, Lincoln's greatness transcended nationality and begged for a dimension of understanding as wide as humanity. The *New York Times* thought that Lincoln remained imperishable because "the fundamental principles for which he fought are still a battleground, testimony that his victory, however great, was incomplete." The nation he had called "the last best hope of earth" would the more surely realize his dream of union, peace, and freedom if his image continued to live in the hearts and minds of the people.[69]

Many groups and organizations contributed to the perpetuation of Lincoln's fame. Despite the demise of the Abraham Lincoln Association, Springfield remained a hub of activity. The Illinois State Historical Library boasted the best Lincoln collection in the country; the Mid-Day Luncheon Club offered a forum for discussion of Lincoln; almost every February 12 the American Legion, the Veterans of Foreign Wars, the Sangamon County Bar Association, the NAACP, and the Young Republicans of Illinois held commemorative exercises; every summer the Abraham Lincoln Players performed *Abe Lincoln in Illinois* at nearby New Salem. The Lincoln Home—ten rooms on two floors—was open to the public. The Old Capitol, where Lincoln had made the "House Divided" speech, was saved from the wrecking-ball in 1961 when the state legislature, at the eleventh hour, voted the money to restore and preserve it as a shrine. Springfield was the terminus of the still somewhat fictitious 425-mile Lincoln National Memorial Highway, the result of the Lincoln Way inquiry begun several decades before. The Gettysburg National Military Park and Cemetery was a center for the interpretation of Lincoln, for the park featured his address as well as the battle. The Old Wills House, where the President had purportedly put the final touches on his speech before going to sleep the night before, became a museum, the bedroom furnished with a wax figure and the address pronounced as the lights dimmed.

In 1940 another historic site, Cooper Union, inaugurated an annual Lincoln Lecture for delivery the Sunday nearest February 12. From this platform T. V. Smith, Lyman Bryson, and other worthies spoke on Lincoln, sometimes with reference to contemporary problems. In 1960, at the time of sit-ins and freedom rides in the South, Ralph McGill, editor of the *Atlanta Constitution*, called for a national dialogue on the race problem. That, after all, was what Lincoln had wanted when he spoke from the same platform a hundred years earlier; but the South would not talk then, and regrettably, McGill said, powerful southern voices rejected dialogue today. The Lincoln Fellowships in major cities were not public forums but, like the Civil War Round Tables, places for thought and discussion. None was more active than the Lincoln Group of Boston under its guiding spirit, F. Lauriston Bullard. A fiftieth anniversary volume in 1988 showed that, in addition to the contributions of guest speakers, its members had presented some 170 papers.[70]

The Lincoln Sesquicentennial Commission encouraged states, localities, schools and colleges, historic sites and museums, libraries, newspapers, radio and television, and appropriate federal agencies to develop their own observances. To this end it distributed 185,000 copies of an informational *Handbook*. 160,000 copies of a sixty-four-page booklet, *Lincoln's Ideals*, being selections from his writings, and a periodic newsletter, *The Intelligencer*. Twenty or more states established commissions or equivalent bodies to commemorate the anniversary. Several of these achieved widespread participation. Kentucky reenacted as a moving historical pageant the 1816 trek of the Lincoln family to Indiana, and opened Farmington, the former home of Joshua Speed, where Lincoln had visited, as a national shrine. The Indiana observance, while centered on the developing system of parks and memorials in and around Lincoln City, covered the state and featured a special mission to Japan headed by Governor William G. Stratton, after thousands of Japanese schoolchildren voted Lincoln the greatest man who ever lived. Illinois's yearlong commemoration kicked off with the visit of Willy Brandt, mayor of Berlin, who spoke on Lincoln's birthday. A Mobile Museum carried an exhibit organized by the Illinois State Historical Society to all parts of the state. In April, Springfield hosted a pilgrimage of ten thousand Eagle Scouts from the fifty states to Lincoln's tomb. In New Hampshire five historians documented Lincoln's all-but-forgotten visit to that state in 1860; in Wisconsin a special exhibit, "Meet Mr. Lincoln," toured the state; in Ohio bronze plaques were placed at sites associated with Lincoln, and 5,925 pupils in 363 schools competed for prizes by submitting to an examination of one hundred questions testing knowledge of Lincoln.[71]

The southern states, except at the fringes of Tennessee, Texas, and Florida, showed little interest in the commemoration. This was somewhat surprising, for most southerners had come to recognize Lincoln as a legitimate national hero. "His greatness is now the Southern boy's heritage," the Old South historian Avery Craven had declared in 1942.

"He belongs to the common folk of the nation." Wasn't this, after all, the point of Ben Ames Williams's prodigious novel? In 1954, for the first time, a representative of the Sons of Confederate Veterans appeared at the Lincoln Memorial on the morning of February 12 to honor its pantheoned hero. "The President's wreath of red, white, and blue carnations was placed along side the wreath of Southern magnolia blossoms and leaves brought by Colonel [John] Virden. Amid the magnolia blossoms were two small Confederate battle flags." There was a gap, however, between acceptance of Lincoln, or even formal honors, and celebration of his memory. For the latter most of the South was not yet ready.[72]

The commission enlisted the aid of the government's international agencies to spread the fame of the country's foremost democrat throughout the world. Copies of the *Collected Works* and of the microfilm of the Robert Todd Lincoln Papers were placed in major libraries abroad. Under State Department auspices a number of Lincoln scholars—Angle, Basler, Current, and others—traveled to distant countries to lecture. The Voice of America beamed programs abroad. Lincoln was honored in strange places, sometimes in strange ways. A bronze bust was dedicated in the civic square of the city named for him in Argentina. Trinidad inaugurated the Lincoln Memorial Walking Race, suggested by the legend—at least in Trinidad—of the young Lincoln who walked nine miles to visit a sick friend. Israel dedicated the Abraham Lincoln Science Building at Bar-Ilam University. In West Germany, United States soldiers built a replica of the birthplace log cabin, which became a permanent exhibit; and authorities sponsored an essay contest entered by five thousand young people and won by eighteen-year-old Christa Reineka of Hamburg, whose prize was a month's tour of the United States. In Ceylon sixty-two thousand people in fourteen cities saw the film *Abe Lincoln in Illinois* during two weeks of February.[73]

Lincoln's international fame had grown steadily in the years since the Second World War. The universal aspiration for human freedom and dignity, as embodied in the United Nations Declaration of Human Rights, was in some sense his legacy. Latin American successors to the Martís and Nabucos of an earlier time continued to be inspired by the Great Emancipator. In 1956 Jawaharlal Nehru, Prime Minister of India, on a state visit to the United States, said that for the last several years he had kept on his desk a treasured memento, the bronze cast of Lincoln's hand. "It is a beautiful hand, strong and firm, and yet gentle. . . . I look at it every day and it gives me strength." Also on Nehru's desk, according to his biographer, was a gold statuette of Gandhi. Lincoln stood for strength and unity, Gandhi for compassion. "The heart of Gandhi and the hand of Lincoln—a fair symbol of Nehru's philosophy of power."[74]

Both at home and abroad Lincoln was honored by commemorative stamps. In Ghana a series of three stamps showed the Prime Minister, Kwame Nkrumah, standing before the statue in the Lincoln Memorial. A series of twelve stamps in Honduras featured a portrait of Lincoln, the birthplace cabin, Lincoln at Gettysburg, Lincoln and the Emancipation Proclamation, and so on. The Republic of China revived the pairing of Lincoln and Sun Yat-sen. Haiti, Nicaragua, Columbia, Argentina, and San Marino issued commemorative stamps. The United States commemorative series began with a four-cent stamp featuring Lincoln in one of the debates against Douglas. G. P. A. Healy's portrait appeared on a green one-cent stamp; Gutzon Borglum's marble head and a drawing after the Memorial statue appeared on other denominations. The image of the Memorial had become inseparable from the image of Lincoln in American memory. As if to recognize this fact the United States Mint redesigned the Lincoln penny with a view of the Memorial impressed on the reverse. The obverse, with the familiar profile portrait by Victor Brenner—said to be the most reproduced portrait in the world—was left undisturbed. In January 1959, forty million of the new pennies were trucked to the nation's Federal Reserve Banks. For thirty years those same banks had been engraving Lincoln's portrait on the five-dollar bill.[75]

The commission focused its educational efforts on junior high and high school youth. The National Education Association, joining in the endeavor, developed a program to stimulate creative thinking and writing about Lincoln. A program pamphlet with other materials went to approximately twenty-nine thousand schools. Students were encouraged to write essays, poems, plays, and stories. "I find," remarked a Michigan teacher, "constant comment by the students that their reverence for Lincoln is as great as mine has always been." The best work submitted by July 1 was published in an unusual book, *Abraham Lincoln Through the Eyes of High School Youth*. The students wrote on the big themes; and nothing seemed more compelling, even in 1959, than the theme of Lincoln's rise to fame and glory in the face of adversity. "The Lincoln family wasn't rich nor influential," wrote one young essayist. "Abe got to be President by hard work. He walked to school barefoot and often toiled in the fields all day." The programs of libraries, colleges, and universities ran to exhibits, lectures, and symposia. From its own rich collections the Library of Congress mounted probably the largest exhibition of Lincolniana ever assembled. Allan Nevins chaired a symposium at Occidental College on "Lincoln's Meaning for Americans." A two-day program at the University of Illinois offered, among other things, three contrasting views of the political Lincoln: "Conservative Statesman," by Norman

Graebner; "Pragmatic Democrat," by T. Harry Williams; "Whig in the White House," by David Donald. Whether by accident or design the three professors made manifest that, while everybody knew something about Lincoln, he was still the man *nobody* knows.[76]

From the beginning the commission aimed to make an enduring contribution to Lincoln scholarship, some literary monument that would survive the avalanche of trivia. And it early decided to make this work the revision and completion of the chronology, *Lincoln Day by Day,* begun by Paul Angle thirty years before. It had been carried forward (actually, backward in time) by his successors at the helm of the Abraham Lincoln Association, Benjamin Thomas and Harry Pratt, until the entire chronology from birth to the presidency was completed in 1941. Baringer was well acquainted with this work. The opening of the Lincoln Papers meant that it could be completed to the President's death and have some of the gaps in the earlier years filled in. Baringer's presence as executive director of the commission undoubtedly contributed to the decision of the ALA to surrender the copyright on the four published volumes; it contributed as well to the commission's willingness to appropriate $60,000 of its first-year budget to undertake the work. After July 1, 1959 Baringer devoted full time to research and assumed editorial responsibility for the first volume, 1809–48, of the new three-volume series. The harder part, covering the war years, was completed by Baringer's associate C. Percy Powell. Published before the commission went out of business in 1960, *Lincoln Day by Day: A Chronology,* under the general editorship of Earl Schenk Miers, was indeed a monument. To celebrate the work a symposium on "The Current State of Lincoln Scholarship" was held at the Library of Congress. Richard Current delivered the keynote address, "Lincoln Unexhausted and Inexhaustible." The scholarly host congratulated themselves on all that had been accomplished since James G. Randall sounded the trumpet a quarter-century ago, and pointed to all that remained to be done.[77]

Of the private corporations that contributed to the commemoration, mention should be made of Broadcast Music, Inc. Its president, Carl Havelin, was a Lincoln buff. BMI commissioned seventy-four radio scripts for broadcast in fifteen-minute segments as *The Abraham Lincoln Story.* Most of these found their way into a volume edited by Ralph Newman, *Lincoln for the Ages.* A similar but shorter volume edited by Henry B. Krantz, *Abraham Lincoln: A New Portrait,* collected twenty-two essays originally developed by the United States Information Agency for translation and distribution abroad. Several of the symposia yielded volumes of lectures or addresses. An occasional lecture rose above the humdrum.

Some useful monographs and compilations appeared. But except for the *Chronology*, the Sesquicentennial added no major work to the Lincoln literature.[78]

However, two brilliant works deeply enriched understanding of the war and yielded new insights on Lincoln. One was Edmund Wilson's *Patriotic Gore*. Described as "a personal encounter with history," these fascinating chapters on some thirty men and women for whom the Civil War was an intellectually penetrating experience related to its centennial rather than to Lincoln's anniversary, though of course the President was one of the subjects. Wilson, a literary critic turned cultural historian, said that Lincoln alone of American presidents might, in a different milieu, have become "a distinguished writer of a not merely political kind." But Wilson was less interested in Lincoln as a writer than in Lincoln's religion, which, he maintained, led the President to see the war in "a more and more apocalyptic aspect." This view would be developed by some of Lincoln's latter-day critics. Wilson also adumbrated the conception of Lincoln as a self-made myth, one whose passion for greatness led him to see himself as a new founder of the nation, and who was properly compared to those iron-willed builders of nations Bismarck and Lenin. They also presided over the consolidation of disparate peoples. This conception was destined to have a remarkable influence on the interpretation of Lincoln the man and the statesman. The following year, 1963, saw the publication of the middle volume of Shelby Foote's *The Civil War: A Narrative*. Scholars had generally dismissed the first volume as just another novelist's flawed attempt to write history; but now they were beginning to pay attention, and some even cheered. Here was an epic on the Homeric scale. Foote framed his narrative on the two protagonists of the conflict, Lincoln and Davis; and although himself a Mississippian, he much preferred Lincoln's "music" to Davis's. Lincoln emerged as the towering genius of the war because he was an artist to the depths of his being as well as a statesman.[79]

The main event of the Sesquicentennial was the joint session of Congress in Lincoln's honor on February 12, 1959. At the "kickoff dinner" the evening before, jointly sponsored by the commission and the Lincoln Group of Washington, two former presidents, Herbert Hoover and Harry S. Truman, sat at the head table with President Eisenhower. In his remarks Eisenhower saluted Lincoln as a world figure and prayed that his spirit be near as Americans, with freedom's sentinels everywhere, met each new challenge. Congressman Schwengel, who wrote the resolution authorizing the joint session, also chaired the committee that arranged it. Carl Sandburg was the committee's automatic choice for featured speaker. In the twenty years since the publication of *Lincoln: The*

War Years, Sandburg had himself become a folk hero. The fact that he was the first private citizen invited to address a joint session of Congress since George Bancroft delivered his eulogy of Lincoln exactly ninety-three years before added to the historic significance of the occasion. The hall of the House of Representatives had seldom, if ever, been more densely populated. The justices of the Supreme Court, cabinet officers, diplomatic corps, and special guests, as well as the membership of the Lincoln Sesquicentennial Commission and the Civil War Centennial Commission, all found a place in the crowded chamber. Following the invocation, the United States Army Band and a choral group from the Coast Guard Academy performed. Frederic March, the celebrated actor, read the Gettysburg Address.

Speaker of the House Sam Rayburn then introduced Sandburg. "Not often in the story of mankind," the white-haired poet began, "does a man arrive on earth who is both steel and velvet, who is as hard as rock and soft as drifting fog, who holds in his heart and mind the paradox of terrible storm and peace unspeakable and perfect." For the next twenty minutes Sandburg held the audience spellbound. Nothing he said was original or profound, but as he wove a delicate tapestry with Lincoln's own words, touching the chords of memory of his audience, he conveyed the majesty of the man. Concluding, he said that Lincoln's most enduring memorial was the invisible one in "the hearts of those who love liberty unselfishly for all men." The audience gave the speaker a standing ovation. Taped for broadcast nationally and sent overseas by shortwave radio, Sandburg's address had great impact. He became a fixture of commemorative programs during the next several years; never, however, did he equal his performance before Congress.[80]

The Lincoln Sesquicentennial flowed into the Civil War Centennial without skipping a beat. The juncture was the anniversary observance of the First Inaugural Address on the steps of the Capitol, March 4, 1961. Sandburg's speech was followed by reenactment of the inauguration before twenty thousand spectators. The anniversary of the Second Inaugural would later be observed in the same fashion. The Emancipation Proclamation, as earlier noted, remained a ticklish subject after one hundred years. To W. E. B. Du Bois, the one Negro leader who remembered the pitifully neglected fiftieth anniversary of emancipation, it was the same old story. "O dark Potomac where looms the glow of the Lincoln Memorial. Father Abraham," Du Bois implored, "unlimber those great limbs; let the bronze blaze with blood and the eyes of sorrow again see."

In 1961 an assembly of the commemorative bodies of the various states met in Charleston under the auspices of the federal commission. The chairman of that body, Ulysses S. Grant III, seemed bent on turning the commemoration into a festival of sectional reconciliation and national patriotism to the exclusion of civil rights for the blacks. In other words, to repeat the error of Reconstruction. When Madaline Williams, a black delegate of the New Jersey State Commission, was refused accommodations at the headquarters hotel, Grant and his executive director proved their incompetence and resigned under fire. The new leadership under Allan Nevins was more sensitive to black concerns, though Nevins in his final report on the Centennial complained of the resistance he had encountered on matters touching civil rights.

Again, as on the fiftieth anniversary, Congress targeted no funds for the commemoration of the Emancipation Proclamation. The commission sponsored a ceremony at the Lincoln Memorial on September 22, 1962, the anniversary of the preliminary proclamation, one hundred days before the ultimate decree. This was the event to which President Kennedy sent a perfunctory message. Negro leaders threatened to boycott; the last-minute addition of Thurgood Marshall, then a federal judge, to the program that featured Adlai Stevenson, Archibald MacLeish, and Mahalia Jackson appeased some, but only about five hundred blacks attended. The NAACP held commemorative meetings in its chapters across the country on a choice of dates—January 1, February 12, July 4—in 1963.[81]

The Commonwealth of Pennsylvania had preempted the National Cemetery at Gettysburg for rites honoring Lincoln's address. Former President Eisenhower, now a neighbor, was a featured speaker; more memorable, however, was the address by E. Washington Rhodes, a descendant of slaves and a Philadelphia newspaper editor and publisher, the first of his race to be accorded this honor at Gettysburg. The commission's tribute took place in Washington. Papers were read by distinguished authors and public figures. The poet Robert Lowell penetrated to the truth of Lincoln's address: "It is part of the battle. . . . By his words, he gave the field of battle a significance that it had lacked. For us and our country, he left Jefferson's ideals of freedom and equality joined to our Christian sacrificial act of death and rebirth. I believe this is a meaning that goes beyond sect or religion and beyond peace and war, and is now part of our lives as a challenge, obstacle, and hope." The Civil War Centennial wound down in 1965. President Johnson hosted a White House Lincoln Birthday luncheon for scholars and members of the commission. One year earlier he had called Lincoln's words "the common covenant of our public life" and vowed "to get on with his work." Now he invoked Lincoln's ideals on behalf of the administration's misadventure in Vietnam. The last event of the tortured history of the

Civil War was Lincoln's assassination and funeral. Appropriately, the Civil War Centennial Commission of Illinois commemorated the one hundredth anniversary of Lincoln's burial at the tomb in Springfield.[82]

Paul Angle, one of the speakers on this occasion, confessed, "The Lincoln theme, if not exhausted, is becoming very, very tired." He had just emerged from a bout with the new books churned up by the Centennial. There had been so many of them that Dwight Macdonald, in the *New Yorker,* had divided all literature into three parts: fiction, nonfiction, and Civil War. Robert Penn Warren, in his long meditation on the one-hundred-year legacy, said the Civil War was "our only *felt* history." and this was why it continued to fill a large space in American historical consciousness. He dwelt on the "maiming liabilities" of the legacy as embodied in two fraudulent traditions. In the South the war became the Great Alibi for all that was false, twisted, and unjust in its history, while in the North the war became the Treasury of Virtue, forgetful of all that was cruel and sordid. Southerners felt trapped by this history; northerners felt redeemed by it. And between the two, Warren concluded, the nation had still failed to achieve the freedom and justice and the deeper union of "one people" to which it had been dedicated in 1776 and for which Lincoln struggled.

The military historian Bruce Catton took a happier view of the legacy. "The memory of our Civil War," he said at the end of the Centennial in Springfield, "has not been a divisive force in this country, it has been a source of unity—something that ties us together and gives us a new depth of understanding. It has given us a common tradition, shared memories that go to the very roots of our existence as a people." Where the balance of truth lay between the legacy as a divide and the legacy as a unifier it is difficult to say; but the American people's absorption with this most singular event in their history was at the same time an absorption with its dominant symbol, Abraham Lincoln. In a Lincoln Day editorial the next year, after the smoke of the sham battles had cleared and the voices of the celebrants had faded, John B. Oates of the *New York Times* wrote that Lincoln's fame alone abided. His vision and his message to the peoples of the world were as clear and sufficient as they were a century past. Tolstoy had said it all in 1909: "He was the only real giant in depth of feeling and in certain moral power. . . . He was one who wanted to be great through his smallness. . . . He wanted to be himself in the world, not the world in himself."[83]

8

Lincoln Everlasting

AS THE TWENTIETH CENTURY drew to a close, the currency of Lincoln in the American mind showed signs of fading. Yet there was no diminution in the amount of thought and expression generated about him both within and without the academy. Frank J. Williams, a leading bibliophile, counted some sixteen thousand Lincoln titles, compared to fewer than four thousand contained in the half-century-old bibliography compiled by Jay Monaghan. The vast redundancy of Lincolnography was conceded; still it continued to yield enough fresh insights and interpretations to sustain a lively dialogue. The more interesting questions were why so many Americans continued to search in Lincoln's life and thought for the meaning of the national experience, why they continued to believe in him so long after his death, and why man and myth, which had become virtually indistinguishable, lived in American memory.[1]

As before, the Lincolnology of the scholars tended to support rather than to undermine the myth. It remained an affront to nay-sayers. By implication, at least, the idealized Lincoln transcended history and was thus beyond desanctification and demythicization. Scholars maintained, of course, that their Lincoln was the historical Lincoln—the *real* Lincoln—and they could not help it if he resembled the idealized Lincoln. Some demurred. Professor Robert W. Johannsen, the biographer of Stephen A. Douglas at the University of Illinois, doubted whether all the scholarship produced since James G. Randall's call had uncovered the real Lincoln. Myth and preconception barred the way to understanding him in the context of his time. Lincoln was a hero, a religion, an industry. "Lincoln, in fact," Johannsen wrote in 1968, "has become one of America's natural resources and one of its great export commodities, watched over and protected by a Lincoln cult or, as one writer recently put it, 'the Lincoln establishment.' " But for all his dissatisfaction with the cult, Johannsen recognized a "new maturity" in Lincoln studies and easily named a number that had, in fact, taken scholars closer to the real Lincoln.[2]

No other famous American had such a large scholarly complement as Lincoln, so many organizations, publications, and activities devoted to cultivating this resource. Washington and Jefferson, although they occupied other cardinal points on the compass of grand memorials in the nation's capital, and although their historic homes, Mount Vernon and Monticello, outshone the humble shrine on Eighth and Jackson in Springfield, were not cultivated and celebrated on the same scale. Standing at the head of the enterprise was the Abraham Lincoln Association, resurrected and embarked on a prosperous third life. It renewed its scholarly commitment with an annual birthday symposium in 1974, to which it added an annual series of *Papers* that evolved into the *Journal of the Abraham Lincoln Association*. In 1986 the ALA joined with the Historic Preservation Agency of Illinois in a project to collect and edit Lincoln's law papers. The "Lincoln Legals"—an estimated fifty-thousand documents in some four thousand law cases—was the last big job to complete the documentary record. At Fort Wayne, Indiana, the Lincoln National Life Foundation was renamed, for its founding director, the Louis A. Warren Library and Museum. (That name has since been changed to the Lincoln Museum.) It continued to publish a bulletin, *Lincoln Lore;* and in 1978 it inaugurated an annual lecture named for Warren's successor, R. Gerald McMurtry. His successor in 1972, Mark E. Neely, Jr., quickly established himself as a major player in the field. In 1982 he edited a new and improved *Lincoln Encyclopedia.**

Lincoln belonged to the nation's cultural heritage as well as to its civil history. Historical museums sought to restage his life for the edification of the general public. In this period, when museums burst upon the American landscape, quite a number were interested in Lincoln. Downtown Springfield virtually became a Lincoln museum, with a New Salem annex. "We sell the Lincoln legend," said the city's tourism director. The place, and the people, to which Lincoln said in his famous Farewell he owed everything, now seemed to owe a good deal to him. Impersonators of Lincoln were frequently encountered at museums and historic sites. There was even an Association of Lincoln Presenters, which in

* Unfortunately, the financial difficulties of Lincoln National Corporation in 1992 cast a cloud over the museum's future.

1992 boasted a hundred members. Museums were found at Fort Wayne, at the Lincoln Shrine in Redlands, California, at Hodgenville, at Lincoln Memorial University (thanks to the generosity of Colonel Harlan Sanders, the Kentucky Fried Chicken magnate), and, of course, at Gettysburg. Ford's Theater, splendidly restored and reopened as a playhouse in 1968, was also a museum, with a focus on Lincoln's assassination. Even Hildene, the Robert Todd Lincoln estate in Vermont, opened its doors to the public after extensive renovation in 1978. (R. T. L. Beckwith, the last of the line, died three years later.) New Lincoln Fellowships formed in Florida and in New York City. In 1990 an annual $50,000 prize for the best work, scholarly or otherwise, on Lincoln and the Civil War was established at Gettysburg College. The first recipient, in 1991, was Ken Burns for his heralded documentary on the war. Thanks to the dynamic leadership of Professor Gabor Boritt, the college now vied with the town and the park in service to Lincoln.[3]

No great collections remained in private hands, but the enterprise of of buying and selling Lincolniana flourished. As before, it was a barometer, always rising, of Lincoln's fame. Harold Holzer, an independent scholar particularly interested in Lincoln portraiture, tracked this activity annually in the magazine *The Antique Trader*. The auction market was dominated by Malcolm S. Forbes, the magazine mogul. In 1984 he paid $231,000 for the manuscript of Lincoln's last speech and considered it a bargain at about $130 a word. (The speech of April 11, 1865, had been found in the secret compartment of an antique table owned by the heiress Jean Payson Whitney.) The brief reply Lincoln wrote to eleven-year-old Grace Bedell, who suggested in October 1860 that he grow a beard to improve his appearance, was sold for $20,000. Because he acted on the suggestion, she, like the letter, was touched with fame. In the same year, 1966, a monument was dedicated to her memory at the place that had become her home, Delphos, Kansas. (A quarter-century later, the letter was offered for sale for $1 million.) Previously unknown or unpublished letters surfaced from time to time; and discoveries, such as Robert Lincoln's "Insanity File" on his mother, found at Hildene, became grist for the scholars' mills. Among the more interesting finds, in 1991, was a single page of manuscript believed by the collector and owner, Lloyd Ostendorf, to be part of the missing sixth copy—the copy the President made for his host, Judge David Wills—of the Gettysburg Address. Experts greeted the claim with skepticism, however; and the manuscript remained to be authenticated. An autographed copy of only the last paragraph of the Second Inaugural Address, previously authenticated, sold at auction on November 20, 1992, for $1.32 million, and one month later a holograph fragment of the "House Divided" speech went for $1.5 million, the most ever paid for an American manuscript.[4] *

With articles and monographs, scholars filled the gaps in the Lincoln story, offered more authoritative studies of old subjects, and tackled some new ones. David A. Nichols, in *Lincoln and the Indians: Civil War Policy and Politics* (1978), carefully examined a neglected subject. In Indian affairs, said the author, the President's toughness and single-minded determination to preserve the Union curbed his humanitarian impulses. Gabor Boritt, in *Lincoln and the Economics of the American Dream* (1978), not only isolated a neglected theme, the political economy of this western Whig who became the sixteenth President, but treated it in a way that resonated with the war's purpose and with the mythopoeic Lincoln. In *The Fate of Liberty* (1991), Mark Neely took a new approach to an old subject. With impressive research, he showed that although the number of arbitrary arrests and confinements had been greater than previously thought, their general significance for civil liberties was much less. The idea of Lincoln as dictator was exploded.

The thesis put forth by two medical scientists in the 1960s that Lincoln suffered from a rare genetic disease, Marfan's Syndrome, unknown in his time, seemed to defy historical analysis. But Boritt, who had inferred from a photograph of Lincoln on his deathbed that he had gigantic big toes, which fit the semiology of Marfan's, set out to find diagrams of the President's feet, and after seven years' investigation succeeded. Alas, the toes were of normal size. Such was Boritt's conclusion in *How Big Was Lincoln's Toe?* (1989), which weakened the Marfan's hypothesis. A seemingly trivial pursuit had a serious purpose after all. Although historians generally pooh-poohed the retrospective diagnosis, in 1991 medical scientists actually proposed DNA testing of Lincoln's hair and tissue fragments to settle the question. This became mixed up in public discussion with the notion of genetic cloning of Lincoln. The political humorist Russell Baker suggested that the scientists be directed to identify Lincoln's disease and, like General Grant's whiskey, send it to all our presidents. The project was laughed out of court.[5]

For many years full-length biography had ceased to engage serious students of Lincoln. The last to appear, by Reinhard Luthin, in 1960,

* In 1993 Ostendorf was a principal in a project to publish a purported memoir of Lincoln's laundress in Springfield, Mariah Vance. The title, "A House Divided," suggested its picture of the Lincoln household. When scholars greeted the reminiscences from a distant past with disbelief, the publisher, William Morrow & Company, backed away from publication (*Washington Post*, Aug. 24, 1993).

was not much read. In 1977 Stephen B. Oates, mindful of the maxim that each generation should reassess the heroes of the past, published *With Malice Toward None*, a one-volume life that tended to replace Benjamin Thomas's of a quarter-century earlier, just as his *Lincoln* had replaced Lord Charnwood's. The biography was basically a synthesis of the scholarship of the last thirty years or so. Oates drew a portrait which, although meant to be "fair and unflinching in its realism," was entirely partial to Lincoln. Above all, he emphasized the President's commitment to liberal (that is, Radical) Republicanism, including the civil rights of the freedmen. The fact that Oates's biography of Lincoln was preceded by one of John Brown and followed by one of Martin Luther King suggested the author's political commitment. Richard Current, in an appraisal of the biography, wondered if Oates was not engaging in a new form of mythmaking by appropriating Lincoln for the civil rights revolution. Remarkably, Oates shunned the new fashion of psychoanalytic interpretation in historical biography. By the standard of several psychobiographical studies soon to appear, *With Malice Toward None* seemed old-fashioned, though not less persuasive for that.[6]

The year 1984 marked the 175th anniversary of Lincoln's birth. The public was largely oblivious to this fact, but Lincolnian precincts buzzed with activity. Especially notable were two conferences of scholars, one at Gettysburg College, the other at Brown University. The former, under Boritt's guidance, ranged over a broad field, from Lincoln's humor and religion to Lincoln as a subject of psychobiography. No better introduction to recent scholarship could be found than *The Historian's Lincoln* (1988), the volume that collected the conference papers. Represented in it were most of the new generation of Lincoln scholars—Boritt himself, Neely, Oates, Jean Baker, Charles Strozier, and others—along with some of the older generation brought up under the Randall's tutelage. An exhibit, "The Lincoln Image," hung in the college's museum. This was a companion to the fine book of that title authored by Holzer, Boritt, and Neely, which demonstrated through popular prints how swiftly Lincoln had been transported from history to myth.

The conference at Brown, jointly sponsored by the Lincoln Group of Boston, focused on a single theme, "Lincoln and the American Political Tradition." The papers by six political historians were published in a book with that title. The authors were not, strictly speaking, Lincoln scholars; the hidden agenda of the conveners was to entice leading historians of the war era to put their minds on the dominant figure. The program also featured an exhibit drawn from the university's great Lincoln collection and a series of workshops on various aspects of Lincolniana. Certain historiographical tendencies, for instance the revolt against "the great white fathers," of whom Lincoln was indubitably one, the shift to quantitative or behavioral methods in political history, and the rising prominence of social history, might have been expected to erode Lincoln's place, but such was not the case.[7]

"As long as we believe in America," Stephen Oates wrote in 1984, "we will have towering Father Abraham as our greatest mythical hero. And as long as he is that hero, he will remain a powerful presence to be reckoned with." There were signs, however, that regardless of the health of the enterprise among scholars, Lincoln was a faltering presence in the hearts and minds of the American people. In part, this could be attributed to waning resonance between the Lincoln symbol and the critical human problems of the late twentieth century. What, after all, did Lincoln have to teach about global overpopulation, depleted natural resources, and decay of the environment? And in a society scarred by violence, poverty, and despair, how could Lincoln continue to serve as an inspirational example to youth? The age of heroes had passed, it was commonly said. More than that, the moral and emotional bonds previous generations had felt with Lincoln no longer held. "His Truth Is Not Marching On" was the provocative title of a Lincoln Day speech by a black educator, Benjamin H. Alexander, in 1986. It was *not* marching on in the schools, in the work place, in the family, or in race relations. A book of photographs by George A. Tice, in 1984, testified poignantly to the debasement of the Lincoln icon. A view from behind of the beloved Borglum bronze in Newark, New Jersey, revealed a figure tarnished with graffiti and gazing upon a blighted urban landscape dominated by a Burger King restaurant.

Curiously, Lincoln was also becoming the butt of humor. Perhaps this was part of a celebrity-crazed generation's effort to "humanize" him, but Lincoln had always been, as Johnny Carson was wont to remark, the one personage living or dead off-limits to humorists. Now Bob Newhart cracked jokes about him, and political columnists Russell Baker and William Safire kicked him back and forth for laughs. Thomas Meehan wrote a funny piece, "Abraham Lincoln: Lawyer, Statesman, and Golf Nut," in the *New Yorker*. The President, it seems, like some of his successors, had a passion for golf, which explained why the Gettysburg Address was so short. He wanted to get in a few holes on the local links before returning to Washington. Presumably with a view to extending Lincoln's appeal to the younger generation, advertisements for Illinois tourism portrayed the sad-eyed emancipator in shades.

Speechmaking about Lincoln, especially by politicians, fell off dramatically. "Getting right with Lincoln" was no longer important. The generic President's Day, which the nation began to observe in 1971, proved a flat failure as far as it concerned commemoration of Lincoln, or Washington, or anybody. Instead of a day of remembrance the three-day weekend was a time of fun. Editors of major newspapers, like the *Los Angeles Times*, voiced displeasure and asked for reconsideration of a national holiday for Lincoln. By 1990 even visitation at the Lincoln Memorial was on the decline; the following year the Vietnam War Memorial surpassed it as the national capital's leading tourist attraction. And in Boston, where a replica of the Freedman's Monument had stood in Park Square since 1879, there was agitation to remove it because, it was said, "some people now think the freed slave looks like he's shining Lincoln's shoes." At last report the statue stood undisturbed. Another, and quite astonishing, victory for Father Abraham occurred at Disneyland in 1991. After the Disney Corporation announced that it would discontinue "Great Moments with Mr. Lincoln," an animated figure delivering excerpts from his speeches, which had debuted at the New York World's Fair in 1964, and replace it with an attraction featuring Kermit the Frog, there was such an outcry from patrons and employees alike that the company reversed the decision. After all, as a twelve-year-old protested, "Lincoln was president, Kermit is a frog."[8]

THE THEMES AND IMAGES of the apotheosis that sent the Martyr President on his amazing career in American thought and imagination were still present in the collective memory 125 years or more after his death, though fractured and revised, blurred and attenuated. They may be recapitulated in this final reckoning.

But first a word about the macabre interest in the assassination itself. On August 3, 1977, the syndicated columnists Jack Anderson and Les Whitten headlined their story "FBI Probes Lincoln Assassination." Allegedly, eighteen missing pages of John Wilkes Booth's diary had been discovered, and they implicated Edwin Stanton in the murder conspiracy. There was, in fact, no FBI probe, but it was a time when self-appointed investigators claimed to have uncovered the awful truth of the assassination. A best-selling book, *The Lincoln Conspiracy*, by David Balsinger and Charles Sellier, and the popular Sun Classics motion picture based upon it, appeared in 1977. An assassination newsletter, the *Lincoln Log*, was followed some years later by the *Journal of the Lincoln Assassination*. The Booth home in Maryland was a shrine of sorts; a brisk business went on in relics and mementoes; in 1985 ABC aired a drama on Booth's escape and second life as John St. Helen.

The dramatic possibilities of the assassination made it almost irresistible to playwrights. Lincoln Kirstein, known as a ballet impresario rather than a playwright, had long been interested in the man after whom he was named; and he wrote a play, *White House Happening*, that premiered in Cambridge in 1967. The action takes place in the White House in the hours before the fateful departure for the theater. The Radicals, it seems, are behind the plot; but the shocker of this drama is that the President knew about it and by going to the theater in effect committed suicide. The psychobiographers might have explained the act as Lincoln's ultimate bid for immortality; Kirstein, however, attributed it to the blood-guilt of the war together with feelings of disgust and despair rising, in part, from his wife's insanity. Evidently the play was not better as theater than it was as history, for it died in Cambridge.

The reopened Ford's Theater was nervous about staging plays treating the assassination, but they naturally gravitated to the historic site. (The actor James Whitmore said that whenever he performed there he looked up to the President's box, and once saw Lincoln's chair rocking.) In 1969 Paul Shyre's *The President Is Dead*, which exploited the parallel to the Kennedy assassination, was abruptly canceled by the theater, causing the author to charge censorship. The National Park Service management, however, said its objection was to the false depiction of Stanton as the archvillain. In 1984 another brouhaha occurred, over a one-man show about Booth by Chris Dickerson. It re-created Booth's last visit to Taltavul's saloon across the alley from the theater. In this instance Ford's director, Frankie Hewitt, intervened to halt production of a play approved by the management. "You've got to be kidding!" she exclaimed when her action was questioned. "Would you go see a play about Lee Harvey Oswald in the Texas Book Depository? It's sick."

The obsession with Lincoln's assassination was probably incurable; but it was a case, as Mark Neely pointed out, where professional historians had abdicated responsibility. They had left the field to frauds, dramaturgists, sleuths, and sensation-mongers. At last, in 1983, William Hanchett, professor of history at San Diego State University, produced a scholarly study, *The Lincoln Murder Conspiracies*. It provided the first authoritative study of the assassination as well as of its historiography before and after Otto Eisenschiml's unfortunate book. Speculating on the popularity of the thesis that traced the conspiracy to the War Department, Hanchett compared it to the Orwellian "Big Lie," which generated its own credibility. "For many people, the thesis is simply too audacious to be disbelieved."[9]

The image of Lincoln as Savior of the Union lay at the heart of the apotheosis. It coupled his fame with Washington's. It crowned him the hero of American nationhood triumphant over the once-vaunted rights of the states and the petty loyalties men rendered to them. It identified him with a strong national government empowered not only to preserve itself but also, with the war amendments, to secure the fundamental rights of all the citizens. John G. Nicolay and John Hay embodied the theme in their great *History*. The values of the image were specific to the American Union and Constitution. They were neither universal nor timeless. In the twentieth century several important changes occurred in this conception.

Lincoln's fundamental cause and purpose in the Civil War, many came to believe, was not to save the Union but to advance human freedom. The Union and the Constitution, as he himself put it in a biblical metaphor, composed the frame around the picture; the picture was the apple, freedom, which the frame was meant to preserve and adorn. The Civil War came to be understood less as a war to preserve the old Union than as a revolutionary refounding. James McPherson, whose epic history of the war, in 1988, was entitled *Battle Cry of Freedom*, developed the idea of the "Second American Revolution" in a way entirely different from Charles A. Beard, who first employed that language. It was not that the war was an economic revolution, as Beard had maintained, but that the war effected a revolutionary change in southern society and in the condition of the Negro people and placed the enhanced authority of the national government behind fundamental civil rights. Preservation was not conservation for Lincoln; he sought a Union refounded on a more liberal as well as more national basis. "To understand the Constitution as Abraham Lincoln did," Harry Jaffa wrote, "must mean, primarily, and essentially, to understand the Constitution as an expression of the principles of the Declaration of Independence." And that was to understand it, with the Union, in a way none of the founders had envisioned.[10]

A different reinterpretation of Lincoln and the Union took flight from Edmund Wilson's key insight in his 1953 article "Abraham Lincoln: The Union as Religious Mysticism," which was the basis of the chapter in *Patriotic Gore*. Wilson was startled by a passage in Lincoln's Lyceum Address of 1838, entitled "The Perpetuation of Our Political Institutions." A hymn to the legacy of the Founding Fathers coupled with a lawyerly denunciation of "the mobocratic spirit," the address took an unexpected turn when young Lincoln warned that ambitious men among the sons and grandsons of the fathers could not be satisfied with being mere legatees and guardians, occupying a governor's seat here or a senator's chair there. "What! think you these places would satisfy an Alexander, a Caesar, or a Napoleon? Never! Towering genius disdains a beaten path." And he went on to say that if the passion for fame could not be gratified by building up, then lofty genius would set boldly to the task of tearing down. "The effect of this," Wilson said, "is curiously ambiguous:

it is evident that Lincoln has projected himself into the role against which he is warning them [the audience]." Indeed, not only did he imagine himself the "towering genius," he also prefigured an assassin, a Brutus, as the necessary denouement of the tragic drama. Here, in sum, was the germ of an interpretation of Lincoln and the Union rooted in the strivings of the inmost self. Psychobiographers seized upon this neglected and, to them, suddenly prophetic passage. None of Lincoln's writings has received more critical attention during the last quarter-century than the Lyceum Address. Although it could be interpreted, as it was by Jaffa, as a brilliant yet characteristically Whig protest against certain tendencies of Jacksonian Democracy, many scholars, following Wilson, read deeper meanings into it.[11]

The eclipse of the image of the Great Emancipator coincided with the rise of Jim Crow. It was saved, in part, by transformation into the conception of Lincoln as a universal liberator; and although it entered into the civil rights revolution, it was attacked from both left and right, then virtually vacated by the new myth of a black Moses in the person of Martin Luther King. Increasingly among Afro-American historians the slaves were seen as their own emancipators, which, of course, diminished the importance of the Emancipation Proclamation. Yet to an extent, the currents of Lincoln scholarship validated the old image. If the war was fundamentally about slavery and freedom, as historians from Allan Nevins to James McPherson maintained, then it was fundamentally about emancipation. That was not to say that the President placed emancipation ahead of saving the Union, for he clearly did not. Without the frame the picture would be torn or broken or lost. He hated slavery from the depths of his being, but he had a high official duty to the Union. His predicament, Hannah Arendt suggested, was like Machiavelli's when he declared, "I love my native city more than my own soul."

Nevertheless, emancipation underlay the "new birth of freedom" resolved at Gettysburg. Lincoln, like Jefferson in 1776, was committed to the proposition of human equality, and unlike Jefferson he was capable of acting upon it with regard to the bondsmen. In the eyes of conservative intellectuals like M. E. Bradford, this was Lincoln's great heresy, and he assailed the idea of a second founding as the sentimental residue of "dying god" martyrdom. During these years Bradford, professor of English at the University of Dallas, became "the symbolic adversary" of what he named "the secular religion of Lincoln." The plaint was too much for conservative columnist George Will, who stigmatized Bradford "the nostalgic Confederate remnant of the conservative movement," and in 1981 added his voice to the liberals' campaign to defeat Bradford's prospective nomination by President Reagan to head the National Endowment for the Humanities.[12]

Even among believers, however, there remained nagging doubts about the fullness of Lincoln's commitment to racial equality. To John Hope Franklin, the black historian, Lincoln was, unfortunately, a flawed hero. "The fight for union that also became a fight for freedom never became a fight for equality or the creation of one racial world." And a younger black historian, Armstead Robinson, said that Lincoln practiced "an ideology of racial gradualism which could move with such deliberate speed as to the slow the pace of progress toward universal citizenship virtually to nil." He was responding to a paper at the Gettysburg conference by La Wanda Cox, author of a prizewinning study, *Lincoln and Black Freedom*, which, along with the work of Peyton McCrary on the Reconstruction experiment in Louisiana, held that the President took radical steps toward black freedom and equality before his death. Of course, he was limited by what was attainable, but his record on the rights of the freedmen needed no apology.

These works, with others, hinted at scholarly anxiety to preserve the image of the Great Emancipator. Historians proved resourceful in finding new explanations to acquit the President of fault or blame in race relations. Thus if he had been slow to back limited black suffrage in Louisiana, it was because he was waiting for public opinion to catch up with him. Thus if he adhered too long to the chimera of colonization, the reason was psychological: colonization served as a defense mechanism against thinking about the tough problems of racial accommodation under conditions of freedom. Although the brightness of Lincoln's image had faded among blacks, his name would always be associated with emancipation. Emancipation Day actually enjoyed a modest revival; in 1979 Juneteenth (June 19) became a holiday in Texas; and by the logic of history even Martin Luther King Day was a salute to his memory.[13]

Lincoln the hero of democracy found embodiment in the image of Man of the People. Elected President as the Railsplitter, he was from the first reckoned one of the common people, and at his death "the gentlemen who jeered," as the poet said, joined the chorus of praise; but time was required for the mind to take in the significance of his achievement for American democracy. "Genius is no snob," President Wilson declared at Lincoln's birthplace memorial in 1916, and his tribute bears repeating. "Here is proof of it. This little hut was the cradle of one of the great sons of men, a man of singular, delightful, vital genius who presently emerged upon the great stage of the nation's history, gaunt, shy, ungainly, but dominant and majestic, a natural ruler of men, himself inevitably the central figure of the great plot. . . . Such are the authentic proofs of the validity and vitality of democracy." It was never a question of Lincoln being ordinary. He was the common man writ large, the common man become master, his commonness lifted to "consummate personal nobility," in Herbert Croly's words. This was the paradox and the mystery. In the late twentieth century, democracy had become more than a form of government; in the West it had become an ethical ideal of universal humanity, and, as Tolstoy understood, Lincoln personified it.[14]

The image developed the theme of Lincoln's commitment to democratic government and his masterly leadership. The war had a third purpose, beyond preserving the Union and emancipating the slaves; it was about saving the democratic experiment in America. In his first message to Congress, July 4, 1861, Lincoln declared, "This is essentially a People's contest. On the side of the Union, it is a struggle for maintaining in the world, that form, and substance of government, whose leading object is, to elevate the condition of men—to lift artificial weights from all shoulders—to clear the paths of laudable pursuit for all—to afford all, an unfettered start, and fair chance, in the race of life." The war threatened that experiment in the most fundamental sense, for the secession and rebellion of the Confederate states struck at the very process of free elections and majority rule through which the people governed. This was "the central idea" of the contest, Lincoln had told John Hay, who recorded it in his diary. "We must settle this question now, whether in a free government the minority have the right to break up the government whenever they choose. If we fail it will go far to prove the incapability of the people to govern themselves." At Gettysburg he viewed the war as a test of the viability of government of, by, and for the people. Somehow, in the vastness of the Lincoln literature, the theme had been obscured. But A *People's Contest* was the title Phillip S. Paludan gave to his 1991 book on the war years in the Union states. And William S. Safire's mammoth historical novel *Freedom*, in 1987, was written in the conviction that Lincoln's great purpose was neither Union nor emancipation but the preservation of democratic government. Safire contrived to have Lincoln declare the passage from Hay's diary to Benjamin F. Wade, who wrongly supposed the war was about the abolition of slavery.[15]

Although politicians loved to appeal to Lincoln, they had experienced difficulty making him a touchstone for their special causes and agendas. In the late twentieth century, his name was invoked less and less in ongoing political debate. New York Governor Mario Cuomo, for whom Lincoln was a personal hero, was one politician who continued to do so, however. Speaking at Springfield on "Abraham Lincoln and Our Unfinished Business," in 1986, he insisted that his hero would never have tolerated the "new slavery" of poverty, homelessness, drugs, and despair. In his time one in seven Americans was enslaved. He did not falter but gave his life to end that injustice. Today, when one in seven Americans lives in poverty, said Cuomo, we should not rest until this injustice, too, is righted. Some efforts to project Lincoln into contemporary politics were so forced as to be ludicrous. George Will, for instance, looking ahead to the sure-to-be-controversial "pro-life" plank in the Republican platform of 1992, thought that the opposition might be assuaged by borrowing Lincoln's strategy toward slavery: don't attack abortion head on but control it at the periphery, as Lincoln barred slavery in the territories, by, for instance, limiting the right to the first eight weeks of pregnancy and withholding public funds.

The Lincoln scholar Harold Holzer suggested that the difference between the Republican candidate in 1860 and present-day presidential politics was not as great as generally imagined. Indeed, Lincoln was "the first media President." Sitting so often for his own portrait, crafting his own image, he instinctively understood the place of mass communications in politics. But was the greater truth not in the difference? Would not Lincoln, who wrote his own speeches and engaged in reasoned discourse, have been an anachronism in the age of the "sound bite"? The political cartoonist Tony Auth showed Lincoln flunking the test of political viability administered by "Candidate Images, Inc.," while Jack Ohman imagined him offering a sound bite, à la President Bush, at Gettysburg.[16]

The image of Lincoln as the First American proceeded from James Russell Lowell through Walt Whitman to Carl Sandburg. It belonged to the poets rather than to the preachers, historians, and politicians, though, of course, it radiated everywhere. Lowell had discerned the archetype of a new national character, one long imagined but only now realized in a man who looked at things, who related to people, who bore himself in ways indigenous to the American continent. The President who had spoken of "a new birth of freedom" was himself "new birth of our soil," and in that sense "the first American" authenticated the refounded nation. For it was only in the mirror of Lincoln, as Sandburg would say, that the American people finally came to see themselves. In *The Prairie Years* he accented the vulgar, folksy, frontier side of Lincoln's character. The portrait filled out the Americanist conception of Lincoln. It had immense influence during the era when Americans searched out the roots of their culture. That time had passed, but this conception of Lincoln was so completely assimilated, and its cultural implications were so well understood, that hardly anyone needed to think about it any longer. The utter individuality of Lincoln, which had impressed all who knew him, entered into the definition of his Americanness, to the point that his physiognomy blended into the representation of Uncle Sam.

But Lincoln as a representative American was more than a folk type. From the beginning, delineations of his character had recognized two sides, the vulgar and the saintly, and had sought to balance or reconcile the two lives they implied. The enigma of that character was another part of its Americanness. Scholars traced the tradition of the folklore Lincoln to the work of William H. Herndon. Although Herndon's Lincoln was actually too complex to be defined in those terms, the eclipse of the village Boswell's reputation coincided with the eclipse of this mythic representation of Lincoln. The older formulas for modeling his character no longer interested biographers. Nor were they interested in old questions of parentage and ancestry. When, at last, Adin Baber sorted out the genealogy of the Hanks family, nobody cared. Even children's biographies began to approach Lincoln from the point of view of the White House instead of the log cabin. Russell Freedman's prize-winning *Lincoln: A Photobiography*, in 1987, for instance, was distinguished not alone for its creative use of photographs and other images but also for its straightforward, unsentimental, non-didactic characterization.[17]

The image of the Self-made Man, reverberating with the American and the democratic hero, had been the dominant pedagogical model, known to every schoolboy and to every schoolgirl. Governor Cuomo, the son of Italian immigrants, attested to its importance in his own life and in the American ideal. Lincoln's belief in the promise of equal opportunity, in a society and a government that afforded all, regardless of condition, as he had said, "an unfettered start, and a fair chance, in the race of life," arose from his own experience. "His personal mythology," said Cuomo, "became our national mythology." And because of Lincoln Americans "are still reaching up, for a better job, a better education, a better society." Boritt's book *Lincoln and the Economics of the American Dream*, in a sense, provided the historical documentation for this conception. Pursuing a lead thrown out by Richard Hofstadter, Boritt showed how Lincoln's belief in the self-made man lay at the center of his life and his politics. His Whig economics esteemed free labor against slave labor and advocated a positive government helping in the development of the country for the greater opportunity of all the people. Another scholar went so far as to argue that Lincoln, "the self-made myth," as Hofstadter had called him, actually invented the myth for personal political purposes. Holding with Louis Warren that Lincoln's family background was respectably middle-class, hence a source of pride and strength for the son, he could have had no cause other than political ambition to furnish himself with a legend of poor and humble beginnings.[18]

The ideal of the self-made man has suffered some hard knocks since the Great Depression. Viewing American history from that perspective, Hofstadter had said that Lincoln ironically presided over the economic revolution that destroyed it. Boritt, too, speculated on Lincoln's responsibility for the Gilded Age. Yet the image, with the ideal, had survived. The speeches of the Governor of New York, in which it is associated with a liberal and activist government rather than a laissez-faire ideology, offer proof of that. Still one wonders if it lifts the hearts and extends the horizons of American youth. At Lincoln Memorial University the Lincoln story had long been part of the curriculum. His example, his virtues and teachings, said the president, Robert L. Kincaid, in 1950, should be "a conscious and compelling part of the educational heritage of every American youth." Only the shell of that idea survived forty years later. A professor at the Appalachian university that had grown to eighteen hundred students, when asked if the life of its spiritual father was still taught and upheld as an ideal, replied soberly, "No, Lincoln doesn't cut it with these kids any more."[19]

ON FEBRUARY 12, 1976, Daniel J. Boorstin, Librarian of Congress, emptied Lincoln's pockets. Thirty-nine years earlier, Mary Lincoln Isham, the granddaughter, had given to the library a box containing the things in the President's pockets the night he was murdered. For reasons nobody could explain, the box had reposed undisturbed in a vault until now. Gingerly, Boorstin removed the contents: two pairs of eyeglasses, a penknife, a watch fob, a cufflink, a monogrammed handkerchief, and a wallet. The wallet contained eight newspaper clippings about his leadership, all of them congratulatory, but no money except, on a second look, a Confederate five-dollar bill. "This must have been insurance," Boorstin quipped. One pair of glasses had been broken and crudely, heartbreakingly, repaired with twine. So many people asked about the eyeglasses that Boorstin called in the experts. "Will our knowledge of these glasses help us understand why he married that lady?" The answer was no; they showed, however, that the President suffered from presbyopia, or "old eye-sight," meaning that he grew far-sighted with age, and one pair was more powerful than the other. Among the oculists present were representatives of the Washington firm, Franklin and Co., that fitted the President for new glasses in 1864 and, amazingly, had the $2.50 check to prove it. The contents of Lincoln's pockets were placed on exhibit in the Great Hall of the Library. They suggested that Lincoln was much like the rest of us. Having an immediacy that transcended time, they should, Boorstin said, help to humanize a man "mythologically engulfed."[20]

In the poem by Stanley Kunitz, "the Lincoln relics" evoke a lost father figure and a dream betrayed.

> He is slipping away from us
> into his legend and his fame,
> having relinquished, piece by piece,
> what he carried next to his skin,
> what rocked to his angular stride,
> partook of his man smell,
> shared his intimacy and his needs.
> Mr. President,
> in this Imperial City
> awash in gossip and power,
> where marble lots marble
> and your office has been defiled,
> I saw the piranhas darting
> between rose-veined columns,
> avid to strip the flesh
> from the Republic's bones.
> Has no one told you
> how the slow blood leaks
> from your secret wound?

Kunitz recalls happier years, for himself and the republic, and yearns to be shone upon again by the "prairie star." The poem ends with a glimpse of "a rawboned, warty, gangling fellow" who mysteriously steps out of the crowd and disappears.[21]

It was sometimes said that Lincoln must be liberated from the wooden prisons of the scholars, on the one hand, and the pious keepers of his shrine, on the other hand, in order to humanize him. A humanized Lincoln would be a flesh-and-blood Lincoln rather than one who was a story, a model, and a memory. Dramatists responded to the challenge, but without much success. No play remotely equaled Robert Sherwood's. It was reported in 1990 that Lincoln had been portrayed in 133 motion pictures—three times as many as his nearest competitor among historical figures, General Grant—and almost always Lincoln the immortal smothered the palpable human being. A partial exception was David Wolper's six-part dramatization for television, *Sandburg's Lincoln*, based upon *The War Years*, which starred Hal Holbrook, in 1974. Some television productions were a positive disservice, for instance ABC's dramatization of *The Perfect Tribute,* which perpetuated a hoary myth about the Gettysburg Address. Herbert Mitgang's one-man stage play *Mister Lincoln,* with an eloquent performance by Roy Dotrice, a British actor, failed on Broadway but in the more intimate medium of television offered a believable portrait of Lincoln through his talk. Robert Wilson, the idiosyncratic impresario of monumental theatrical events, introduced Lincoln into his five-part *The CIVIL Wars: A Tree Is Best Measured When It Is Down.* (The President is shot while watching the motion picture *Gone with the Wind.*) But the gigantic opera had little to do with Lincoln or the Civil War. The palpable human being, if found anywhere, was found at last in Ken Burns's historical documentary film *The Civil War.*[22]*

Psychoanalysts had been summoning Lincoln to the couch since the 1920s. L. Pierce Clark, it may be recalled, traced much of what is puzzling in Lincoln's behavior to a mother fixation. Two decades later the psychiatrist Edward J. Kempf maintained that the young Lincoln had suffered brain damage from the kick of a horse, and this explained his chronic melancholy. In a sense, Herndon, in his intuitive fashion, had been the first to probe Lincoln's personality at levels that could not be reached by the historian's usual methods. American historians had been slow to introduce modern or Freudian psychoanalytic theory into their researches. In 1957, however, William L. Langer, in his presidential address before the American Historical Association, called upon the profession to make the exploitation of the findings of modern depth psychology "the next assignment." Many scholars rose to the challenge. With respect to Lincoln, three important psychobiographical studies appeared between 1979 and 1982. All were perhaps as much indebted to Edmund Wilson as to Sigmund Freud. For they took the literary critic's insightful reading of the Lyceum Address, which he had been content to leave at that, and combined it with the theory of subconscious Oedipal wishes in Freudian psychology, so that the Caesarism Lincoln prophesied became a revelation of his true self as well as a key to understanding the mythical Lincoln in American history.[23]

The first of these books, *Patricide in the House Divided,* by George B. Forgie, was based upon the author's doctoral dissertation at Stanford University. Lincoln reenacts in public life the rejection of his own father by revolting against the domination of the Founding Fathers. Unable to

*Another documentary, *Lincoln,* aired on ABC on December 26, 1992, and, published as a book as well, was especially important not alone for its accomplishment but also for its auspices. The film was developed by Philip and Peter Kunhardt, who were third-generation descendants of Frederick Hill Meserve, the pioneer collector.

recognize his resentments, he does not himself become the tyrant who imaginatively slays the father, but projects this role onto his rival, Stephen A. Douglas. Douglas, not Lincoln, becomes "the bad son." By slaying him politically, Lincoln discharges his filial obligation to the fathers and at the same time realizes his own ambition. Fratricide resolves the guilt of patricide. This was a bold thesis, and in some parts, for instance the light shed upon it by the author's reading of Lincoln's reading of Shakespeare, brilliantly argued. But the evidence was flimsy, the reasoning tortuous. Political historians objected to a thesis that reduced Lincoln's opposition to Douglas to the overt expression of his repressed desires. Moreover, by fantasizing a "house divided," according to Forgie, Lincoln set the stage for the calamity of the Civil War, all because he wished to be master of a house of his own. As one reviewer observed, the psychobiographical interpretation of Lincoln and the Civil War came to the same conclusion as the "needless war" thesis of the discredited revisionists.[24]

Dwight G. Anderson's *Abraham Lincoln: The Quest for Immortality* began as a doctoral dissertation in political science at the University of California, Berkeley. He portrayed a downright demoniacal Lincoln, reminiscent of the hateful images of last-ditch defenders of the Confederacy. Although his methods were not unlike Forgie's, he dismissed Forgie's thesis as "preposterous." Lincoln, not Douglas, was "the bad son," and the patricide was his. Anderson made much of Lincoln's early reading, and constant re-reading, he said, of Weems's *Life of Washington*. Washington became his imaginary father. Moreover, from an early age Lincoln developed abnormal fears of death, as evidenced by his favorite poem, and corresponding passions for immortality. Like Forgie, Anderson read Lincoln's psyche through Shakespeare's plays. He *was* Macbeth; he *was* Richard III; he *was* Claudius. At first, he sought to realize his ambition by faithfulness to Washington, but after the humiliating failure of his career in Congress and fading prospects within the system, Lincoln turned to tearing it down, killing his imaginary father, and setting himself up as the tyrant-father of a new nation complete with a civil religion in which he was God's anointed for his "almost chosen people." This was fantasy masquerading as history. As Robert Bruce observed in response to Anderson's summation of the book at the Gettysburg conference, the argument, even where it agreed with the known facts, passed over the more obvious, plausible, and reasonable explanations for explanations suited to the thesis; as for the thesis, it was ludicrous to portray Lincoln, rather than Jefferson Davis, as presiding over the destruction of George Washington's Union.[25]

A better book, both as history and as psychology, was Charles B. Strozier's *Lincoln's Quest for Union*, in 1982. A seasoned Lincoln scholar, with credentials in psychoanalysis, Strozier sought to understand how Lincoln achieved a strong and cohesive personal identity despite the wrenching emotional disturbances and insecurities of his life. He suffered damaging guilts about both his parents; he eventually overcame fears of intimacy by marriage to Mary Todd, yet in the end the marriage was not fulfilling and a void formed between them; he suffered from recurrent bouts of depression. Yet none of this, Strozier acknowledged, interfered with Lincoln's effectiveness as a person and a leader. By humor, empathy, creativeness, and wisdom, he transcended his psychological problems. But Strozier, like Forgie and Anderson, was also interested in the public outcomes of Lincoln's inner strivings. (The subtitle of the book was *Public and Private Meanings*.) Lincoln ultimately resolved his personal conflicts in the consecration of the Union, Strozier argued. "His own ambivalent quest for union—with his dead mother, his bride, his alienated father—gave meaning to the nation's turbulence as it hurtled toward civil war. . . . The image he shaped [of a united nation] dissolved struggles over father, fatherhood, and founders. It put him in touch with God. And it gave America its greatest hero." This was a leap of the imagination sure to raise frowns on the brows of sober historians. Was the course of political history to be so lightly discarded? Did saving the Union serve a similar private purpose in the lives of Stanton or Seward or Sumner? And why did Strozier assume that the Union, rather than freedom, was Lincoln's great concern?[26]

The psychobiographers, with their insights, seemed likely to stimulate renewed interest on the part of more conventional biographers and historians in Lincoln's personal growth and development. In 1990 there were signs of a revival of Herndon's reputation. He had, of course, written about the same kind of problems in Lincoln's personal life that interested the psychobiographers. But they had declined to use Herndon's pivotal disclosure, the Ann Rutledge romance, for it had long since been dismissed as a fable. But now two Lincoln scholars, John Y. Simon and Douglas L. Wilson, gave the love affair a rehearing and concluded it had, indeed, occurred and should be restored to its proper place in Lincoln's biography. And since the story rested upon the recollective evidence of Herndon's informants, its return to credibility reopened the larger question of the truthfulness of the mass of undocumented reminiscence about Lincoln. Whatever the influence the psychobiographers may have had on these developments, and whatever final assessment is placed upon their work, that work, far from humanizing Lincoln, only elaborated the myth.[27]

The resemblance, unacknowledged, between the psychobiographers and the historical novelists is suggested by Gore Vidal's blockbuster best-seller *Lincoln*, published in 1984. Nothing ever written about the sixteenth President was more widely read. Random House's first printing was two hundred thousand copies; the first printing of the paperback edition a year later ran to one million copies. Vidal, a celebrated writer who had distinguished himself in historical fiction, tipped his hand on the novel in a newspaper article three years prior to publication. He spoke contemptuously of Sandburg and the hagiographers who pretended to be historians. He admired Herndon as a guide to the real Lincoln, even trumpeting the truth of the churlish story that Lincoln had contracted syphilis, which Herndon had never trusted to print, adding his own belief that he communicated the disease to Mary and the children, thus accounting for her insanity and the early death of three of them. This was nasty. In the novel the luridness of the tale is enhanced by an inebrious Herndon telling it to young John Hay in a Washington brothel. Vidal got the attention he was seeking. The article was "a cheap shot," one outraged reader wrote, while another declared, "Vidal had just done the impossible! He has once again assassinated Abraham Lincoln!" Vidal, too, had felt "the shock of recognition" communicated by Wilson's reading of the Lyceum Address. He kept hearing the prophesy of a new Caesar as he wrote the book. It was this Lincoln of "the family of lion," driven to surpass the fame of the fathers and rear his own pantheon, Vidal sought to portray. "Nothing that Sandburg ever invented was equal to Lincoln's invention of himself and, in the process, us."[28]

Although all the action in the novel revolves around Lincoln, he is never himself an actor in it. Nor does Vidal ever enter his head. The President—the novel begins with Lincoln's inauguration and ends with his death—is seen in the multiple perspective of those around him. The narrative is carried mainly through Hay. The Lincoln thus portrayed is a towering figure—Nicolay and Hay's Tycoon—admired by the author, but neither loved nor liked by him. The lore of anecdote and reminiscence is drawn upon to reveal his amiable and generous side. (Despite his contempt for Sandburg, Vidal pays him the the compliment of borrowing freely from him.) But the dominant impression is of a man smart, cold, calculating, devious, and dictatorial. It was not a portrait pleasing to the fraternity of Lincoln scholars. And in 1984 not many American historians approved the implication that the war was waged mainly to gratify Lincoln's passion for fame. In the last sentence of the book, Vidal has Hay reflecting on the Tycoon's place in history, comparing him to Bismarck, even speculating like Kirstein "that Lincoln, in some mysterious fashion, had willed his own murder as a form of atonement for the great and terrible thing he had done by giving so bloody and absolute a rebirth to his nation."

The main issue between Vidal and the scholars, however, was the old one of the novelist's literary license with history. Certain characters, certain incidents, and virtually all the dialogue were invented. Events were exaggerated for dramatic effect. Vidal claimed that he stuck to "the agreed upon facts" and that the principal characters "did pretty much what I have them saying and doing." How much was "pretty much"? The historian-critics thought that Vidal blurred the line between fact and fiction, intermingling the two so artfully that the average reader could not know where one ended and the other began. Even if the facts were true, the critics said, Vidal's use of them was unfaithful to the larger historical truth of his subject. William Safire, in *Freedom*, which took similar, and even greater liberties with the historical record, acknowledged the problem and tried to meet it, cleansing his conscience, as he said, by compiling an "Underbook" of 133 pages detailing his sources, his inventions, and his departures from fact "to get at truth."[29]

NBC's four-hour dramatization of Vidal's *Lincoln* in 1988 revived the issue between the novelist and the historians. Defending the authority of the book and the film, for which he was not directly responsible, Vidal targeted his criticism on Richard Current, by now the dean of Lincoln scholars, who had blasted the novel in a review in the *Journal of Southern History*. "Current," Vidal wrote, "has fallen victim to the scholar-squirrels' delusion that there is a Final Truth revealed only to the tenured in their footnote maze; in this he is simply naive." The scholar-squirrels seethed with resentment, Vidal continued, because his own hagiography was better than theirs. As to falsifying and distorting the truth, he defended assertions in the novel that Lincoln clung to his belief in Negro inferiority and advocated colonization two years after the Emancipation Proclamation. He scored a direct hit on Current by citing a passage in his 1958 book, *The Lincoln Nobody Knows*, which seemed to take the same view.* Current replied that his statement was carefully hedged, and he cautioned the reader about General Butler's veracity. But Vidal had made his point: the professionals were revising not only Lincoln but themselves to keep up with changing times. The "agreed

---

*"He still clung to his old ideas of postponing final emancipation, compensating slaveholders, and colonizing freedmen. Or so it would appear. As late as March of 1865, if the somewhat dubious Ben Butler is to be believed, Lincoln summoned him to the White House to discuss the feasibility of removing the colored population of the United States." (p. 230).

upon facts," in Vidal's slightly cynical opinion, were not hard and fixed but malleable and mutable, and one man's myth was as good as another's.[30]

Curiously, Vidal returned to the attack two years later in a communication to the *American Historical Review*. This time his quarrel was with that "very high priest" of the Lincoln guild Don E. Fehrenbacher. His brief critical report on "Vidal's Lincoln" to the Gettysburg conference, when the novel had only just appeared, had been published in the conference proceedings, *The Historian's Lincoln*, in 1988. What Fehrenbacher had labeled "errors" Vidal dismissed as "quibbles." Moreover, since everything that he related about Lincoln was from the point of view of one of the six historical characters around him, the so-called errors were theirs, not his. Such an assertion, as Fehrenbacher wrote in reply, exploded Vidal's claim for the historical soundness of his work. "As argument, this can only mean that Vidal disclaims responsibility for anything erroneous he may have put into the minds of his characters." By some strange quirk the author seemed to want to have his cake and eat it at the same time, that is, to indulge his literary imagination and to be historically authentic as well; and that was impossible. In a twist of fate Vidal got the last kick in this encounter. In 1989, out of his deep knowledge of Lincoln, Fehrenbacher edited for the Library of America a sixteen-hundred page selection of his *Speeches and Writings*. When the Library, showing its Madison Avenue colors, decided to abridge the work in paperback for a larger audience, who should be asked to write the Introduction—his name splashed upon the cover—but Gore Vidal![31]

Lincoln, the natural resource, still appeared inexhaustible in the last decade of the twentieth century. His memory had lost some of its clarity, warmth, and power among Americans, but it exerted more appeal than that of any other national saint or hero. The reasons for this influence and the courses through which it flowed have been charted in these pages. There was, first of all, the wonderful story of his rise from poverty and neglect to unsurpassed fame. There was the belief that he represented, in his character as in his politics, the democratic ideal. There was the fact that his greatness lay at the very core of that huge ganglion of American history, the Civil War, which, as William James had said, was "a sacred spiritual possession worth more than all the blood poured out." As everything about the war lived in memory, so did Lincoln. And his assassination in the hour of victory stabbed at the nation's heart. He was a man of mystery and paradox: raw and folksy, yet sturdy and dignified; a laughing friend, a melancholy stranger; "hard as rock and soft as drifting fog," as Sandburg said. There was so much of him, he seemed fated to remain forever unknown.[32]

Lincoln lived, above all, off the spiritual capital of his words. Were everything else about him forgotten, and only his writings together with some of his talk to survive, he would stand revealed as an extraordinary human being, as one who in the terrible crisis of the nation's history re-created the promise of its origins and thereby secured its destiny, as one who understood so well the values of freedom and the fundamentals of democratic government that even if he should cease to instruct Americans he might still instruct and inspire the peoples of other lands. In the aftermath of the collapse of communism, leaders in the states of the former Soviet Union and in eastern Europe, wishing to establish free and democratic governments, seemed eager to learn from the history of the American experiment. They knew Lincoln, as Tolstoy had known him, as a prophet of human freedom and brotherhood. Now they wished to know more, not only of Lincoln but of Jefferson and the founders. In July 1989, a visiting delegation of educators from Poland asked Governor Cuomo to help in building a Polish archive of American writings on democracy. Told that not a single volume of Lincoln's writings translated into Polish existed in their country, the Governor undertook, with the assistance of Harold Holzer, the "Lincoln on Democracy" project. It led to a book with that title, published simultaneously in English and Polish in 1990. The four-hundred page book consisting of selections of Lincoln's writings organized chronologically, with commentary by seven Lincoln scholars, was an instant success in both countries, and seemed likely to be translated into other languages.

Indicative of Lincoln's eminence as a man of letters was his inclusion, like Jefferson before him, in the aforementioned Library of America. Everybody agreed that the sixteenth President was the preeminent master of the English language ever to occupy the White House, unless one excepted Jefferson; moreover, like Jefferson, but unlike modern Presidents, he wrote his own speeches. In a charming essay, "How Lincoln Won the War with Metaphors," James McPherson contrasted his rich figurative language with Jefferson Davis's wooden and platitudinous prose, which called to mind David Potter's observation that if the two presidents had changed places the Confederacy might have won the war.

The gem of Lincoln's literary genius, of course, was the Gettysburg Address. It was "the Sermon on the Mount of politics" in *Safire's Political Dictionary*, the ultimate text of *The American Testament* as expounded by Mortimer J. Adler and William Gorman, and in Robert Lowell's words "a symbolic and sacramental act." In *Lincoln at Gettysburg* (1992) Garry Wills offered the finest of all appreciations of the address as literature, as politics, and as sacrament. In 272 words Lincoln completed the work of armies and cannon; he "revolutionized the Revolution" by giving the nation a new birth of freedom in the name of the fundamental proposition of the Declaration of Independence. The book, in this sense, put the capstone on the work of the generation of Lincoln scholars that overturned revisionism and restored the iron nerve of ideology to Lincoln's thought and politics. More than that, it rejected the psychobiographers with their labored explanations of Lincoln's quest for union or immortality, and it rebuked the conservatives who contended that equality was no part of the American ideal. By Wills's accounting, Jefferson, not Washington, was Lincoln's imaginary father.[33]

"Fellow-citizens, *we* cannot escape history," the President intoned at the close of his annual message to Congress in 1862. "We of this Congress and this administration will be remembered in spite of ourselves. No personal significance, or insignificance, can spare one or the other of us. The fiery trial through which we pass, will light us down, in honor or dishonor, to the latest generation." These words, indeed the entire peroration, are among the most eloquent he ever uttered. They show a profound sense of his own responsibility as an actor on the stage of history. They show as well, as Robert Bruce has suggested, that Lincoln's quest for immortality had nothing to do with deep psychological compulsions, nor had it anything to do with the afterlife of Christian believers; rather it expressed the wish for honorable remembrance in the mind of posterity. If so, the wish was lovingly gratified through all the generations of American memory.[34]

NOTES

1 *Apotheosis*

1. Moorfield Story, "Dickens, Stanton, Sumner, and Story," *Atlantic Monthly*, v. 145 (1930), 463–65; Charles A. Leale, "Lincoln's Last Hours," *Harper's Weekly*, v. 53 (1909), 7–10, 27; James Tanner, "At the Deathbed of Lincoln," *American Historical Review*, v. 29 (1924), 514–17; Maunsell B. Field, *Memories of Many Men* (London, 1874), 523–26. The traditional version of Stanton's statement is given. Some heard it as "He belongs to the ages now," and his stenographer thought he said, "He belongs to the angels now." See Dorothy M. Kunhardt and Philip B. Kunhardt, Jr., *Twenty Days* (New York, 1965), 80–81.

2. L. A. Gobright, *Recollection of Men and Things at Washington* (Philadelphia, 1869), Ch. 18; Louis L. Snyder and Richard B. Morris, eds., *A Treasury of Great Reporting* (New York, 1962), 151–53; Lawrence Greene, *America Goes to Press* (Freeport, N.Y., 1970), 179; Richard D. Brown, *Knowledge Is Power* (New York, 1989), 258–65.

3. George Haven Putnam, *Memories of My Youth, 1844–1865* (New York, 1914), 431; *The Civil War Notebook of Daniel Chisholm*, W. Springer Menge and J. A. Shimrok, eds. (New York, 1989), 81–82; Mary Logan, *Reminiscences of the Civil War and Reconstruction* (Carbondale, Ill., 1970), 122.

4. *A Philadelphia Perspective: The Diary of Sidney George Fisher, 1834–1871*, Nicholas B. Wainwright, ed. (Philadelphia, 1967), 492; Jane Addams, *Twenty Years at Hull House* (New York, 1912), 23; Samuel Gompers, *Seventy Years of Life and Labor* (New York, 1925), I, 27, 517; Katharine Lee Bates, *America the Dream* (New York, 1930), 107–9; Elbert Hubbard, *Little Journeys to the Homes of American Statesmen* (New York, 1898), 432; Sherman Grant Bonney, *Calvin Fairbanks Bonney, Harriet Cheney Bonney: A Tribute* (Concord, N.H., 1930), 117.

5. *Chicago Tribune*, April 17, 1865; *Illinois Journal*, May 16, 1865; Edmund C. Stedman verse, in *The Lincoln Memorial: A Record . . .* , (New York, 1865), 54; *Confederate Veteran*, v. 6 (1898), 77.

6. George Templeton Strong, *Diary*, Allan Nevins and M. H. Thomas, eds. (New York, 1952), III, 383–84; *New York Times*, April 16, 1865; James A. Garfield, *Works* (New York, 1882), I, 383–84.

7. Chauncy M. Depew, *My Memories of Eighty Years* (New York, 1922),

108; *Autobiography of Oliver Otis Howard* (New York, 1908), II, 325; John Eaton, *Grant, Lincoln, and the Freedmen* (New York, 1907), 234–35.

8. *Lincolniana: In Memoriam* (Boston, 1865), 59, 207–9; E. E. Adams, in *Philadelphia Evening Bulletin*, April 17, 1865. Lloyd Lewis, *Myths After Lincoln* (New York, 1929), Ch. 8, emphasizes the revengeful motif almost to the exclusion of the peaceful one.

9. [Charles G. Halpine], *Baked Meats of the Funeral* (New York, 1866), 100; *Boston Daily Advertiser*, April 17, 1865; Shea, *Lincoln Memorial*, 98; *Philadelphia Evening Bulletin*, April 17, 1865; *Lincolniana*, 214.

10. The whole is covered in Kunhardt, *Twenty Days*. See also, Victor Searcher, *The Farewell to Lincoln* (Nashville, 1965), and Carl Sandburg, *The War Years* (New York, 1939), IV, Ch. 36.

11. See the principal campaign biographies: John Locke Scripps, *Life of Abraham Lincoln*, Roy P. Basler and L. A. Dunlop, eds. (Bloomington, 1961); William Dean Howells, *Life of Abraham Lincoln* (Bloomington, 1960); Joseph H. Barrett, *Life of Abraham Lincoln* (Cincinnati, 1860).

12. Charles Francis Adams, *Charles Francis Adams* (Boston, 1900), 145–46; John Lothrop Motley, *Correspondence*, George W. Curtis, ed. (New York, 1889), II, 78; John Bigelow, *Retrospections of an Active Life* (New York, 1909), I, 366.

13. Quoted in Isaac N. Arnold, *The History of Abraham Lincoln, and the Overthrow of Slavery* (Chicago, 1866), 274–75. The verse is from A. J. K. Duganne, "The Statue of Lincoln," in *Utterances* (New York, 1865), 28–30.

14. "The President's Policy," *North American Review*, v. 98 (1864), 244. This is also in Lowell's *Political Essays* (Boston, 1899), 179–209.

15. Quoted in William H. Richardson, *The "Makings" of the Lincoln Association of Jersey City* (Jersey City, 1919); *Congressional Globe*, 38 Cong., 1 Sess., Appendix, 2038.

16. *The Civil War in Song and Story, 1860–1865*, Frank Moore, ed. (n.p., 1865), 382; Noah Brooks, *Washington in Lincoln's Time* (New York, 1896), 74; *Independent*, March 9, 1865; *A Cycle of Adams Letters, 1861–65*, W. C. Ford, ed. (Boston, 1920), II, 257; *Living Age*, v. 84 (1865), 426–30, v. 85 (1865), 86–87.

17. The most authoritative account is Admiral David D. Porter's *Incidents and Anecdotes of the Civil War* (New York, 1886), 295–98. This was first published in the *New York Tribune*, January 11, 1885. An important earlier account is "Lincoln in Richmond," *Atlantic Monthly*, v. 15 (1865), 753–55. See also, *Liberator*, April 21, 1865; Daniel Raymond, *Life of Abraham Lincoln* (New York, 1866), 682–83; Donald C. Pfanz, *The Petersburg Campaign* (Lynchburg, 1989), Ch. 9. None of these accounts mentions Lincoln sitting in Davis's chair. For Nasby, see David R. Locke, *The Nasby Letters* (Toledo, 1893), 33; see also the report in the *New York Times*, February 13, 1883. Among authors crediting the legend are Helen Nicolay, *Personal Traits of Abraham Lincoln* (New York, 1912), 224, and Lewis, *Myths*, 38. The latter says Lincoln was "joyous as a boy" when he sat in the chair.

18. *National Intelligencer,* April 16, 1865; Lewis, *Myths,* 41–42; Johnson Brigham, *James Harlan* (Iowa City, 1913), 338.

19. Willard L. King, *Lincoln's Manager: David Davis* (Cambridge, 1960), 227; David C. Mearns, *The Lincoln Papers* (New York, 1948), I, 15; *Chicago Tribune,* April 17, 1865.

20. *New York Times,* April 19, 1865; Merrill D. Peterson, *The Great Triumvirate* (New York, 1987), 487–88.

21. Shea, *Lincoln Memorial,* 126; *National Intelligencer,* April 19, 20, 21, 1865; Lewis, *Myths,* Ch. 11.

22. *National Intelligencer,* April 20, 21, 1865; *New York Tribune,* April 20, 1865; Mary Adams French, in *When I Was a Child,* Edward Wagenknecht, ed. (New York, 1946), 161–66.

23. Alexander V. Allen, *Life and Letters of Phillips Brooks* (London, 1901), II, 13–17; Ralph Waldo Emerson, *Works: Miscellanies* (Boston, 1891), 265–66, and *Journals and Miscellaneous Notebooks,* Ronald A. Bosco and Glen Jackson, eds. (Cambridge, 1982), XIV, 5; *New York Times,* June 6, 1884, reprinting the *Boston Congregationalist.*

24. John Carroll Power, *Abraham Lincoln . . . History and Description of the National Lincoln Monument* (Springfield, 1875), 128; *Baltimore American,* April 21, 1865; *New York Times,* April 22, 23 1865; Lewis, *Myths,* 135.

25. *Philadelphia Evening Bulletin,* April 20, 24, 1865; *Collected Works of Abraham Lincoln,* Roy P. Basler, ed. (New Brunswick, 1953), IV, 240; *New York Times,* April 24, 1965; Phillips Brooks, *Addresses* (New York, 1900), 163–73.

26. Henry Ward Beecher, *Lectures and Orations,* N. D. Hillis, ed. (New York, 1970), IX, 282–83; *New York Times,* April 25, 26, 1865; *New York Tribune,* April 26, 1865; William Cullen Bryant, *Poetical Works* (New York, 1890), 316–17. (The second of four stanzas is quoted.)

27. Lewis, *Myths,* 142, 146; Strong, *Diary,* III, 597; *Chicago Tribune,* May 2, 3, 1865; *New York Times,* May 2, 1865.

28. *Collected Works,* IV, 190; Power, *Lincoln,* 221–24; Jean H. Baker, *Mary Todd Lincoln: A Biography* (New York, 1987), 250–52; *New York Times,* June 16, May 4, 5, 1865; Simpson's hymn is in Kenneth A. Bernard, *Lincoln and the Music of the Civil War* (Caldwell, Idaho, 1966), 311.

29. "A Lincoln Legend," *Survey,* v. 34 (1915), 43; *Current History,* v. 51 (1940), 48, on the clock story, first reported in *New York World,* August 2, 1891; *New York Tribune,* January 25, 1880, extracting *Hartford Courant; New York Times,* April 19, 1865; James L. Vallandingham, *A Life of Clement L. Vallandingham* (Baltimore, 1872), 406; *New York World* editorial, in Shea, *Memorial,* 76–81.

30. Clara Barrus, *The Life and Letters of John Burroughs* (Boston, 1925), I, 100. Whitman wrote several poems in memory of Lincoln. They are grouped together in *Leaves of Grass and Selected Prose,* Sculley Bradley, ed. (New York, 1949).

31. Joshua Chamberlain, *The Passing of the Armies* (New York, 1915), 342; William Howard Brownell, "Abraham Lincoln," *War Lyrics and Other Poems* (Boston, 1866), 111–38. (The poem was first published in the *Atlantic Monthly*, October 1865.)

32. *Boston Daily Advertiser*, June 3, 1865; *New York Times*, June 1, 2, 1865; Holmes's "For the Services in Memory of Abraham Lincoln" is in A. Dallas Williams, ed., *The Praise of Abraham Lincoln* (Indianapolis, 1911), 89; Charles Sumner, *Works* (Boston, 1870–83), IX, 369–428; Sumner to George Bancroft, July 5, 1865, Bancroft Papers, Massachusetts Historical Society.

33. *Living Age*, v. 84 (1865), 426–30, v. 85 (1865), 353–69; George M. Trevelyn, *Life of John Bright* (Boston, 1913), 326; Shea, *Lincoln Memorial*, 279–80. See also, Joseph H. Parks, "Lincoln and Contemporary English Periodicals," *Dalhousie Review*, v. 6 (1926), 297–311.

34. Quoted in Bigelow, *Retrospections*, II, 523; Jean A. A. J. Jusserand, *With Americans of Past and Present Ages* (New York, 1916), 299–300; *Appendix to the Diplomatic Correspondence of 1865: The Assassination of Abraham Lincoln . . . Expressions of Condolence and Sympathy . . .* (Washington, 1866); Andrew C. McLaughlin, in *Journal of Illinois State Historical Society*, v. 17 (1924), 299.

35. For other formulations, see Dixon Wecter, *The Hero in America* (New York, 1941), Ch. 10; L. Gerald Bursey, Ch. 4 of William C. Spragens, ed., *Popular Images of American Presidents* (Westport, Conn., 1988); David Donald, *Lincoln Reconsidered* (New York, 1956), Ch. 8.

36. To Horace Greeley, August 22, 1862, *Collected Works*, V, 388; *North American Review* quoted in Merrill D. Peterson, *The Jefferson Image in the American Mind* (New York, 1960), 218; Bushnell, "Our Obligation to the Dead," in *Democratic Vistas, 1860–1880*, Alan Trachtenberg, ed. (New York, 1970), 46; Joseph T. Thompson, "Abraham Lincoln: His Life and Lessons . . . ," in *Union Pamphlets of the Civil War*, Frank Freidel, ed. (Cambridge, 1967), II, 1172; *Philadelphia Evening Bulletin*, April 15, 1865.

37. Harold Holzer, Gabor Boritt, and Mark E. Neely, Jr., *The Lincoln Image: Abraham Lincoln and the Popular Print* (New York, 1984), 192–205; *New York Times*, April 18, 1865; Charles R. Wilson, *Life of Rutherford B. Hayes* (Boston, 1914), I, 272n.

38. Charles Godfrey Leland, "The Proclamation," first published in *Continental Monthly* in 1862, and included in *Abraham Lincoln and the Abolition of Slavery in the United States* (London, 1879), 129; Douglass quoted in Benjamin Quarles, *Lincoln and the Negro* (New York, 1962), 131; Fred B. Perkins, *The Picture and the Men* (Chicago, 1867), and *Cleveland Leader*, May 1, 1865, on the Carpenter painting.

39. Raymond, *Life of Lincoln*, 637; Roi Ottley, *Black Odyssey* (New York, 1948), 187; *Cleveland Leader*, January 4, 1865; *Nation*, August 24, 1865; *National Anti-Slavery Standard*, May 20, 1865.

40. *National Intelligencer*, April 16, 1865; Rev. Benjamin Watson, in *Lincolniana*, 228; *New York Tribune*, April 19, 1865.

41. *Ibid.*; Lowell, *Works*, VI, 271–72; Boker's Phi Beta Kappa poem is in his *Our Heroic Themes* (Boston, 1865).

42. Lowell, *Works*, X, 22–24; Horace E. Scudder, *James Russell Lowell: A Biography* (Boston, 1901), II, 73; Caroline Ticknor, *Hawthorne and His Publisher* (Boston, 1913), 267–74; Brooks, *Addresses*, 166; *New York Herald*, January 17, 1865.

43. Sumner, *Works*, IX, 410; *New York Times*, January 24, 1865.

44. Josiah G. Holland, *The Life of Abraham Lincoln* (Springfield, Mass., 1866), 58; *Collected Works*, VII, 512; William M. Thayer, *The Pioneer Boy and How He Became President* (Boston, 1864), Preface and *passim*; *Lincoln Lore*, no. 687 (1942); John Cawelti, *Apostles of the Self-made Man* (Chicago, 1965); M. L. Weems, *Life of George Washington* (Philadelphia, 1900), 9.

45. *New York Times*, April 19, 1865; Brooks, *Addresses*, 175.

2 *Shapings in the Postwar Years*

1. Quoted in William H. Herndon and Jesse W. Weik, *Herndon's Lincoln* (Chicago, 1889), III, 517. The oration was frequently reprinted, for instance, *Abraham Lincoln: A Tribute* (New York, 1908). See also, Russel B. Nye, *George Bancroft: Brahmin Rebel* (New York, 1944), 225–34; Ben: Perley Poore, *Perley's Reminiscences* (Philadelphia, 1886), II, 222–23; *New York Times*, February 11, 13, 1866; *Chicago Tribune*, February 24, 1866; *Nation*, February 22, 1866.

2. Parke Godwin, *A Biography of William Cullen Bryant* (New York, 1883), II, 230–31; Norton quoted in *Lincoln and the Gettysburg Address*, Allan Nevins, ed. (Urbana, 1964), 14; David C. Mearns, *The Lincoln Papers* (New York, 1948), Ch. 4.

3. See, in general, Eric Foner, *Reconstruction, 1863–1877* (New York, 1988), and Peyton McCrary, *Abraham Lincoln and Reconstruction: The Louisiana Experiment* (Princeton, 1978).

4. *The Collected Works of Abraham Lincoln*, Roy P. Bassler, ed. (New Brunswick, 1953), VIII, 332–33. I am especially indebted to the exegesis by William Lee Miller, *Lincoln's Second Inaugural: A Study in Political Ethics* (Bloomington, 1980).

5. *Collected Works*, VIII, 399–405. On Louisiana, see McCrary, *Lincoln and Reconstruction*, and La Wanda Cox, *Lincoln and Black Freedom: A Study in Presidential Leadership* (Columbia, S.C., 1981).

6. "George W. Julian's Journal—The Assassination of Lincoln," *Indiana Magazine of History*, v. 11 (1915), 335; Lloyd Lewis, *Myths After Lincoln* (New York, 1929), 72; Hans Trefousse, *The Radical Republicans: Lincoln's Vanguard for Racial Justice* (New York, 1969), 305–8; *Messages and Papers of the Presidents*, James D. Richardson, ed. (Washington, 1897), V, 3505 and *passim*; James G. Blaine, *Twenty Years of Congress* (Norwich, Conn., 1884), II, 10–11; *New York Tribune*, April 19, 1865; Horace Greeley, *Recollections of a Busy Life* (New York, 1868), Ch. 51.

7. *Moses Colt Tyler*, Jessica Tyler Austen, ed. (Garden City, N.Y., 1911),

51; Blaine, *Twenty Years*, II, 123; Carl Schurz, *Speeches, Correspondence, and Political Papers*, Frederic Bancroft, ed. (New York, 1913), I, 383, and *Reminiscences* (New York, 1908), III, 221–24; *Nation*, May 8, 1866. See several later concurrent judgments: A. M. Simons, *Social Forces in American History* (New York, 1914), 299; Lloyd Lewis, in *If Lincoln Had Lived*, McKendree L. Raney, ed. (Chicago, 1935), 16–35; David Donald, "The Radicals and Lincoln," in *Lincoln Reconsidered* (New York, 1956); and Trefousse, *Radical Republicans*.

8. John W. Burgess, *Reconstruction and the Constitution, 1866–1876* (New York, 1902), 36–37; *Dictionary of American Biography*, Allen Johnson and Dumas Malone, eds. (New York, 1931), III, 375; *Congressional Globe*, 39 Cong., 1 Sess., 266–67, and 40 Cong., 2 Sess., Appendix, 244; W. M. Robbins quoted, *ibid.*, 43 Cong., 3 Sess., Appendix, 36. See also the discussion of Henry Watterson's views in *New York Tribune*, November 28, 1903, and James G. Randall, *Lincoln and the South* (Baton Rouge, 1946), 148–49.

9. *New York Times*, May 7, July 4, 1865; Alexander H. Stephens, *A Constitutional View of the Late War Between the States* (Philadelphia, 1868), II, 634n; Ulysses S. Grant, *Personal Memoirs* (New York, 1886), II, 522, 511. Hugh McCulloch, *Men and Measures of Half a Century* (New York, 1889), 408, said Lincoln told the story at his last cabinet meeting.

10. Stephens, *Constitutional View*, II, 589–622; Henry Cleveland, *Alexander H. Stephens in Public and Private* (Philadelphia, 1866), 197–200; R. M. T. Hunter, "The Peace Conference of 1865," *Southern Historical Society Papers*, v. 3 (1877), 174–75; John H. Reagan, *Memoirs* (New York, 1901), 172–79; Julian S. Carr, "The Hampton Roads Conference," *Confederate Veteran*, v. 15 (1917), 57–61; Alexander K. McClure, *Recollections of Half a Century* (Salem, Mass., 1902), 14; Henry Watterson, "The Hampton Roads Conference," in *The Compromises of Life* (New York, 1903), 363–69; see also speech by Hugh Gordon Miller, *Congressional Record*, 75 Cong., 1 Sess., Appendix, 374–77.

11. Allan B. Magruder, "Lincoln's Plan of Reconstruction," *Atlantic Monthly*, v. 37 (1876), 21–24; Robert Stites, "Lincoln's Restoration Policy for Virginia," *Magazine of American History*, v. 22 (1889), 209–23; Anna Pierpont Siviter, *Recollections of War and Peace, 1861–1868* (New York, 1938), Ch. 9.

12. *Collected Works*, VII, 243, 101, VIII, 403; Foner, *Reconstruction*, 49; *New York Times*, June 23, 1865; Cox, *Lincoln and Black Freedom*, 117–19; Blaine, *Twenty Years*, II, 123. For the Wadsworth letter: *New York Tribune*, September 26, 1865; *Nation*, October 5, 1865. Lincoln is quoted in Charles R. Williams, *The Life of Rutherford B. Hayes* (Boston, 1914), II, 274–75.

13. See, in general, Michael Davis, *The Image of Lincoln in the South* (Knoxville, 1971), Chs. 2 and 3; and Mark E. Neely, Jr., and others, *The Confederate Image: Prints of the Lost Cause* (Chapel Hill, 1987). Mary Boykin Chesnut, *A Diary from Dixie*, Isabelle D. Martin and M. L. Avary, eds. (Gloucester, Mass., 1961), 19, 270; Frederick S. Daniel, *The Richmond Enquirer During the War* (New York, 1868), 7–13; John H. Hewitt, *King Linkum the First* (Atlanta, 1947).

14. Emma Le Conte, *When the World Ended*, Earl S. Miers, ed. (New York,

1957), 91; John S. Wise, *The End of an Era* (Boston, 1899), 454–55; Ralph S. Steen, "Texas Newspapers and Lincoln," *Southwestern Historical Quarterly*, v. 51 (1948), 201, 203; Thomas R. Turner, *Beware the People Weeping* (Baton Rouge, 1982), 97; John Smith Dye, *History of the Plots and Crimes of the Great Conspiracy . . .* (New York, 1866), 320; *Diary of Edmund Ruffin*, William K. Scarborough, ed. (Baton Rouge, 1989), III, 859, 949; *Selected Letters of Paul Hamilton Hayne*, R. S. Moore, ed. (Baton Rouge, 1982), 93; *Old Guard*, v. 3 (1865), 243, 326, 571, v. 4 (1866), 343, 574, v. 5 (1867), 79, 207–17, 645–50, 844–47; Pomeroy quoted in George Clemenceau, *American Reconstruction* (New York, 1928), 257; E. Merton Coulter, *The South During Reconstruction, 1865–1877* (Baton Rouge, 1947), 179.

15. Lewis F. Bates, quoted in *The Pictorial Book of Anecdotes and Incidents of the War of the Rebellion* (Hartford, 1866), 641–42; McClure, *Recollections*, 244,17; *Jefferson Davis, Constitutionalist: His Letters, Papers, and Speeches*, Dunbar Rowland, ed. (Jackson, Miss., 1923), VII, 2, 513–14; Jefferson Davis, *The Rise and Fall of the Confederate Government* (New York, 1881), II, 683; quoted in Hudson Strode, *Jefferson Davis* (New York, 1964), III, 193–94.

16. "Abraham Lincoln and Jefferson Davis: A Comparison," *A Land We Love*, v. 4 (1868), 391–92; T. C. DeLeon, "The Real Jefferson Davis," *Southern Historical Society Papers*, v. 36 (1908), 80; Varina Howell Davis, *Life of Jefferson Davis* (New York, 1890), I, 132; Davis, *Lincoln in South*, 141.

17. Basil L. Gildersleeve, *The Creed of the Old South* (Baltimore, 1915), 49; Hampton quoted in Thomas J. Pressly, *Americans Interpret Their Civil War* (New York, 1954), 104.

18. Edward A. Pollard, *The Lost Cause: A New Southern History of the War* (New York, 1866), 101–2, 108–12, 713; Richard Weaver, *The Southern Tradition at Bay* (New Rochelle, N.Y., 1968), 114; Davis, *Lincoln in South* 113–17; *Southern Review*, v. 13 (1873), 328; Stephens, *Constitutional View*, I, 28, 444–46, 520, II, 448 (quoted), 35–36, 40, 354–55, 408, 444; Davis, *Rise and Fall*, 297; Cleveland, *Stephens*, 721.

19. Thomas L. Connelly, *The Marble Man: Robert E. Lee and His Image* (New York, 1977), 36; Davis, *Lincoln in South*, 122–23; Bessie L. Pierce, *Public Opinion and the Teaching of History* (New York, 1926), Ch. 6.

20. *Congressional Record*, 45 Cong., 2 Sess., 970–71; Watterson, *Compromises*, 285–87, 289; Wilson quoted in Pressly, *Americans Interpret*, 199; *New York Tribune*, December 23, 1886; Paul M. Gaston, *The New South Creed* (New York, 1970), 87–89.

21. Benjamin Brown French, *Witness to the Young Republic*, Donald B. Cole and John J. McDonough, eds. (Hanover, N.H., 1989), 497, 410–11; *New York Times*, June 16, 1865; Jean H. Baker, *Mary Todd Lincoln: A Biography* (New York, 1987), 250–52; Willard L. King, *David Davis: Lincoln's Manager* (Cambridge, 1960), Ch. 19; Mary Todd Lincoln to Elihu Washburne, November 29, December 9, 15, 1865, Washburne Papers, Library of Congress; *Congressional Globe*, 39 Cong., 1 Sess., 71–72, 107.

22. Baker, *Mary Lincoln*, 275–77; *New York Times*, October 3, 4, 5, 1867;

Chicago Tribune, October 6, 8, 1867; Poore, *Reminiscences*, II, 277–79; French, *Witness*, 497, 615; Elizabeth Keckley, *Behind the Scenes* (New York, 1868), Ch. 15.

23. Keckley, *Behind the Scenes*, 131, 182, and *passim*; Baker, *Mary Lincoln*, 280. On the matter of authorship, see John E. Washington, *They Knew Lincoln* (New York, 1942), 221–36, and *Lincoln Lore*, no. 364 (1936).

24. Mrs. John A. Logan, *Thirty Years in Washington* (Hartford, 1901), 644–55; see also, Mary Clemmer Ames, *Ten Years in Washington* (St. Louis, 1874), Ch. 24. On the pension, see *Congressional Record*, 41 Cong., 2 Sess., 3151–52, 557–60. For a scorching private estimate, see Alexander K. McClure to Alonzo Rothschild, May 9, 1907, Lincoln Museum (Fort Wayne).

25. *National Lincoln Monument Association* (Washington, 1867), and *Organization and Design* (Washington, 1870); *Congressional Record*, 40 Cong., 1 Sess., 345, 385; Henry W. Bellows, "The National Lincoln Monument," *Old and New*, v. 1 (1870), 9–12; *New York Times*, March 12, 1869. See also, typescript, "The Inception and History of the Lincoln Memorial," v. 10, Records of the Lincoln Memorial Commission, National Archives.

26. Papers of the Lincoln National Monument Association, Illinois State Library; John Carroll Power, *Abraham Lincoln . . . with a History and Description of the National Lincoln Monument* (Springfield, 1875); "Recollections of Robert C. Keyes," ca. 1932, typescript, Illinois State Library; *Chicago Tribune*, January 8, 1867.

27. Larkin G. Mead, Jr., to Richard Oglesby, August 9, 1865; John Hay to Oglesby, August 3, 1865; R. E. Fenton to Oglesby, May 25, 1866, all in Lincoln Monument Association Papers.

28. *New York Times*, October 16, 1874; *New York Tribune*, October 15, 16, 1874; F. Lauriston Bullard, *Lincoln in Marble and Bronze* (New Brunswick, 1952), Ch. 6; Lorado Taft, *History of American Sculpture*, new ed. (New York, 1930), 238–41.

29. Bullard, *Marble and Bronze*, Ch. 6; *New York Times*, November 9, 10, 1876, July 12, 19, 1884.

30. Benjamin Quarles, *Lincoln and the Negro* (New York, 1962), Ch. 1; Joseph T. Wilson, *The Black Phalanx: A History of Negro Soldiers* (Hartford, 1888), 510–14.

31. Robert C. Morris, ed., *Freedmen's Schools and Textbooks* (New York, 1980), 25; *Proceedings of Black State Conventions*, 1840–65, Philip S. Foner and George E. Walker, eds. (Philadephia, 1979), I, 203–4, II, 273–74, 302; *National Anti-Slavery Standard*, July 22, 1865; Frank A. Rollin, *Life and Public Services of Martin R. Delany* (New York, 1969), Ch. 24; National Lincoln Monument Association, *Celebration by the Colored People's Educational Monument Association* (Washington, 1865), 26, 31.

32. Harriet Hosmer, *Letters and Memories* (New York, 1912), 282–88; "Easy Chair," *Harper's Monthly Magazine*, v. 34 (1867), 389–90; Wayland Crow to David Davis, June 17, 1868, Lincoln Monument Association Papers.

33. Thomas Ball, *My Three Score Years and Ten: An Autobiography* (Boston,

1891), 249–53; *New York Times*, July 8, 1873; Bullard, *Marble and Bronze*, 67–68; Quarles, *Lincoln*, 5–6.

34. *Ibid.*, 7–11; *Louisville Courier-Journal*, April 15, 1876; *Life and Writings of Frederick Douglass*, Philip Foner, ed. (New York, 1915), IV, 98–99; *New York Times*, April 15, 1876; Taft quoted in Bullard, *Marble and Bronze*, 69.

35. Frederick Douglass, *Life and Times* (New York, 1890), 383–84, 394–95, 396, 401–4; *National Anti-Slavery Standard*, June 10, 1865; *Washington Post*, February 13, 1888; John Eaton, *Grant, Lincoln, and the Freedmen* (New York, 1907), 174–76.

36. Douglass, *Life and Times*, Appendix.

37. Bullard, *Marble and Bronze*, Ch. 1; Truman Bartlett, "The Physiognomy of Lincoln," *McClure's Magazine*, v. 29 (1907), 311–467; Piatt quoted in Charles Hamilton and Lloyd Ostendorf, *Lincoln in Photographs* (Norman, Okla., 1963), 61; Rufus R. Wilson, *Lincoln in Portraiture* (New York, 1935), 111–12; and see letter of Thomas B. Jones, *New York Times*, February 10, 1957.

38. Mrs. Lincoln quoted in Hamilton and Ostendorf, *Lincoln in Photographs*, 139; John G. Nicolay, "Lincoln's Personal Appearance," *Century*, v. 42 (1891), 933; Chambrun, *Impressions*, 100–104; Marchant quoted in Holzer and others, *Lincoln Image*, 108.

39. On Healy, see Rufus R. Wilson, *Intimate Memories of Lincoln* (Elmira, N.Y., 1945), 306, and William T. Sherman to Isaac N. Arnold, November 28, 1872, Chicago Historical Society; on Carpenter, Fred B. Perkins, *The Picture and the Men* (Chicago, 1867); Holzer and others, *Lincoln Image*, 112–28, 175–77, 211–13; *Congressional Record*, 45 Cong., 2 Sess., 968–71; *New York Times*, February 12, 1929; *Nation*, v. 3 (1866), 455, v. 4 (1867), 16; W. R. Emerson, "Marshall's Portrait of Abraham Lincoln," *Atlantic Monthly*, v. 18 (1866), 643–44; Winfred B. Truesdell, *Engraved and Lithographed Portraits of Abraham Lincoln* (Champlain, N.Y., 1933).

40. Richard L. Hoxie, *Vinnie Ream* (St. Louis, 1908); Bullard, *Marble and Bronze*, Ch. 4; Vinnie Ream Hoxie Papers, Library of Congress; *New York Times*, January 11, 1971; Logan, *Thirty Years*, 93.

41. Homer Saint-Gaudens, ed., *The Reminiscences of Augustus Saint-Gaudens* (London, 1913), I, 352–55; Bullard, *Marble and Bronze*, Ch. 8; *Letters of Richard Watson Gilder*, Rosamund Gilder, ed. (Boston, 1916), 149–50, 480–81; R. W. Gilder, *Complete Poems* (Boston, 1909), 117–18; for Keller, Laurence Hutton, *Talks in a Library* (New York, 1905), 395; Leonard W. Volk, "The Lincoln Life Mask and How I Made It," *Century*, v. 23 (1881), 223–26; *New York Times*, June 21, 1881, June 26, October 22, 23, 1887; *Chicago Tribune*, October 22, 23, 1887; and Donald Charles Durman, *He Belongs to the Ages: The Statues of Abraham Lincoln* (Ann Arbor, 1951), 3–5.

42. *Ibid.*; Gilder, *Letters*, 150; Taft, *History*, 291.

43. *Letters of Franklin K. Lane*, Anne W. Lane and L. H. Wall, eds. (Boston, 1922), 368–69; *Arena*, v. 38 (1907), 427–28; Carl Holliday, *Abraham Lincoln in Heaven* (New York, 1936), 93–94.

44. Joseph H. Barrett, *Life of Abraham Lincoln* (New York, 1865); Henry J.

Raymond, *The Life and Public Services of Abraham Lincoln* (New York, 1865); Francis Brown, *Raymond of the Times* (New York, 1951), 249, 274–75; *New York Times*, December 11, 1865.

45. *New York Evening Post*, April 18, 22, 1865; *Independent*, April 27–August 10, 1865; Francis B. Carpenter to William H. Herndon, December 4, 1866, Herndon-Weik Papers, Library of Congress; *Six Months in the White House* (New York, 1866), 50–52, 114–15, 81, and *passim*.

46. Jay Monaghan, *Lincoln Bibliography, 1839–1939* (Springfield, 1943), I, 217; H. M. Plunkett, *Josiah Gilbert Holland* (New York, 1894), Ch. 6; Holland, *Life of Abraham Lincoln* (Boston, 1866), 58, 30, 83–85, 198–99, and *passim*.

47. Quoted in David Donald, *Lincoln's Herndon* (New York, 1948), 212–13; Holland, *Life*, 236–42, 542, 455, 544.

48. Isaac N. Arnold, *The History of Abraham Lincoln, and the Overthrow of Slavery* (Chicago, 1866), Ch. 3 and *passim*; *New York Times*, October 23, 1867; *Dial*, v. 5 (1885), 261–63; "Abraham Lincoln," *Transactions of the Royal Historical Society*, v. 10 (1882), 313–42.

49. See, in general, Donald, *Herndon*; Benjamin P. Thomas, *Portrait for Posterity* (New Brunswick, 1947), Ch. 1; and Joseph Fort Newton, *Lincoln and Herndon* (Cedar Rapids, Iowa, 1910).

50. To Caroline Dall, May 26, 1865, September 3, 1866, Caroline H. Dall Papers, Massachusetts Historical Society; Horace White to Herndon, May 22, 1865, Herndon-Weik Papers; to Isaac N. Arnold, November 20, 1866, in Emanuel Hertz, ed., *The Hidden Lincoln, from the Papers of William H. Herndon* (New York, 1938), 38.

51. Quoted in Donald, *Herndon*, 184.

52. *Abraham Lincoln Quarterly*, v. 1 (1941), 343–83, 403–41; abridged in Carpenter, *Six Months*, 323–50; substantially repeated in William H. Herndon and Jesse W. Weik, *Herndon's Lincoln*, Ch. 20; Herndon to Richard Oglesby, January 8, 1866, Oglesby Papers, Illinois State Historical Library.

53. *Abraham Lincoln Quarterly*, v. 3 (1943), 178–203.

54. *Lincoln and Ann Rutledge and the Pioneers of New Salem* (Herrin, Ill., 1945); *Chicago Tribune*, November 28, 1866; Donald, *Herndon*, 223; Holland, *Life*, 24; John McNamar to G. W. Niles, May 5, 1866, Herndon-Weik Papers. On the poem, see Douglas L. Wilson, "Abraham Lincoln's Indiana and the Spirit of Mortal," *Indiana Magazine of History*, v. 87 (1991), 155–70.

55. Dall Papers; Herndon to Owen Clark, October 4, 1866, Illinois State Historical Library; *New York Times*, December 16, 1866.

56. "Pioneering," *Atlantic Monthly*, v. 19 (1867), 403–16; "Gath," *New York Tribune*, February 15, 1867; Francis Carpenter to Herndon, December 4, 1866, Herndon to Hickman, December 6, 1866, Herndon-Weik Papers.

57. King, *Davis*, 240; Donald, *Herndon*, 191; Hertz, *Hidden Lincoln*, 68, 82; Newman Hall, *From Liverpool to St. Louis* (London, 1870), Ch. 6; Robert Todd Lincoln to Herndon, December 13, 1866, Herndon-Weik Papers.

58. *Chicago Tribune*, March 6, 7, 1867; James Smith to Herndon, January

24, 1866, Herndon-Weik Papers; *New York Times*, March 9, 10, 17, 1867; Robert Dale Owen to Herndon, July 27, 1867, Herndon-Weik Papers.

59. Herndon to Lamon, February 17, 26, March 17, 1869, in Hertz, *Hidden Lincoln*, 59–61; Donald, *Herndon*, 246–65.

60. Lamon to Herndon, February 18, 1870, Herndon-Weik Papers; Donald, *Herndon*, 256; Black to Lamon, March 8, 1870, Carpenter to Herndon, December 4, 1866, Herndon-Weik Papers; Thomas, *Portrait*, 37.

61. *Ibid.*, 43–45; *Diary of Orville H. Browning*, 1850–81, T. C. Pease and J. G. Randall, eds. (Springfield, 1931), II, 351; Black to Osgood, May 4, 6, 1872, Black to Lamon, May 6, 13, 1872, Lamon to Black, May 15, 1872, and Black to Herndon, January 9, 1873, in Jeremiah S. Black Papers, Library of Congress.

62. Herndon to Black, January 4, 1873, Black Papers; Lamon, *The Life of Abraham Lincoln* (Boston, 1872), 144, 237, 504, and *passim*.

63. "Lamon's Life of Lincoln," *Southern Review*, v. 13 (1873), 328–68, 364. See also the review by W. H. Browne in *Southern Magazine*, v. 11 (1872), 367–74.

64. *Atlantic Monthly*, v. 30 (1872), 364–71; *Scribner's Monthly*, v. 4 (1872), 506–10; Lowell to James A. Manson, May 11, 1881, Lincoln Museum.

65. *Scribner's Monthly*, v. 6 (1873), 333–43; Thomas, *Portrait*, Ch. 3; Black to Herndon, February 5, June 23, 1873, Black Papers; R. Gerald McMurtry, ed., *Lincoln's Religion: The Text of the Addresses Delivered by William H. Herndon and Reverend James A. Reed . . .* (Chicago, 1936); *New York Herald*, December 15, 1873; Herndon to Dall, January 24, 1873, Dall Papers.

3 *Filling Up The Image*

1. *New York Herald*, April 19, 1873.

2. *Ibid.*; *New York Times*, April 22, 1873; *Cleveland Leader*, May 17, 1873; *Galaxy*, v. 16 (1873), 518–30, 687–700, 793–804; William E. Smith, *The Franklin Preston Blair Family in Politics* (New York, 1933), II, 466–67.

3. Joseph Fort Newton, *Lincoln and Herndon* (Cedar Rapids, Iowa, 1910), 45; Louis A. Warren quoted by B. A. Botkin in *New York Folklore Quarterly*, v. 14 (1958), 66; Carl Schurz, *Abraham Lincoln: An Essay* (Cambridge, 1891), 31. In general on history and memory, see David Lowenthal, *The Past Is a Foreign Country* (Cambridge, 1985), Michael Kammen, *Mystic Chords of Memory* (New York, 1991), and special issue of *Journal of American History*, v. 17 (1989).

4. Frank Crosby, *Life of Abraham Lincoln* (Philadelphia, 1865); [Charles G. Halpine], *New York Herald*, April 17, 1865; Carl Becker, *Everyman His Own Historian* (New York, 1935).

5. *Independent*, September 1, 1864; Josiah Holland, *Life of Abraham Lincoln* (Boston, 1866), 213–15.

6. Moncure Conway, "Personal Recollections of President Lincoln," *Fortnightly Review*, v. 1 (1865), 56–65, and *Autobiography* (London, 1904), II, 85;

Horace White, *The Life of Lyman Trumbull* (Boston, 1913), 428–29; Gideon Welles, *Diary* (New York, 1960), II 323; Horace Greeley, *Recollections of a Busy Life* (New York, 1868), Ch. 51; Don C. Seitz, *Horace Greeley* (Indianapolis, 1926), Ch. 10; *Century*, v. 42 (1891), 371–82.

7. *The Lincoln Memorial: Album-Immortelles*, Osborn H. Oldroyd, ed. (New York, 1882); William B. Benham, *Life of Osborn H. Oldroyd* (Washington, 1927).

8. *Reminiscences of Abraham Lincoln*, Allen T. Rice, ed. (New York, 1885); *The Every-day Life of Abraham Lincoln*, Francis F. Browne, ed. (New York, 1886); *Abraham Lincoln: Tributes from His Associates*, William Hayes Ward, ed. (New York, 1895).

9. Eleanor Atkinson, *The Boyhood of Lincoln* (New York, 1908), Ch. 1, which is based on an 1889 interview, also quoted in Carl Sandburg, *The Prairie Years* (New York, 1926), I, 16; Crawford quoted in Bess V. Ehrmann, *Missing Chapter in the Life of Abraham Lincoln* (Chicago, 1938), 64; Arnold, *Life of Abraham Lincoln* (Chicago, 1885), 31n; Ida M. Tarbell, *Life of Abraham Lincoln* (New York, 1900), I, 57–58; Della C. Miller, *Abraham Lincoln: A Biographical Trilogy* (Boston, 1965), II, 27; Onstot *Pioneers of Menard and Mason Counties* (Forest City, Ill. 1902), 25, 28; Joseph Gillespie, quoted in *Intimate Memories of Lincoln*, Rufus R. Wilson, ed. (Elmira, N.Y., 1945), 333–36 (this was first published in the *Cincinnati Commercial Gazette* in 1888).

10. Joshua F. Speed, *Reminiscences of Abraham Lincoln* (Louisville, 1896), 10, substantially as in Oldroyd, *Album*, 145–46, and Arnold, *Lincoln*, 53–54; Oglesby, "Origins of the Rail-splitter Legend," in *Intimate Memories of Lincoln*, 192–93; "When Joe Cannon Met Lincoln," *ibid.*, 186; *Reminiscences*, 402; David H. Bates, *Lincoln in the Telegraph Office* (New York, 1907), 157; Rush C. Hawkins memorandum, August 17, 1872, Brown University.

11. John L. Scripps, *Life of Abraham Lincoln*, Roy P. Basler and L. A. Dunlop, eds. (Bloomington, 1961), 147; Henry J. Raymond, *Life and Public Services of Abraham Lincoln* (New York, 1865), 66; Holland, *Life*, 189; Ward Hill Lamon, *Life of Abraham Lincoln* (Boston, 1872), 415; Joseph Medill, in *Chicago Tribune*, May 9, 1895; Don E. Fehrenbacher, *Prelude to Greatness: Lincoln in the 1850s* (Stanford, 1962), Ch. 6; Joseph Logsdon, *Horace White, A Nineteenth-Century Liberal* (Westport, Conn., 1971), 53–55. Regarding Douglas and Lincoln's hat (footnote), see: Holland, *Lincoln*, 278; Arnold, *History of Abraham Lincoln and the Overthrow of Slavery* (Chicago, 1866), 278; Watterson, "*Marse Henry*": *An Autobiography* (New York, 1919), I, 78; and Frank Maloy Anderson, *The Mystery of "A Public Man"* (Minneapolis, 1948), 101–2 and *passim*.

12. Stephen Fiske, in *Lincoln Among His Friends*, Rufus R. Wilson, ed. (Caldwell, Idaho, 1942), Ch. 33.

13. James R. Gilmore, *Personal Recollections of Abraham Lincoln and the Civil War* (London, 1899), Ch. 16; "Peace Mission to Richmond," *Atlantic Monthly*, v. 14 (1864), 372–83; Lucius E. Chittenden, "The Faith of President

Lincoln," *Harper's*, v. 82 (1891), 385–91; John Hays Hammond, *Autobiography* (New York, 1935), II, 707.

14. Conant, "Personal Recollections of Abraham Lincoln," *New York Times*, December 4, 1888, reprinted in Alexander K. McClure, *Lincoln's Yarns and Stories* (Chicago, 1901), 44; Mrs. General Pickett, "President Lincoln: Intimate Personal Recollections," *Lippincott's*, v. 77 (1906), 560; Freeman to F. Lauriston Bullard, July 12, 1938, and Nevins to Bullard, October 18, 1932, Bullard Papers, Boston University; *Butler's Book: Autobiography . . .* (Boston, 1892), 903–8, and see Mark Neely, Jr., "Abraham Lincoln and Black Colonization: Benjamin Butler's Spurious Testimony," *Civil War History*, v. 25 (1979), 77–83.

15. Charles Chiniquy, *The Assassination of President Lincoln* (Cleveland, 1886); Burke McCarty, *The Suppressed Truth About the Assassination of Abraham Lincoln* (Philadelphia, 1924); *New York Times*, December 12, 1921, and December 12, 1922; Carl R. Fish, in *American Historical Review*, v. 29 (1924), 723–24.

16. Donn Piatt, *Memories of the Men Who Saved the Union* (New York, 1887), Preface and Ch. 1 (original article in *New York Tribune*, March 22, 1885, and *Reminiscences*, 477–500); Henry Villard, *Memoirs* (London, 1904), I, 96–97; Swisshelm, in Oldroyd, *Album*, 413–15; Charles A. Dana, *Recollections of the Civil War* (New York, 1899), Ch. 12.

17. Noah Brooks, "Personal Recollections of President Lincoln," *Harper's New Monthly Magazine*, v. 31 (1865), 222–30; *Washington in Lincoln's Time* (New York, 1895), 76–80, 241, 296–98, 302.

18. Lamon, *Life*, 329; James L. King, "Lincoln's Skill as a Lawyer," *American Historical Review*, v. 166 (1898), 186–195; Abram Bergen, "When Lincoln Defended Duff Armstrong," in *Lincoln Among Friends*, Ch. 15; Ida Tarbell, *In the Footsteps of Lincoln* (Cambridge, 1923), 320; Edward Eggleston, *The Graysons: A Story of Illinois* (New York, 1887); Armstrong, in *Among Friends*, Ch. 16, and *New York Times*, June 28, 1896.

19. Alexander K. McClure to Ward H. Lamon, July 18, 1891, Lamon Papers, Huntington Library; *New York Times*, July 10, 1891; Charles Hamlin to John G. Nicolay, July 14, 1891, and *passim*, Nicolay Papers, Library of Congress; Hay quoted in Helen Nicolay, *Lincoln's Secretary: A Biography of John G. Nicolay* (New York, 1949), 325; H. Draper Hunt, *Hannibal Hamlin of Maine* (Syracuse, 1969), 188–89.

20. "The Final Estimate of Lincoln," *New York Times*, February 12, 1898.

21. Don C. Seitz, *Artemus Ward* (New York, 1919), 113–14.

22. Keith W. Jennison, *The Humorous Mr. Lincoln* (New York, 1965), Foreword; P. M. Zall, *Abe Lincoln Laughing* (Berkeley, 1983), Introduction; Constance Rourke, *American Humor* (New York, 1931), 152–55; Charles Sumner, *Works* (Boston, 1874), IX, 411. See also, Walter Blair, *Horse Sense in American Humor* (New York, 1962), Ch. 6.

23. *New York Times*, April 18, 1881.

24. See the hint in Kenneth Lynn, *Mark Twain and Southwest Humor* (New York, 1972), 263n.

25. Davis quoted in Lamon, *Life*, 479; Francis Grierson, "Abraham Lincoln," in *The Humour of the Underman and Other Essays* (London, 1911); Louis A. Warren, "The Mystery of Lincoln's Melancholy," *Indiana History Bulletin*, v. 3 (1925), 53–60.

26. Greeley, *Recollections*, 405; Douglass in *New York Tribune*, July 5, 1885; Silas W. Burt, "Lincoln on His Own Story-telling," *Century*, v. 73 (1907), 502; William H. Seward, *Autobiography*, F. W. Seward, ed. (New York, 1891), III, 209; Holland, *Life*, 74–75; Depew, in Rice, *Reminiscences*, 428.

27. Albert D. Richardson, *The Secret Service . . .* (Hartford, 1865), 314–15; *New York Tribune*, March 5, 1893; Horace Porter, "Lincoln and Grant," *Century*, v. 30 (1885), 940, and *Scribner's Monthly*, v. 15 (1878), 885–86; Titian J. Coffee, in Rice, *Reminiscences*, 235–36; Alexander H. Stephens, *A Constitutional View of the Late War Between the States* (Philadelphia, 1868), II, 615n.

28. Quoted in Dorothy M. Kunhardt and Philip B. Kunhardt, Jr., *Twenty Days* (New York, 1965), 63; Piatt, in *New York Tribune*, March 22, 1885; Villard, *Memoirs*, I, 143; and see Ellery Sedgwick, *The Happy Profession* (Boston, 1946), 165; Brooks, "Recollections," 228; Conwell, *Lincoln Laughing*, 122, and George W. Julian, in Rice, *Reminiscences*, 541.

29. *New York Times*, January 14, 1866; *Collected Works of Abraham Lincoln*, Roy P. Basler, ed. (New Brunswick, 1953), III, 279, VII, 384; Hay's diary, December 23, 1863, quoted in William R. Thayer, *The Life and Letters of John Hay* (Boston, 1915), I, 209; McClure, *Lincoln's Own Yarns*, 124, and John Bartlett, *Familiar Quotations*, 11th ed. (New York, 1938), 457.

30. Raymond, *Life*, 663n; Theodore L. Cuyler, *Recollections of a Long Life* (New York, 1902), 154; Locke, in Rice, *Reminiscences*, 442; Carl Schurz, *Reminiscences* (New York, 1908), II 340–41; Paul Selby, *Lincoln's Life, Stories, and Speeches* (Chicago, 1902), 199, and John Eaton, *Grant, Lincoln, and the Freedmen* (New York, 1907), 90; Zall, *Lincoln Laughing*, 41, 22; Charles H. Zane, "Lincoln as I Knew Him," *Journal of Illinois State Historical Society* v. 14 (1921), 76.

31. Roy P. Basler, *The Lincoln Legend* (Boston, 1935), 125; Frank Crosby, *Life of Abraham Lincoln* (Philadelphia, 1865), 389; Herman Melville, "The Martyr," in *Battle Pieces and Aspects of the War* (Gainesville, Fla., 1960), 141; George H. Boker, in *Our Heroic Themes* (Boston, 1865).

32. Francis De Haas Janvier, *The Sleeping Sentinel* (Philadelphia, 1863); James E. Murdoch, ed., *Patriotism in Poetry and Prose* (Philadelphia, 1866), 103–7; Chittenden, *Recollections*, Ch. 32.

33. Hall quoted in Holland, *Life*, 432; Mary A. Livermore, *My Story of the War* (Hartford, 1889), 557; Eaton, *Grant*, 91, 180.

34. G. Wayne Smith, *Nathan Goff, Jr. : A Biography* (Charleston, W. Va., 1959), 48–49; *Chicago Tribune*, February 7, 1909, from the book by Thomas H. Livermore; John M. Bullock, "President Lincoln's Visiting Card," *Century*, v. 55 (1898), 565–71; Anna Pierpont Siviter, *Recollections of War and Peace* (New York, 1938), 295–303.

35. Ward Hill Lamon, *Recollections of Abraham Lincoln*, 2d ed. (Washington, 1911), 86; Dana, *Recollections*, Ch. 14.

36. J. T. Doris, "President Lincoln's Clemency," *Journal of Illinois State Historical Society* v. 20 (1928), 553; *New York Times*, February 9, 1880; Livermore, *My Story*, 568; *Reminiscences*, 583.

37. Cordelia A. P. Harvey, "A Wisconsin Woman's Picture of President Lincoln," *Wisconsin Magazine of History*, v. 1 (1918), 233–55. This reproduces a lecture Harvey gave after the war; portions of it were used in Holland's *Life*, 443–53, on which Mrs. Lincoln commented in a letter to Harvey, December 4, 1865, University of Chicago.

38. To Erastus Corning and others, June 12, 1863, *Collected Works*, VI, 266; Harold Hyman and William Wiecek, *Equal Justice Under Law*, 1835–1875 (New York, 1982), 233–37, 265; John A. Marshall, *American Bastille* (Philadelphia, 1869), 528 and *passim*.

39. [Daniel B. Lucas], *Memoir of John Yates Beall* (Montreal, 1865); *New York Times*, April 30, 1867; *Confederate Veteran*, v. 7 (1899), 67–69, v. 9 (1901), 3–4, v. 19 (1911), 343; *Christian Observer*, v. 92 (1904), 3; Isaac Markens, *President Lincoln and the Case of John Y. Beall* (New York, 1911); Markens Papers, Virginia Historical Society. Regarding Belle Boyd (footnote), see: Louis L. Sigaud, *Belle Boyd, Confederate Spy* (Richmond, 1944); Ruth Scarborough, *Belle Boyd, Siren of the South* (Mercer, GA., 1983); J. V. Ryals, *Yankee Doodle Dandy* (Richmond, 1891), 467.

40. Ralph Waldo Emerson, "Abraham Lincoln," *Works: Miscellany* (Boston, 1891), 258.

41. *The President's Words*, Edward Everett Hale, ed. (Boston, 1865); John Earle, *English Prose* (New York, 1891), 477–80; Daniel K. Dodge, *Abraham Lincoln: The Evolution of His Literary Style*, in *University of Illinois Studies*, v. 1 (Urbana, 1900); *The Words of Abraham Lincoln for Use in the Schools*, Isaac Thomas, ed. (Chicago, 1898).

42. For an overview, see Roy P. Basler, "Lincoln's Development as a Writer," which is the Introduction to his *Abraham Lincoln: His Speeches and Writings* (Cleveland, 1946), 1–49; *New York Times*, January 2, 1885; Shelby M. Cullom, *Fifty Years of Public Service* (Chicago, 1911), 88; Holland, *Life*, 81; Robert G. Ingersoll, *Works* (New York, 1915), III, 129, and *Washington Post*, January 15, 1883; Walter B. Stevens, *A Reporter's Lincoln* (St. Louis, 1916), 62.

43. John G. Nicolay, "Lincoln's Literary Experiments," *Century*, v. 57 (1894), 823–32; Joseph Jefferson, *Autobiography* (New York, 1890), 28–30; John Hay, "Lincoln in the White House," *Century*, v. 41 (1890), 36; *Richard II*, Act III, Scene 1.

44. Joseph H. Barrett, *Life of Abraham Lincoln* (New York, 1888), 838–39; Milton Hay, in *Intimate Memories*, 48; Andrew Carnegie, *Miscellaneous Writings*, Burton Hendrick, ed. (Garden City, N.Y., 1933), I, 181; Hamilton Wright Mabie, "Lincoln the Literary Man," *Outlook*, v. 58 (1898), 327; Helen Nicolay, *Personal Traits of Abraham Lincoln* (New York, 1912), 367.

45. *New York Times*, May 16, 1885; Caroline Dall, "Pioneering," *Atlantic Monthly*, v. 19 (1867), 413; Brooks, *Washington*, 79–80, 297–98.

46. Douglass, in *Reminiscences*, 194–95; Richard Hofstadter, *The American Political Tradition* (New York, 1948), 131.

47. Hamilton Wright Mabie, "The Education of Lincoln," *Outlook*, v. 76 (1904), 456; Daniel K. Dodge, *Abraham Lincoln: Master of Words* (New York, 1924), 65 and *passim*; Richard Watson Gilder, *Lincoln the Leader and Lincoln's Genius of Expression* (Boston, 1909), 91 and *passim*; Gustave Koerner, *Memoirs*, Thomas J. McCormack, ed. (Cedar Rapids, Iowa, 1909), II, 33.

48. See James R. Perry, "The Poetry of Lincoln," *North American Review*, v. 193 (1911), 213–14. Regarding the poem quoted in the footnote, see: Lydia Avery Coonley, "Lincoln's Wish," *New England Magazine*, n.s., v. 20 (1899), 95; *Herndon's Lincoln*, 425.

49. John G. Nicolay, "Lincoln's Gettysburg Address," *Century*, v. 57 (1894), 596–608; Henry S. Burrage, *Gettysburg and Lincoln* (New York, 1906); *Boston Daily Advertiser*, November 23, 1863; *New York Tribune*, August 13, 1893; Isaac Markens, "Lincoln Masterpiece," in *Abraham Lincoln: The Tribute of the Synagogue*, Emanuel Hertz, ed. (New York, 1927), 424–36; Lamon, *Recollections*, 174–75; *Collected Works*, VII, 17–23.

50. Arnold, *Lincoln and the Overthrow*, 423; Poore, in Rice, *Reminiscence*, 228; Usher, in *New York Tribune*, May 7, 1882.

51. "The Perfect Tribute," *Scribner's*, v. 40 (1906), 17–24; Mary R. Shipman Andrews to Mrs. Bigelow, n.d., Lincoln Museum; Louis A. Warren, *Lincoln's Gettysburg Declaration: "A New Birth of Freedom"* (Fort Wayne, 1964), xviii.

52. Sumner, *Works*, IX, 405; Bayard Taylor, *Poetical Works* (Boston, 1884), 219; Richard Watson Gilder, *Complete Poems* (Boston, 1908), 163.

53. David C. Mearns, *The Lincoln Papers* (New York, 1948), 71; Emerson, "Lincoln," 259; Nicolay and Hay, *Abraham Lincoln: A History* (New York, 1890), I, x. On the *History*, see Benjamin P. Thomas, *Portrait for Posterity*, Ch. 4, which is excellent but for the misleading label "romantic" applied to the work.

54. Nicholay, *Lincoln's Secretary, passim*; Thayer, *Hay*, I, *passim*.

55. Nicolay, *Lincoln's Secretary*, 277, 268; John Hay, *Abraham Lincoln* (n.p., n.d.), which is a facsimile of Hay's letter to Herndon, September 5, 1866; John S. Goff, *Robert Todd Lincoln: A Man in His Own Right* (Norman, 1969), 179–80.

56. Hay to Charles Francis Adams, December 19, 1903, in Tyler Dennett, *John Hay* (New York, 1934), 137; Nicolay to Isaac Arnold, September 17, 1874, Nicolay Papers.

57. R. W. Gilder to Nicolay, March 7, 1882, Nicolay to Gilder, March 21, 1882, Roswell Smith to Hay, March 19, 1885, Nicolay Papers; L. Frank Tooker, *The Joys and Tribulations of an Editor* (New York, 1924), 305–6; Gilder to Edmund Gosse, November 2, 1885, in *Letters of Richard Watson Guilder*, Rosamund Gilder, ed. (Boston, 1916), 174–76.

58. Tooker, *Joys*, 305–6; Nicolay, *Lincoln's Secretary*, 297–99; R. W. Gilder to Nicolay, January 18, 1887, Nicolay Papers; Hay to Robert Todd Lincoln, January 27, 1884, in Hay, *Letters and Extracts from Diary* (New York, 1969), II, 87; Thomas, *Portrait*, 110–12.

59. In addition to *History*, I, 189–90, 202, 104–5, and *passim*, see Clarence King, "The Biographers of Lincoln," *Century*, v. 32 (1886), 861–69.

60. *History*, II, 381, III 247–52, 207–8, and *passim*; Nicolay, *Lincoln's Secretary*, 55; to James T. Hale, January 11, 1861, *Collected Works*, IV, 172.

61. *History*, III, 374, 371, 319–23, 446–49; R. T. Lincoln to Nicolay, December 16, 1873, July 10, 1874, Nicolay to Lincoln, July 17, 1874, Nicolay Papers; Nicolay, *Lincoln's Secretary*, 275–78; *History*, IV, 269–76, VI, 254–71.

62. *History*, IV, 257, 45, 445–46, 468–69; Thayer, *Hay*, II, 29; *Lincoln and the Civil War Diaries and Letters of John Hay*, Tyler Dennett, ed. (New York, 1939), 34–36; *History*, V, 101, 357, VI, 25–27; George B. McClellan, "The Peninsular Campaign," *Century*, v. 30 (1885), 136–50; *McClellan's Own Story* (New York, 1887), 154 and *passim*; Comte de Paris, *History of the Civil War* (New York, 1875–88), I, 573–74, II, 557–58; George T. Curtis, *Life and Services of McClellan* (Philadelphia, 1886), 85–95.

63. *History*, IV, 423–25, VI, 120–23, Ch. 18; to Albert G. Hodges, April 4, 1864.

64. *Collected Works*, VII, 282–83.

65. *History*, VIII, 316–17, IX, 250–51; *Lincoln and the Civil War*, 237–38; *Collected Works*, VII, 514–15.

66. Thayer, *Hay*, I, 48–49; Hay quoted in Thomas, *Portrait*, 127.

67. For Schurz, *Atlantic Monthly*, v. 67 (1891), 721–50; [Jacob D. Cox], in *Nation*, v. 52 (1891), 13–14, 35–36; Bancroft, in *Political Science Quarterly*, v. 5 (1891), 711–16; "Editor's Study," *Harper's*, v. 82 (1891), 478–82.

68. Quoted in *The Hidden Lincoln, from the Papers of William H. Herndon*, Emanuel Hertz, ed. (New York, 1938), 15, 147–57; Thomas, *Portrait*, 123.

69. David Donald, *Lincoln's Herndon* (New York, 1948), 287–88; *A Card and Correction* (1882), Handbill (1882), and "A Statement," January 8, 1886, in Herndon-Weik Papers, Library of Congress; Jean H. Baker, *Mary Todd Lincoln: A Biography* (New York, 1987), 323–26.

70. Donald, *Herndon*, Ch. 14; *Hidden Lincoln*, 91–140 *passim*, especially 105, 124, 134, 135; Herndon to Weik, October 9 and November 11, 1886, Herndon-Weik Papers.

71. *Ibid.*, September 10 and October 27, 1887; Donald, *Herndon*, 320–21; *Herndon's Life of Lincoln* (New York, 1983), 2–7. All citations are to this reprint edition, with Introduction and notes by Paul M. Angle. The first edition was published in three volumes by Belford, Clarke & Company, Chicago, in 1889; a second edition, corrected and revised, with important additions, was published in two volumes by Appleton, New York, in 1892.

72. Donald, *Herndon*, 368; *Herndon's Lincoln*, 55, 47–48, 86, 94.

73. *Ibid.*, Chs. 6–7, pp. 107, 389, 181–82, 350; McClure, *Anecdotes*, 20–22, 234–36.

74. *Herndon's Lincoln*, 252, 282–83, 304, 325–26; Donald, *Herndon*, 119n.

75. *Herndon's Lincoln*, 344, 248, 249, 331.

76. Donald, *Herndon*, Ch. 21; quoted in Thomas, *Portrait*, 150–51; Herndon to Weik, December 5, 1890; [Jacob Cox], in *Nation*, v. 49 (1889), 173–74; *Atlantic Monthly*, v. 64 (1889), 711–14.

77. Donald, *Herndon*, Ch. 21; White to Weik, February 11, 21, March 12, September 30, 1890, October 16, November 2, 1891, January 10, 1892, Herndon-Weik Papers.

78. Donald, *Herndon*, 363.

79. Greeley in *Century*, v. 42 (1891), 371–82; John T. Morse, Jr., *Abraham Lincoln* (Boston, 1893), 1, 31–33; Schurz, *Abraham Lincoln: An Essay* (Boston, 1891), 92.

80. William M. Thayer, *The Pioneer Boy and How He Became President* (Boston, 1864), Preface, Ch. 5; the revision was *From Pioneer Home to the White House* (New York, 1882); *Lincoln Lore*, no. 687 (1942).

81. Z. A. Mudge, *The Forest Boy* (New York, 1867), 201–2; Horatio Alger, Jr., *Abraham Lincoln, the Backwoods Boy* (New York, 1883), 65, 107, 98; Hezekiah Butterworth, *In the Boyhood of Lincoln* (New York, 1892), 57; Noah Brooks, *Abraham Lincoln* (New York, 1888), 53; Brooks, "A Boy in the White House," *St. Nicholas*, v. 10 (1882), 57–65; Bayard Taylor, *Ballad of Abraham Lincoln* (Boston, 1870).

82. Walt Whitman, *Prose Works*, Floyd Stovall, ed. (New York, 1963), II, 509; Joel Chandler Harris, *The Kidnapping of President Lincoln and Other War Detective Stories* (New York, 1909).

83. Colfax to John G. Nicolay, July 17, 1875, Nicolay Papers; A. Y. Moore, *Life of Schuyler Colfax* (Philadelphia, 1868), Appendix; Donald, *Herndon*, 294; Robert Ingersoll, *Abraham Lincoln* (New York, 1894), also in *Works*, III, 123–73; *New York Times*, February 14, 1888, *Washington Post*, January 15, 1883; Henry Watterson, *The Compromises of Life* (New York, 1903); "Abraham Lincoln," *Cosmopolitan*, v. 46 (1909), 363–75; *The Editorials of Henry Watterson*, Arthur Krock, ed. (New York, 1923), *passim*; Joseph F. Wall, *Henry Watterson* (New York, 1956), 220–21, 293; *Chicago Tribune*, February 13, 1895.

84. Horace Traubel, *With Walt Whitman in Camden* (Boston, 1906), I, 38; *Reminiscences*, Ch. 27; William E. Barton, *Abraham Lincoln and Walt Whitman* (Indianapolis, 1928), 91–92, 96, 105; *The Correspondence of Walt Whitman*, Edward H. Miller, ed. (New York, 1964), III, 108, 149, 178; "Death of President Lincoln," *Prose Works*, II, 495–509; Walter Lowenfels, *Walt Whitman's Civil War* (New York, 1961), 266; *New York Times*, April 18, 1881, April 15, 1887; Martí quoted in *Walt Whitman Abroad*, Gay Wilson Allen, ed. (Syracuse, 1955), 203.

85. Edmund C. Stedman, "Walt Whitman," *Scribner's Monthly*, V. 21 (1880), 63; Oppenheim, in William W. Betts, Jr., *Lincoln and the Poets: An Anthology* (Pittsburgh, 1965), 123–27.

4 To The Afterwar Generation

1. "Lincoln in 1854," *Putnam's Monthly*, v. 5 (1909), 729; Robinson, *Collected Poems* (New York, 1922), 317–19; Fletcher, "Lincoln" (1916), in *Breakers and Granite* (New York, 1921), 158–63, and Amy Lowell, *Tendencies in Modern American Poetry* (Boston, 1917), 327. On Lincoln as a twentieth-century symbol, see *Dial*, v. 46 (1909), 101–3, and Ralph H. Gabriel, *The Course of American Democratic Thought*, 2d ed. (New York, 1956), Ch. 30.

2. William Ellery Leonard, *The Vaunt of Man and Other Poems* (New York, 1915), 115–20; Communication from Francis N. Thorpe, *Century*, v. 50 (1895), 476.

3. George R. Stewart, *Names on the Land* (New York, 1945), 299, 311, 320; William H. Richardson, *The "Makings" of the Lincoln Association of Jersey City* (Jersey City, 1919); *New York Tribune*, February 13, 1891; *Cleveland Leader*, February 4, 1875; *New York Times*, February 13, 1896.

4. *American Missionary*, v. 53 (1899), 29; Allison R. Ensor, "The House United: Mark Twain and Henry Watterson Celebrate Lincoln's Birthday, 1901," *South Atlantic Quarterly*, v. 74 (1975), 259–68; *Brooklyn Eagle*, January 28, 1898, January 31, 1900; *New York Times*, February 12, 1901, February 11, 1917; *Springfield Republican*, February 16, 1903. On LMU, see also, Joseph E. Suppinger, *Phoenix of the Mountains: The Story of Lincoln Memorial University* (Harrogate, Tenn., 1977).

5. Richard Oglesby quoted in *Intimate Memories of Abraham Lincoln*, Rufus R. Wilson, ed. (Elmira, N.Y., 1945), 192–93; *Lincoln Lore*, no. 1449 (1958); *New York Tribune*, January 9, 1898.

6. William B. Benham, *Life of Osborn H. Oldroyd* (Washington, 1927), and Kathleen R. Coontz, "A Boy Who Loved Lincoln," *St. Nicholas*, v. 53 (1926), 256–61; Robert Todd Lincoln to Mary E. Brown, June 24, 1921, Robert Todd Lincoln Papers, Illinois State Historical Library; Oldroyd to Jesse W. Fell, April 3, 1882, Fell Papers, University of Illinois; *New York Times*, October 22, 1893, April 26, 1896; *Chicago Tribune*, February 10, 1895.

7. *Lincoln Lore*, no. 932 (1947); David H. Donald, *Lincoln's Herndon* (New York, 1948), 314–15; *New York Tribune*, January 8, 1894, and *New York Times*, December 10, 1894.

8. Andrew C. Zabriski, A *Descriptive Catalogue of the Political and Memorial Medals Struck in Honor of Abraham Lincoln . . .* (New York, 1873); *New York Times*, December 18, 1892.

9. [Charles Henry Hart], *Catalogue of a Collection of Engraved and Other Portraits of Lincoln . . .* (New York, [1899]); *New York Tribune*, April 16, 1899; Truman H. Bartlett, "The Portraits of Lincoln," included in a new edition of Carl Schurz, *Abraham Lincoln: A Biographical Essay* (Boston, 1907); Frederick Hill Meserve, *The Photographs of Abraham Lincoln* (New York, 1911).

10. In general, see *Lincoln Bibliography*, 1839–1939, Jay Monaghan, ed., in *Collections of Illinois State Historical Library*, v. 31–32 (Springfield, 1943), xvii, xxiv, and *passim*; Donald, *Herndon*, 207–12.

11. Daniel Fish, "Lincoln Collectors and Lincoln Bibliography," in *Proceedings and Papers of the Bibliographical Society of America*, v. 3 (1980), 49–64; White, "Lincoln in 1854," 726–27; Albert H. Griffith, "Lincoln Literature, Lincoln Collectors, and Lincoln Collections," *Wisconsin Magazine of History*, v. 15 (1915), 148–67; *Lincoln Bibliography*, xix; *Philadelphia Evening Bulletin*, February 8–9, 1909, and *New York Evening Post*, February 11, 1909; Charles W. McLellan Papers, Brown University. For a recent and comprehensive account, see Ralph G. Newman, *Preserving Lincoln for the Ages: Collectors, Collections, and Our Sixteenth President* (Fort Wayne, 1989).

12. See, in general, Simon Gratz, A *Book About Autographs* (Philadelphia, 1920), Thomas F. Madigan, *Word Shadows of the Great* (New York, 1930), and Mary A. Benjamin, *Autographs: A Key to Collecting* (New York, 1986); George S. Hellman, *Lanes of Memory* (New York, 1927), 47–48; *Collector*, v. 27 (1914), 38–39. A price-annotated copy of *The Library of the Late Major William H. Lambert* (New York, 1914) is in the Lincoln Museum, Fort Wayne.

13. Charles Hamilton, *Collecting Autographs and Manuscripts* (Norman, 1961), 35–40; *New York Tribune*, December 6, 1891, and *Complete Works*, national ed. (New York, 1906), VII, 270; John Kobler, "Yrs. Truly, A. Lincoln," *New Yorker*, v. 32 (1956), 38–39; *Uncollected Letters of Abraham Lincoln*, Gilbert A. Tracy, ed. (Boston, 1917); Worthington C. Ford, in *Proceedings of the Massachusetts Historical Society*, v. 61 (1928), 183–95; Madigan, *Word Shadows*, 72–73.

14. William H. Townsend to Ida Tarbell,February 18, 1933, Tarbell Papers, Allegheny College. On Tarbell generally, see her autobiography, *All in the Day's Work* (New York, 1939), Kathleen Brady, *Ida Tarbell: Portrait of a Muckraker* (New York, 1984), and Benjamin P. Thomas, *Portrait for Posterity* (New Brunswick, 1947), Ch. 7.

15. *Day's Work*, 11.

16. *Ibid.*, 118–22, 146–53; Brady, *Tarbell*, Ch. 6.

17. *Day's Work*, 161–64.

18. *Ibid.*, 165–68; Charles Hamilton and Lloyd Ostendorf, *Lincoln in Photographs* (Norman, 1963), 4–7. The daguerreotype was by N. H. Shepherd, Springfield, 1846.

19. Whitney to Jesse Weik, undated [1895], and April 28, 1896, Herndon-Weik Papers, Library of Congress; *Day's Work*, 165.

20. *Ibid.*, 170, 167; Halstead quoted in *McClure's*, v. 4 (1896), 214; *Lincoln Bibliography*, I, 308; on *McClure's* circulation, see Brady, *Tarbell*, Ch. 6, and Peter Lyon, *Success Story: The Life and Times of S. S. McClure* (New York, 1963), 142, and for Gilder comment, 135; Ray Stannard Baker, *American Chronicle* (New York, 1945), 78.

21. *Day's Work*, 169.

22. *Life of Abraham Lincoln* (New York, 1900), I, 47, 174–80, II, 89–90, 96–97; Whitney to Weik, April 28 and May 23, 1896, Herndon-Weik Papers.

23. *Life of Lincoln*, I, 93–94, 297–99; Joseph Medill to Jesse Weik, May 28, 1896, Herndon-Weik Papers, and to Tarbell, May 15 and June 9, 1896, Tarbell

Papers; "Lost Speech of Lincoln," by Ida M. Tarbell, newspaper clipping, June 14, 1896, Lincoln Museum; Henry C. Whitney, "Lincoln's Lost Speech," *McClure's*, v. 7 (1897), 319–31, and *Life of Lincoln* (New York, 1908), 327–54; Nicolay to R. Johnson, January 6, 1897, Nicolay Papers, Library of Congress; Paul Angle to Ida Tarbell, December 14, 1936, Tarbell Papers.

24. Quoted in Thomas, *Portrait*, 184–85, 199–200; *Day's Work*, 385; *He Knew Lincoln and Other Billy Brown Stories* (New York, 1927), 8–9, and *Life*, I, 246.

25. George S. Boutwell, *Reminiscences of Sixty Years in Public Affairs* (New York, 1902), II, 300; *Republican Campaign Text-book*, 1904 (n.p., 1904), 458–59; Mary R. Dearing, *Veterans in Politics: The Story of the* GAR (Baton Rouge, 1952), Ch. 1.

26. Carl Schurz, *Speeches, Correspondence, and Political Papers*, Frederick Bancroft, ed. (New York, 1913), III, 295, II, 232.

27. *Text-book*, 49; F. W. Taussig, "Abraham Lincoln on the Tariff: A Myth," *Quarterly Journal of Economics*, v. 28 (1914), 814–20, and "Lincoln and the Tariff: A Sequel," *ibid.*, v. 29 (1915), 426–29, and "The Lincoln Tariff Myth Finally Disposed Of," *ibid.*, v. 35 (1921), 500; *Congressional Record*, 51 Cong., 1 Sess., 878–79, 1006; M. A. DeWolfe Howe, *Portrait of an Independent, Moorfield Storey* (Boston, 1932), 193–94, 239; *Springfield* (Mass.) *Republican*, February 13, 1899, January 22, 1900, March 5, 1902.

28. *Congressional Record*, 56 Cong., 2 Sess., Appendix, 215; *Commoner*, February 13, 22, 1902; Ray Ginger, *Altgeld's America: The Lincoln Ideal Versus Changing Realities* (New York, 1958), 1–10; William J. Bryan, *The First Battle* (Chicago, 1896), 316, 376; *New York Times*, September 28, 1896; *Commoner*, February 13, 22, April 5, 1901,February 12, 26, 1904, February 22, 1907, February 12, 1909.

29. *New York Times*, October 3, 1896; Helen Nicolay, *Personal Traits of Abraham Lincoln* (New York, 1912), 380–87; see also one of the principal sources of the quotation, George H. Sibley, *The Money Question* (Chicago, 1896), 282–83. For other examples: *Lincoln's Words on Living Questions*, H. S. Taylor and D. M. Fulwiler, eds. (Chicago, 1900), 132–33; Charles A. Lindbergh, in *Congressional Record*, 62 Cong., 2 Sess., Appendix, 56; E. Jay Jernegan, *Henry Demarest Lloyd* (New York, 1976), 94; and *Lincoln Encyclopedia*, Archer B. Shaw, ed. (New York, 1950), 40; W. J. Ghent, "Lincoln and the Social Problem," *Collier's Weekly*, v. 35 (1905), 23–24, and "Lincoln and Labor," *Independent*, v. 66 (1909), 301–5; see also, Herman Schlueter, *Lincoln, Labor, and Slavery* (New York, 1913), 174–80.

30. James Russell Lowell, *Literary Essays* (Boston, 1890), II, 280, and *Works*, VI, 271–72; John L. Motley to Mary Todd Lincoln, May 1, 1865, Lincoln Museum; Charles A. Dana, *Recollections of the Civil War* (New York, 1899), Ch. 12; Cullom, in *New York Times*, March 22, 1908; Tarbell, in *New York Times Magazine*, February 11, 1917.

31. Alonzo Rothschild, *Lincoln, Master of Men: A Study in Character* (Cambridge, 1906), 3, 283, and *passim*.

32. Arthur O. Dillon, "Abraham Lincoln," in *The Ancestors of . . . and His Poems* ([Pomona, Calif.], 1927); see Robert Crunden, Ministers of Reform: The Progressive Achievement in American Civilization, 1889–1920 (New York, 1982) for the argument that the Progressives were conscious legatees of Lincoln; Rose Strunsky, *Abraham Lincoln* (New York, 1914), and "Abraham Lincoln's Social Ideal," *Century*, v. 87 (1914), 588–92.

33. *Letters of Vachel Lindsay*, Marc Chenetier, ed. (New York, 1979), 432; quoted in Margaret H. Carpenter, *Sara Teasdale: A Biography* (New York, 1960), 297–301; *The Poetry of Vachel Lindsay*, Dennis Camp, ed. (Peoria, 1984), I, 169.

34. *Letters of Theodore Roosevelt*, Elting Morison, ed. (Cambridge, 1951), IV, 1049–50, 1132–33; Nicholas Roosevelt, *Theodore Roosevelt: The Man as I Knew Him* (New York, 1967), 43; J. B. Bishop, *Theodore Roosevelt and His Time Shown in His Own Letters* (New York, 1902), I, 352; quoted in Charles T. White, *Lincoln and Prohibition* (New York, 1921), 111.

35. Theodore Roosevelt, *An Autobiography* (New York, 1913), 395–99; *Letters*, VI, 1527; *Works of Theodore Roosevelt*, XVI, 3; Bishop, *Roosevelt*, II, 1150.

36. *Works*, XVII, 7–10, 66, 121, 361; *New York Times*, February 12, 1911, April 28, 29, 30, 1912; John M. Harlan, in *Congressional Record*, 62 Cong., 2 Sess., Appendix, 356–57; *Nation*, v. 94 (1912), 50–51.

37. *The Papers of Woodrow Wilson*, Arthur Link, ed. (Princeton, 1966–92), VIII, 299–300; *Forum*, v. 19 (1895), 544–59; "A Calendar of Great Americans," *Wilson Papers*, VIII, 375, 378.

38. *Ibid.*, XVIII, 602; XIX, 54, 113; XX, 365, 389.

39. *Ibid.*, XXIII, 466; XXIV, 75, 152–53; XXXVIII, 323–34.

40. *Independent*, v. 65 (1908), 529–34; *Freeman* (Indianapolis), August 22, 1908; and for the southern reaction, see *Atlanta Constitution*, August 15, 16, 1908.

41. James McPherson, *The Abolitionist Legacy, from Reconstruction to the NAACP* (Princeton, 1975), 388–89; Oswald Garrison Villard, *Fighting Years* (New York, 1939), 191–94; W. E. B. Du Bois, *Pamphlets and Leaflets*, Herbert Aptheker, ed. (White Plains, N.Y., 1986), 166.

42. In general, see I. A. Newby, *Jim Crow's Defense: Anti-Negro Thought in America*, 1900–1930 (Baton Rouge, 1965), 74–75; for the former view, see Thomas N. Norwood, *A True Vindication of the South* (Savannah, 1917), Ch. 41, and John A. Price, *The Negro: Past, Present, and Future* (New York, 1907); for the latter view, see the writings of Thomas Dixon, and Edward Eggleston, *The Ultimate Solution of the American Negro Problem* (New York, 1913); *Congressional Record*, 57 Cong., 1 Sess., 626–27; *Nation*, v. 64 (1890), 64, 91–92.

43. See, in general, Raymond A. Cook, *Fire from the Flint: The Amazing Career of Thomas Dixon* (Winston-Salem, 1968).

44. *The Clansman* (New York, 1905), 47.

45. For a typical article, see "Booker T. Washington and the Negro," *Saturday Evening Post*, v. 78 (1905), 1–2; *Norfolk Virginian Pilot*, September 23,

1905; *Richmond Times-Dispatch*, September 23–26, 1905; *Colored American*, v. 9 (1905), 599–603; for Hay and Wilson endorsements, see Cook, *Dixon*, 137, 170; Sutton E. Griggs, *The Hindered Hand; or, The Reign of the Repressionist* (Nashville, 1905), contains a Negro's response; on Griffith and *Birth of a Nation*, see Thomas Cripps, *Slow Fade to Black: The Negroes in American Films, 1900–1942* (New York, 1977), 47; and Richard Schickel, *D. W. Griffith: An American Life* (New York, 1984), Ch. 9.

46. *The Southerner* (New York, 1913), 274, 345; for plot analysis, see Don E. Fehrenbacher, *Lincoln in Text and Context* (Stanford, 1987), 132–33. Dixon later wrote a play, *The Man of the People* (New York, 1920), based on the novel; for a similar view see William P. Pickett, *The Negro Problem: Abraham Lincoln's Solution* (New York, 1909).

47. *The Slave Narratives of Texas*, Ronnie C. Tyler and L. R. Murphy, eds. (Austin, 1974), 133.

48. *The American Slave: A Composite Autobiography*, George P. Rawick, ed. (Westport, Conn., 1972), III, 203, VII, 75, 225–26, IX, 357–58, XIV, 60; John E. Washington, *They Knew Lincoln* (New York, 1942), *passim*; Norman R. Yetman, *Voices from Slavery* (New York, 1970), 73, 304; C. T. Corliss, in *Lincoln Memorial*, Osborn H. Oldroyd, ed. (New York, 1882), 551; Walter H. Brooks, *Impressions at the Tomb of Abraham Lincoln* (Washington, 1926); Carter G. Woodson, *Negro Orators and Their Orations* (Washington, 1925), 541–78; Charles Henry Fowler, *Patriotic Orations* (New York, 1910), 3–111.

49. See, in general, William H. Wiggins, Jr., *O Freedom! Afro-American Emancipation Celebrations* (Knoxville, 1987). For examples of celebrations: *Voice of the Negro* (Atlanta), v. 1 (1904), 73, v. 2 (1905), 127–28; and *Baltimore Afro-American*, January 3, 1903, October 1, 1904, January 7, 1905, September 23, 1905.

50. Earl E. Thorpe, *The Mind of the Negro* (Baton Rouge, 1961), 395–96; Maude K. Griffin, "Lincoln and the Simple Life," *Colored American*, v. 12 (1907), 94–95, and "Lincoln the Man of Many Sides," *ibid.*, v. 14 (1908), 188–89; *Voice of the Negro*, v. 2 (1905), 127–28; *The Upward Path: A Reader for Colored Children*, Mary White Ovington and M. T. Pritchard, eds. (New York, 1920); Washington quoted in *Chicago Tribune*, February 7, 1909.

51. *Up from Slavery* (New York, 1901), 7, 263; Centennial tribute, *Chicago Tribune*, February 7, 1909; *Papers of Booker T. Washington*, Louis R. Harlan, ed. (Urbana, 1974–89), III, 30, 93, IV, 514, V, 36, X, 33–39.

52. *Voice of the Negro*, v. 4 (1907), 242–47; George Washington Williams, *History of the Negro Race in America* (New York,1882), II, 241, 254, 270–71; Du Bois, *Book Reviews*, Herbert Aptheker, ed. (Millwood, N.Y., 1977), 50–51; *Selections from "The Crisis,"* Aptheker, ed.(Millwood, N.Y., 1983), I, 51–52; on Grimké and Lincoln, see *Lincoln Lore*, no. 1681 (1978); Rayford Logan, *Betrayal of the Negro*, new ed. (New York, 1954), 369; *New York Times*, October 15, 1912; Albert E. Pillsbury, *Lincoln and Slavery* (Cambridge, 1913), and Henry W. Wilbur, *President Lincoln's Attitude Towards Slavery and Emancipation* (Philadelphia, 1914).

53. Benjamin O. Flower in *Arena*, v. 41 (1909), 480; *The Crusade for Justice: The Autobiography of Ida B. Wells* (Chicago, 1970), 321; Markham, "The Coming of Lincoln," in *The Priase of Lincoln: An Anthology*, A. Dallas Williams, ed. (Indianapolis, 1911), 38; Editorial, *Colored American* v. 16 (1909), 135. On the democracy-nationality link, see especially Herbert Croly, *The Promise of American Life* (New York, 1909), Ch. 4.

54. George L. Knapp, "Lincoln," *Lippincott's*, v. 83 (1909), 207; Editorial, *Dial*, v. 46 (1909), 101–3; Croly, *Promise*, 89–95.

55. *New York Times*, January 25, 1905; *ibid.*, October 8, 1896; Shelby M. Cullom, *Fifty Years of Public Service* (Chicago, 1911), Ch. 32; Papers of the Lincoln Centennial Association, Illinois State Historical Library; *Chicago Tribune*, February 7, 1909; White, "Lincoln's Rise to Greatness," in *Lincoln Among His Friends*, Rufus R. Wilson, ed. (Caldwell, Idaho, 1942).

56. See, in general, Roy Hays, "Is the Lincoln Birthplace Cabin Authentic?," in *Abraham Lincoln Quarterly* (1948); *Lincoln Lore*, nos. 1016, 1019 (1948). Events associated with the birthplace cabin may be followed in newspapers. On the Goose Neck Prairie cabin, see the letters between Eleanor Gridley and Louis A. Warren, September 30, 1929, April 22, 1932, Lincoln Museum, and Gridley's *The Story of Abraham Lincoln* (n.p., 1900).

57. Jones to Ida M. Tarbell, April 2, 1923, Tarbell Papers; Jones, "The Birthday Farm," *Collier's Weekly*, v. 36 (1906), 12–14.

58. *Ibid.*, and "The Lincoln Farm Association," v. 38 (1907), 14–15; *New York Times*, June 24, 1907, February 22, 1906; *House Reports*, 60 Cong., 1 Sess., no. 1641.

59. *Collier's Weekly*, v. 40 (1908), 5; Roosevelt, *Works*, XI, 210–14; Taft quoted in *New York Times*, November 10, 1911.

60. Gloria Peterson, *An Administrative History of Abraham Lincoln Birthplace National Historic Site* (Washington, 1968); James M. Cox, *Journey Through My Years* (New York, 1946), 438; *Wilson Papers*, XXXVIII, 142; Tarbell, *In the Footsteps of Lincoln* (New York, 1924), 96.

61. Robert P. King, "Lincoln in Numismatics," *Numismatist*, v. 37 (1924), 151; *New York Times*, February 12, 1989; Sandburg quoted in Harry Golden, *Carl Sandburg* (Cleveland, 1961), 246; Randle B. Truett, *Lincoln in Philately* (Washington, 1959); *Congressional Record*, 60 Cong., 2 Sess., 856; *Lincoln Tribute Book . . . with a Centenary Medal*, Horatio S. Krans, ed. (New York, 1909); Montgomery Schuyler, "A Medallic History of Lincoln," *Putnam's Monthly*, v. 5 (1909), 678–81.

62. For an overview, see *Abraham Lincoln: The Tribute of a Century, 1809–1909*, Nathan W. MacChesney, ed. (Chicago, 1909); *Chicago Tribune*, February 7–13, 1909.

63. "The Lincoln Centenary," *Journal of Illinois State Historical Society*, v. 2 (1909), 27–33; *Chicago Tribune*, February 11, 12, 1909; *Colored American*, v. 15 (1909), 135.

64. *New York Times*, January 21, February 7, 15, March 1, 1909; *Boston*

Globe, February 7, 12, 1909; on Stahlberg, see *New York World*, February 7, 13, 1909.

65. *Ibid.*, February 7, 1909; *New York Times*, February 12, 1909; Taft, "Abraham Lincoln," *Cosmopolitan*, v. 46 (1909), 361.

66. Whitlock, *Abraham Lincoln* (Boston, 1909), 118–22; David R. Andrews, *Brand Whitlock* (New York, 1968), 61–63; *Washington Post*, February 7, 1909; *Boston Globe*, February 12, 1909. For a compilation of magazine articles, see William Abbatt's *The Lincoln Centenary in Literature*, 2 v. (New York, 1909).

67. For understanding Grierson's interpretation, see *Abraham Lincoln, the Practical Mystic* (New York, 1918); *Valley of the Shadows* (London, 1913), 198. For an appreciative essay, see *Arena*, v. 41 (1909), 498–508.

68. Percy MacKaye, *Ode on the Centenary of Abraham Lincoln* (New York, 1909); Leonard, *Vault of Man* (New York, 1912), 115–20; Thomson, in Williams, *Praise*, 138–44; and *Collier's Weekly*, v. 40 (1908), 8–9. Other poems by Thomson are "We Talked of Lincoln," *ibid.*, v. 42 (1909), 11; and "When Lincoln Died," *Youth's Companion*, v. 83 (1909), 71. These, too, are collected in Williams's anthology.

69. *Lincoln, and Other Poems* (New York, 1903), 1–3. The last quoted phrase is from John Gould Fletcher's poem in *Breakers and Granite* (New York, 1921), 158.

70. *New York Times*, May 11, 1882, November 16, 1883, March 22, 1893; *New York Tribune*, February 14, 1899; *Chicago Tribune*, February 19, 1899; and Brand Whitlock, *Forty Years of It* (New York, 1914), 59. For Cullom's bill, see *Congressional Record*, 49 Cong., 1 Sess., 570, 2986. For the McMillan Plan, see Senate Report 166, 57 Cong., 1 Sess.; and Paul Caemmerer, "Charles Moore and the Plan of Washington," *Records of Columbia Historical Society*, v. 46 (1944), 237–58. For the 1909 action, see House Report 2106, 60 Cong., 2 Sess.; and Cullom's bill, *Congressional Record*, 61 Cong., 3 Sess., 2086–89.

71. See, in general, Michael Davis, *The Image of Lincoln in the South* (Knoxville, 1971), 166–70; see also, *Baltimore Sun*, February 7, 13, 1909; *Chicago Tribune*, February 7, 1909; *Services in Commemoration of the One Hundredth Anniversary of the Birth of Abraham Lincoln Arranged by Union and Confederate Veterans . . .* (Atlanta, 1909).

72. "Lincoln and the South," *Sewanee Review*, v. 17 (1909), 129–38; F. Lauriston Bullard, *Lincoln in Marble and Bronze* (New Brunswick, 1952), 128–29.

73. Thompson, *Lincoln's Grave* (Cambridge, 1894).

74. *Watson's Jeffersonian Magazine*, v. 3 (1909), 3–12; *Baltimore Sun*, February 7, 1909; George L. Christian, *Abraham Lincoln* (Richmond, 1909); *Confederate Veteran*, v. 17 (1909), 153–54; Minor, *The Real Lincoln* (Richmond, 1907); *Boston Journal*, June 27, 1905; George Edwards [Elizabeth Avery Meriwether], *Facts and Falsehoods Concerning the War on the South, 1861–1865* (Memphis, 1904).

75. In *Lee at Appomattox and Other Papers*, 2d ed. (Boston, 1903), Ch. 6.

See also Adams's address on Lee's centennial, in *Studies Military and Diplomatic* 1775–1865 (New York, 1961), 291–413; *Wilson Papers*, XIX, 33–48. On the Lee cult, see Thomas L. Connelly, *The Marble Man: Robert E. Lee and His Image in American Society* (New York, 1977).

5 *Themes and Variations*

1. Joseph B. Oakleaf, ed., *Lincoln Bibliography* (Cedar Rapids, Iowa, 1925); for Morley's parody, *Saturday Review of Literature*, v. 38 (February 12, 1955), 9–10; Bayne, "Tad Lincoln's Father," *Atlantic Monthly*, v. 133 (1924), 660–65, expanded into a small book in 1931; *New York Times*, October 1–4, 1921.

2. H. L. Mencken, *Prejudices: Third Series* (New York, 1922), 174.

3. Quoted in *Woman's Home Companion*, v. 70 (1943), 71; *St. Nicholas*, v. 53 (1926), 410, reprinted from *Sunshine: Florida's Magazine*. (The magazine piece was expanded into a pamphlet, *He Could Take It*, by Arno B. Reincke, which was abridged in *Reader's Digest*, v. 34 [1939], 1–3.) For the prescription, see *Springfield* (Mass.) *Republican*, December 20, 1903; and the advertisement, *New York Times* (Section F), February 7, 1909. For commercialization of the Lincoln name, see the suggestive discussion by Robert J. S. and Kellie O. Gutman, "Lincoln Dollars and Cents: The Commercialization of Abraham Lincoln," in *Books at Brown*, v. 31–32 (1984–85), 97–113.

4. Judge R. M. Wanamaker provided the epigram for Ervin Chapman, *Latest Light on Abraham Lincoln* (New York, 1917), vii; see, for instance, *Selections from the Writings of Abraham Lincoln*, J. G. DeRoulhac Hamilton, ed. (New York, 1922); Charles Evans Hughes to John Wesley Hill, February 12, 1912, Lincoln Museum; on LMU., see Gladys Parker Williamson, "Living Memorial to Abraham Lincoln," *Missionary Review of the World*, v. 46 (1923), 279–82, Robert L. Kincaid, "The Ageless Lincoln," *Vital Speeches*, v. 16 (1950), 302–5, John Wesley Hill to Louis A. Warren, April 19, 1930, Lincoln Museum, and *New York Times*, April 11–16, 1930; Michael Pupin, *From Immigrant to Inventor* (New York, 1923), and "The Revelation of Lincoln to a Serbian Immigrant," *Abraham Lincoln Association Papers*, v. 3 (1926), 65–80; see the file "Abraham Lincoln Foundation," in Ida M. Tarbell Papers, Allegheny College, and *New York Times*, June 15–16, 29, 1929, February 13, 1930.

5. *The Poetry of Vachel Lindsay*, Dennis Camp, ed. (Peoria, 1984), I, 169; on Wilson, see *New York Times* editorial, September 5, 1916, and Tarbell interview in *Collier's*, included in *The Papers of Woodrow Wilson*, Arthur Link, ed. (Princeton, 1966), XXXVIII, 326; *London Spectator*, September 26, 1914, October 21, 1916, March 9, 1918; *New York Times*, August 19–20, 1917; Choate's speech, New York City, April 23, 1917, included in *Modern Eloquence* (New York, 1928), I, 243–44.

6. *New York Times*, May 28, 1917, December 21, 1916, February 11–13, 1917; J. Hugh Edwards, *David Lloyd George* (New York, 1929), I, 48–49; *New York Times*, December 1, 1922, October 7, 19, 22, 1923; Lloyd George's speech in *Journal of Illinois State Historical Society*, v. 17 (1924), 241–45.

7. Lodge, *The Democracy of Abraham Lincoln* (Malden, Mass., 1913), 3; Binns, *Abraham Lincoln* (London, 1907).

8. See Charnwood's address in Springfield, *Journal of Illinois State Historical Society*, v. 12 (1920), 498–502; "Some Further Notes on Abraham Lincoln," reprinted from *Anglo-French Review* in *Living Age*, v. 305–7 (1920), 404–13, 100–108, 532–42; Lady Charnwood, *An Autograph Collection and the Making of It* (New York, 1930), 314.

9. Charnwood to William Roscoe Thayer, February 28, 1918, Houghton-Mifflin Collection, Houghton Library, Harvard University; quotations from *Abraham Lincoln* (New York, 1916), 155, 182, 438.

10. Carl Russell Fish, in *American Historical Review*, v. 22 (1918), 413–15; John T. Morse, in *Massachusetts Historical Society Proceedings*, v. 51 (1918), 90–105; Howells, in "Easy Chair," *Harper's* v. 138 (1918), 134–36; Charnwood on Springfield address, *Journal*, 498–502.

11. On Chapin, see *New York Times*, April 1, 1906; "Lincoln in the Hearts of the People," *Independent*, February 11, 1909; Daniel Fish, "Lincoln Collections and Lincoln Bibliography," *Papers of the Bibliographical Society of America*, v. 3 (1908), 51; *Lincoln Bibliography*, Jay Monaghan, ed. (Springfield, 1945), I, 336, II, 69–70.

12. *Abraham Lincoln: A Play* (London, 1918) was followed by an American edition (Boston, 1919), with an Introduction by Arnold Bennett; "An English View of the Success of Abraham Lincoln," reprinted from *English Review* in *Living Age*, v. 304 (1920), 791.

13. *Ibid.*

14. Burns Mantle, *The Best Plays of 1919–1920* (New York, 1937), v; Frank McGlynn, *Sidelights on Lincoln* (Los Angeles, 1947), especially Chs. 8 and 9.

15. Butler, *Looking Forward* (New York, 1932), 354; Croly, "The Paradox of Lincoln," *New Republic*, v. 21 (1920), 350–53. Croly's essay was reprinted in the *New Republic*, in 1954 and 1961; see also his editorial "The Living Lincoln," *ibid.*, v. 41 (1925), 298–99.

16. For a general discussion, John M. Palmer, "President Lincoln's War Problem," *Journal of Illinois State Historical Society*, v. 21 (1928), 200–217; John J. Nicolay and John Hay, *Abraham Lincoln: A History* (New York, 1890), IV, 159; Comte de Paris, *History of the Civil War* (New York, 1875), I, 573–74; Henderson, *Life of Stonewall Jackson* (London, 1900), I, xii-xv, II, 334.

17. *Wisconsin Historical Society Proceedings*, 1917, 106–40, and M. M. Quaife's comment, *Mississippi Valley Historical Review*, v. 14 (1927), 412; the books by Ballard and Maurice were both published in London, 1926; chapters from the latter's work appeared in *Atlantic Monthly*, v. 138 (1926), 224–36, and *Forum*, v. 75 (1926), 161–69.

18. *New York Times*, November 16, 1926; Lyon Sharman, *Sun Yat-sen: His Life and Its Meaning* (New York, 1934), 92; Randle Bond Truett, *Lincoln in Philately* (Washington, 1959), 10; *New York Times*, February 17, 1929, February 12, 1935; *Lincoln Lore*, no. 1149 (1951).

19. See, in general, Charles Moore, "The Memorial to Abraham Lincoln,"

Art and Archaeology, v. 13 (1922), 247–52; and *Washington Past and Present* (New York, 1929), Chs. 18–19; Hay as quoted by Charles McKim, *Century*, v. 82 (1922), 150, and also *New York Times Magazine*, February 9, 1913; Cannon quoted in Philip C. Jessup, *Elihu Root* (New York, 1938), I, 380; Records of the Lincoln Memorial Commission, National Archives, especially the ten-volume Scrapbook incorporating letters, reports, and related documents.

20. *Ibid.*; *The Lincoln Memorial* (Washington, 1927), Ch. 2; *New York Times*, January 20, 1913; *New York Times Magazine*, February 9, 1913; *Congressional Record*, 62 Cong., 2 Sess., Appendix, 59.

21. *Independent*, v. 72 (1912), 320–22, and v. 74 (1913), 280–82, 693–94.

22. Margaret Cresson French, *Journey into Fame: The Life of Daniel Chester French* (Cambridge, 1947), Ch. 17 and *passim*; Michael Richman, *Daniel Chester French, an American Sculptor* (New York, 1976), 121–29, 171–86; *Lincoln Memorial*, Ch. 2.

23. *Barnard's Lincoln* (Cincinnati, 1917), 21–30 and *passim*; Milton Bronner, "A Sculptor of Democracy," *Independent*, v. 89 (1917), 355.

24. The controversy may be followed in the pages of the *New York Times* from August 26, 1917, the *London Times*, from August 27, 1917, and *Art World*, v. 2 (1917). The Robert Todd Lincoln Papers, Illinois State Historical Library, beginning with his letter to William Howard Taft, March 22, 1917, contain much information and include Barnard to Lincoln, April 14, 1917, and Conkling to Lincoln, April 27, 1917. MacMonnies, in *North American Review*, v. 206 (1917), 837–40.

25. See especially, Gutzon Borglum, "The Beauty of Lincoln," *Everybody's Magazine*, v. 22 (1910), 217–20; Bartlett, "The Portraits of Lincoln," in Carl Schurz, *Life of Lincoln* (Boston, 1907), 5–39; Borglum quoted in F. Lauriston Bullard, *Lincoln in Marble and Bronze* (New Brunswick, 1952), 212, 210–16.

26. Cox, "Barnard's Lincoln as a Noted Painter Sees It," *New York Times Magazine*, October 28, 1917; see editorials in *New York Times*, August 26, October 3, 21, November 15, 1917, and January 2, 1918; *Congressional Record*, 65 Cong., 1 Sess., 7919.

27. *London Times*, September 24, 1917, July 29, 1920; *Philadelphia Evening Telegraph* cited in Bullard, *Lincoln*, 230, and 84–86; *New York Times*, January 1, December 21, 1918, and quoting *Manchester Guardian*, January 3, 1919; American Association for International Conciliation, *Presentation of the Saint-Gaudens Statue of Lincoln* (New York, 1920).

28. *Lincoln Memorial*; "Lincoln Memorial Number," *Art and Archaeology*, v. 13 (1922); Cresson, *French*, Ch. 17; Ralph Adams Cram, "The Lincoln Memorial," *Architectural Record*, v. 53 (1923), 478–508.

29. *Lincoln Memorial*, 81, 82–83, 83–84.

30. R. L. Duffus, "Lincoln Memorial," *New York Times Magazine*, February 9, 1941.

31. Langston Hughes, "Lincoln Monument," *Opportunity*, v. 4 (1927), 85.

32. Charles Olson, in *Lincoln and the Poets: An Anthology*, William W. Betts, Jr., ed. (Pittsburgh, 1965), 139.

33. Davis quoted in Richard Current, *The Lincoln Nobody Knows* (New York, 1958), 55; Paul F. Boller, *George Washington and Religion* (Dallas, 1963); Sherwood Eddy quoted in Edgar DeWitt Jones, *Lincoln and the Preachers* (New York, 1948), 134.

34. Gray, *Life of Abraham Lincoln* (Cincinnati, 1867); *American Missionary*, v. 53 (1899), 29; quotation, *New York Tribune*, February 13, 1899; Silverman, in *New York Times*, February 13, 1910.

35. Hobson, *The Master and His Servant* (Dayton, 1913); Bruce Barton, *The Man Nobody Knows* (Indianapolis, 1925), 15–16, 147–52, 80–81; Halpin Whitney, in *Southern Methodist*, September 1, 1926, clipping in Lincoln Museum.

36. Jones, *Lincoln and the Preachers*, Ch. 15.

37. Joseph Fort Newton, *River of Years: An Autobiography* (Philadelphia, 1946), 122; White to C. H. Howard, February 10, 1921, in *Selected Letters of William Allen White*, Walter Johnson, ed. (New York, 1947), 214–15.

38. *The Autobiography of William E. Barton* (Indianapolis, 1932), 35 and *passim*. The principal collection of Barton papers is at the University of Chicago; there is also a collection in the University of Illinois; and Barton letters may be found in collateral collections, e.g., Tarbell Papers. See the measured assessment in Benjamin P. Thomas, *Portrait for Posterity* (New Brunswick, 1947), Ch. 9.

39. Review in *New Republic*, v. 60 (1929), 107.

40. *The Soul of Abraham Lincoln* (New York, 1920), vii, 244–45; William J. Wolf, *The Religion of Abraham Lincoln* (New York, 1963), 195; reviews, *Boston Transcript*, April 7, 1920, *Springfield* (Mass.) *Republican*, March 15, 1920, and *Outlook*, v. 124 (1920), 656.

41. Barton, *Soul*, 48, 50, 66, 155–57.

42. Henry B. Rankin, *Personal Recollections of Abraham Lincoln* (New York, 1916), 321–26; *Soul*, 241–42, 309–13, 258, 125–26.

43. Francis B. Carpenter quoted in Henry J. Raymond, *Life and Public Services of Abraham Lincoln* (New York, 1865), Appendix, 731–32; Tarbell, *Life of Abraham Lincoln* (New York, 1900), II, 89–90; *Independent*, March 16, 1865; *Collected Works* (New Brunswick, 1953), vi, 196–97, viii, 55–57; Brooks, "Personal Recollections of President Lincoln," *Harper's New Monthly*, v. 31 (1865), 226; Elliott V. Fleckles, *Willie Speaks Out!* (St. Paul, 1974), 21; Murdoch quoted in Fowler, *Patriotic Orations*, 96–97.

44. The Sickles report is known by way of James F. Rusling, first in *Men and Things I Saw in Civil War Days* (New York, 1899), then in an address, *Lincoln and Sickles, Before the Reunion of the Third Army Corps* (n.p., 1910), which is quoted herein.

45. Lincoln's "Reply," September 7, 1964, is in *Collected Works* VII, 542; Chapman, *Latest Light*, 288; David Burrell, in *New York Times*, February 12, 1923, and *Lincoln Lore*, no. 1401 (1956); Butterworth, *Songs of History* (Boston, 1887), 28–33, and Murat Halstead, quoted in Hobson, *Master and Servant*, 85.

46. For the debate, see Collis, *The Religion of Abraham Lincoln* (New York, 1900); for Ingersoll, see in general, "The Religious Belief of Abraham Lincoln," *Works* (New York, 1915), XII, 248–55, and *Letters*, Eva Ingersoll Wakefield, ed. (New York, 1951), 341–46.

47. Remsburg, *Abraham Lincoln* (New York, 1893); see also his *Six Historic Americans* (New York, 1906).

48. Lewis, *Lincoln the Freethinker* (New York, 1924), 29 (italics in original) and *passim*.

49. *Conservator*, v. 1 (1890), 65; *New York Times*, February 28, 1924, February 14, 1927, February 13, 1928, February 14, 1949.

50. See, in general, Jay Monaghan, "Was Abraham Lincoln a Spiritualist?" *Journal of Illinois State Historical Society*, v. 34 (1941), 209–32, and *Lincoln Lore*, no. 888 (1946); quotation in Nettie Colburn Maynard, *Was Abraham Lincoln a Spiritualist?* (Philadelphia, 1891), x-xi.

51. *Ibid.*, 66–74, 89–95, and *passim*; B. F.Austin, *The Religion of Abraham Lincoln* (Los Angeles, 1924), 74–75; J. J. Fitzgerrell, *Lincoln Was a Spiritualist* (Los Angeles, 1924), 15–23; "Lincoln the Mystic," *Chicago Tribune*, February 7, 1909; Doyle, *The History of Spiritualism* (New York, 1975), I, 143–46.

52. *New York Tribune*, April 13, 1896; *New York World*, February 7, 1909; Irene Hale to Librarian of Lincoln Collection, February 3, April 23, 1938, John Hay Library, Brown University; Shelton, *Lincoln Returns* (New York, 1957); Fleckles, *Willie Speaks Out!*

53. See, in general, Isaac Markens, *Abraham Lincoln and the Jews* (New York, 1909), Bertram W. Korn, *American Jewry and the Civil War* (Philadelphia, 1951), *Memoirs of American Jews, 1775–1865*, Jacob R. Marcus, ed. (Philadelphia, 1955), and Emanuel Hertz, *Abraham Lincoln, a New Portrait* (New York, 1931), I, 340–41; on chaplains, see Lee M. Friedman, *Jewish Pioneers and Patriots* (Philadelphia, 1941), Ch. 4; on Grant's order, see Korn, *American Jewry*, Ch. 6, and Markens, *Lincoln and Jews*, 118–20.

54. *Louisville Daily Journal*, April 23, 1865; James G. Heller, *Isaac M. Wise: His Life, Works, and Thought* (New York, 1865), 371.

55. James Yaffe, *The American Jews* (New York, 1968), 245; Hertz, *Abraham Lincoln: The Tribute of the Synagogue* (New York, 1927), 363 and *passim*; Levy quoted in Jones, *Lincoln and the Preachers*, 107; Stanley F. Chyet, *Lives and Voices* (Philadelphia, 1972), 278.

56. Markens and Hertz, both cited in note 53 above; and Hertz, *Abraham Lincoln the Seer* (New York, 1925).

57. *Collected Works*, III, 511; Herndon quoted in Emanuel Hertz, *Lincoln Talks* (New York, 1939), 3; Autobiography, in *Collected Works*, IV, 61; *New England Historical and Genealogical Register*, v. 19 (1865), 268–69; Edward Channing, *History of the United States* (New York, 1925), VI, 227.

58. Snider, *Lincoln at Richmond: A Dramatic Epos* (St. Louis, 1914), 17; see the general discussion in Thomas, *Portrait*, 222–38.

59. *Herndon's Life of Lincoln*, Paul Angle, ed. (New York, 1942), 7; later

editions of Cathey's book—the fourth in 1939—appeared under the title *The Genius of Abraham Lincoln*; Coggins, *Abraham Lincoln a North Carolinian* (Gastonia, 1927) and *The Eugenics of Abraham Lincoln* (Elizabethtown, Tenn., [1940]).

60. King to Hay, September 7, 1896, Hay Papers, John Hay Library, Brown University; Alley, *Random Thoughts and Musings of a Mountaineer* (Salisbury, N.C., 1941), Chs. 21–22; Graves, *The Fighting South* (New York, 1943), 206.

61. A good early summary is in Morse, *Abraham Lincoln* (Boston, 1895), Ch. 1; Lea and Hutchinson, *Ancestry of Lincoln* (Boston, 1909), 139; the Learned title: *Abraham Lincoln: An American Migration* (Philadelphia 1909); see also, F. Lauriston Bullard, "The New England Ancestry of Abraham Lincoln," *New England Magazine*, v. 39 (1909), 685–91, and Ida M. Tarbell, *In the Footsteps of Lincoln* (New York, 1924), Preface.

62. The full title of Barton's book: *The Paternity of Abraham Lincoln. Was He the Son of Thomas Lincoln? An Essay on the Chastity of Nancy Hanks* (Indianapolis, 1920); J. G. DeRoulhac Hamilton, "The Many-sired Lincoln," *American Mercury*, v. 5 (1925), 129–35.

63. Tarbell's correspondence with Hitchcock, in Tarbell Papers, Allegheny College, falls mainly into two periods: 1895–1900 and 1923–26; see also Jesse Weik to Tarbell, November 9, 1899. Two independent and opposing views of Hitchcock's *Nancy Hanks* are Howard W. Jenkins, "The Mother of Lincoln," *Pennsylvania Magazine of History and Biography*, v. 24 (1900), 129–38 (favorable), and Howard E. and Ernestine Briggs, *Nancy Hanks Lincoln* (New York, 1952), 30–38 (unfavorable).

64. Holland, *Life of Abraham Lincoln* (Springfield, 1866), 23; Nicolay and Hay, *History*, I, 24; George Williamson, *The Laws of Heredity*, 2d ed. (San Francisco, 1898), 208.

65. Chapman, *Latest Light*, Ch. 1; the novel is Marie Daviess, *The Matrix* (New York, 1920); on appearance, see Albert J. Beveridge, *Abraham Lincoln* (Boston, 1928), I, 16n; the "Madonna" metaphor is Morse's, in *Lincoln*, I, 9.

66. On the monument, see Louis A. Warren, *Lincoln's Youth* (Indianapolis, 1959), 55; Monroe's poem is collected in *You and I* (New York, 1914), 100–102. There are poems about Nancy Hanks by Vachel Lindsay, Edgar Lee Masters, and Rosemary and Stephen Vincent Benét, among others.

67. Barton to Daniel Fish, December 27, 1922, May 26, August 6, 1921, and Fish to Barton, May 9, 1922, Lincoln Museum.

68. Warren wrote no memoir, nor have his life and work been studied; but see the bound volume "Religious Contacts of Louis A. Warren," in Lincoln Museum, and the personal sketch, *Lincoln Lore*, no. 1000 (1948); Tarbell to Miss O'Dell, October 29, 1922, Tarbell Papers; Thomas, *Portrait*, 225–26; Jesse Weik, *The Real Lincoln: A Portrait* (Boston, 1922), 42.

69. See Barton's *The Lineage of Lincoln* (Indianapolis, 1929) for the newspaper report of the Chicago lecture, 361–65; Tarbell-Hitchcock correspondence, 1923–25, Tarbell Papers.

70. "Story of Lincoln's Lost Grandmothers," *New York Times*, February 8,

1925; *Christian Science Monitor*, May 2, 1925; Lincoln to Barton, July 22, 1925, Barton Papers, University of Chicago.

71. In addition to Warren's book, especially Ch. 2, see his "Lincoln's Honorable Parentage," *Century*, v. 112 (1926), 532–39; Barton to Warren, September 29, 1926, Lincoln Museum.

72. Warren to Tarbell, May 2, 1928, Tarbell to Myra Hank Rudolph, October 10, 1932, Tarbell Papers; Warren to Martha Stephenson, December 20, 1930, Lincoln Museum.

73. Tarbell quoted in Thomas, *Portrait*, 236. The recent work by Adin Baber includes *Nancy Hanks, the Destined Mother of a President* (Kansas, Ill., 1963).

74. Barton, *Life of Lincoln*, II, 250–53; *New York Times*, August 29, 1926, and "An Old Myth Dispelled," *New York Times Magazine*, October 3, 1926; for the principal source, Waldo F. Glover, *Abraham Lincoln and the Sleeping Sentinel* (Montpelier, Vt., 1936).

75. *Ibid.*; F, Lauriston Bullard to Edwin L. Page, December 9, 23, 1939, in Bullard Papers, Boston University Library.

76. Barton, *Life of Lincoln*, II, Ch. 18; *New York Times*, August 3–16, 1925; *Collected Works*, VIII, 116–17.

77. Quoted in *Abraham Lincoln: His Speeches and Writings*, Roy P. Basler, ed. (Cleveland, 1946), 35. See especially, F. Lauriston Bullard, *Abraham Lincoln and the Widow Bixby* (New Brunswick, 1946). On the reproductions, see Charles Hamilton, *Collecting Autographs and Manuscripts* (Norman, 1961), 51–53; *New York Tribune*, February 13, 1906; Barton, *Beautiful Blunder*, 35–36; *New York Times*, July 29, 1918.

78. Oberholtzer, *Abraham Lincoln* (Philadelphia, 1904), 339; Principal to Bullard, December 22, 1906, Bullard Papers; Edward C. Stokes, in *Addresses at National Republican Club* (New York, 1927), 163; Bullard letter, *New York Times*, August 3, 1925. Joseph George, Jr., "F. Lauriston Bullard as a Lincoln Scholar," *Lincoln Herald*, v. 64 (1960), 173–82, is of general interest.

79. See *Lincoln Lore*, nos. 185 (1932), 601 (1941), 1285 and 1286 (1953); Barton, *Beautiful Blunder*, 56; Gerald McMurtry to W. F. Oakeshott, June 3, 1963, Lincoln Museum; Bullard, *Widow Bixby*, Foreword and 143.

80. See, in general, William H. Townsend, *Lincoln and Liquor* (New York, 1934), and Harry M. Lydenberg, "Lincoln and Prohibition: Blazes on a Zigzag," *American Antiquarian Society Proceedings*, n.s., v. 62 (1952), 9–62; Swett, in *Reminiscences of Abraham Lincoln*, Allen T. Rice, ed., 8th ed. (New York, 1889), 462–63; William O. Stoddard, *Inside the White House in War Times* (New York, 1890), 59–61.

81. See the marginal note by Charles W. McLellan on a letter from Truman Bartlett, May 13, 1908, McLellan Collection, Brown University; Ida Tarbell to Langbourne Williams, August 26, 1927, Tarbell Papers; Chapman, *Latest Light*, 153–55; *Lincoln-Douglas Debates*, Robert W. Johannsen, ed. (New York, 1965), 42–43, 53; Burner reminiscence in *Chicago Tribune*, February 10, 1895;

Townsend-Tarbell correspondence, April 13 and May 2, 1932, Tarbell Papers; Townsend, *Lincoln and Liquor*, 32–33, 38.

82. *Ibid.*, Ch. 4; advertisement in *Baltimore Sun*, February 12, 1909; *Collected Works*, I, 271–79, 273; editorial, *New York Times*, September 14, 1928; Charles T. White, *Lincoln and Prohibition* (New York, 1921), *passim*; Barton, *Life*, II, 450, on Merwin; see also, Jonathan T. Hobson, *Footprints of Lincoln* (Dayton, 1909), Chs. 7–8.

83. *Christian Advocate*, February 6, 1919, quoted in Hill, *Lincoln, Man of God*, 148n, and, in part, in White, *Lincoln and Prohibition*, 155; see also, Chapman, *Latest Light*, 174–75.

84. Quoted in Lydenberg, "Lincoln and Prohibition," 9; *New York Times*, March 12, 1922.

85. Louis Albert Banks, *The Lincoln Legion* (New York, 1903), Chs. 2, 11, and *passim*.

86. Virginius Dabney, *Liberalism in the South* (Chapel Hill, 1942), 341–42; Norwood, A *True Vindication of the South* (Savannah, 1917), 321; Donald Davidson, *The Attack on Leviathan* (Chapel Hill, 1938), 217–18; *Richmond Times-Dispatch*, June 22, 1922; *New York Times*, June 22, 23, 26, 1922; Arthur J. Jennings, "The Lincoln Resolution," *Confederate Veteran*, v. 30 (1922), 285–86; *Literary Digest*, v. 74 (July 15, 1922), 30–31.

87. *New York Times*, February 14, 1928; *Richmond Times-Dispatch*, February 12, 1928, *Richmond News Leader*, February 14, 1928; Mary D. Carter to Mr. Griffith, April 19, 1928, Lincoln Museum; *Confederate Leaders and Other Citizens Request the House of Delegates to Repeal the Resolution of Respect to Lincoln, the Barbarian* (Richmond, 1928); Lucy Shelton Stewart, *The Rewards of Patriotism* (New York 1930), Prologue; *Confederate Veteran*, v. 32 (1924), 156, v. 33 (1925), 364.

88. Tyler, "The South in Germany," *William & Mary College Historical Papers*, v. 26 (1917), 1–20, and "The South and Self-determination," *ibid.*, v. 27 (1919), 217–25; [O. W. Blacknall], *Lincoln as the South Should Know Him* (Raleigh, 1915); Landon Bell, *The Old Free State* (Richmond, 1927), II, 481; Robert L. Preston, *Southern Miscellanies*, no. 2 (Leesburg, Va., 1919). For a brief biography of Tyler, see John E. Hobeika, *The Sage of Lion's Den* (New York, 1948).

89. "Propaganda in History," *Tyler's Quarterly*, v. 1 (1920), 217–30; "After Sixty Years," *ibid.*, v. 12 (1930), 84; Blease in *Congressional Record*, 70 Cong., 1 Sess., 1998–2007, 9809–14.

90. See, in general, Bessie Louise Pierce, *Public Opinion in the Teaching of History in the United States* (New York, 1926), Ch. 6; and *Tyler's Quarterly*, v. 10 (1928), 69–70; *Confederate Veteran*, v. 40 (1932), 128–30.

91. *Time*, v. 11 (April 9, 1928), 11–12; Thomas, *Portrait*, 237; *New York Times*, February 14, 1935; Cabell, "To John Wilkes Booth," in *Ladies and Gentlemen* (New York, 1934), 265–67; Dolly Blount Lamar, *When All Is Said and Done* (Athens, Ga., 1942), 12–24.

6 *From Memory to History*

1. *American Historical Review*, v. 41 (1936), 270–94; Benjamin P. Thomas, quoted by T. Harry Williams, *Saturday Review of Literature*, v. 35 (1952), 17.

2. "Lincoln Theme," 282, 288, 292 (my italics).

3. See John S. Goff, *Robert Todd Lincoln: A Man in His Own Right* (Norman, 1969), 234 and *passim;* Bess Beatty, *A Revolution Gone Backward: Black Response to National Politics*, 1876–1896 (New York, 1987), 63, 94, 134; and Jesse T. Moore, *A Search for Equality: The National Urban League*, 1910–1961 (University Park, Pa., 1981), 114; "The Son of Abraham Lincoln," *Literary Digest*, v. 90 (August 14, 1926), 36–42. Randolph (footnote) quoted in Brailsford R. Brazeal, *The Brotherhood of Sleeping Car Porters* (New York, 1946), 43; Jack Sandino, *Miles of Smiles, Years of Struggle* (Urbana, 1989), 154.

4. On the Rothschild matter, see the correspondence with Isaac Markens in *A Portrait of Abraham Lincoln in Letters by His Oldest Son*, Paul Angle, ed. (Chicago, 1968), 24–46; and the Robert Todd Lincoln Papers, Illinois State Historical Library.

5. See especially, David C. Mearns, *The Lincoln Papers* (New York, 1948), I. Ch. 7 and *passim;* see also Mearns's memorandum, March 21, 1947, in his papers, Library of Congress.

6. *Across the Busy Years* (New York, 1939), Ch. 9; Emanuel Hertz, *The Hidden Lincoln* (New York, 1938), 17–19; Mearns, *Papers*, Ch. 8.

7. See, in general, Paul Angle, *A Shelf of Lincoln Books* (Springfield, 1946), 6–11; Tarbell's Introduction to Tracy's *Uncollected Letters* (Boston, 1917); *Books at Brown* (Providence, 1985), 6–7; Hertz, *Abraham Lincoln*, Introduction, v. 2 (New York, 1931); *Lincoln Lore*, no. 499 (1938).

8. See, in general, Ralph Goeffrey Newman, *Preserving Lincoln for the Ages* (Fort Wayne, 1989); Roy P. Basler, *A Touchstone for Greatness* (Westport, Conn., 1973), 36; Carl Sandburg, *Lincoln Collector: The Story of Oliver R. Barrett's Great Private Collection* (New York, 1949), 21–22, 45–46, 72.

9. Thomas B. Littlewood, *Horner of Illinois* (Evanston, 1969), 23–24; the Horner Papers in the Illinois State Historical Library furnish details on collecting; *A Catalog of the Alfred Whital Stern Collection of Lincolniana* (Washington, 1960); *New York Times*, January 26, 1941, May 2, November 19, 1950, February 13, 20, 21, 1952.

10. Interview with Thomas L. Schwartz, September 1990; *Lincoln Lore*, no. 1697 (1979); Lincoln Centennial Association Papers, Illinois State Historical Library.

11. Angle, "Abraham Lincoln: Circuit Lawyer," *Lincoln Centennial Association Papers*, v. 5 (1928), 19–44.

12. Angle, *Shelf*, 70.

13. Rankin, "The First American—Abraham Lincoln," *Journal of Illinois State Historical Society*, v. 8 (1915), 260–67; John J. Duff, *Abraham Lincoln, Prairie Lawyer* (New York, 1960), 181; E. O. Laughlin, in *Literary Digest*, v. 80 (February 1923), 40; [Charles M. Thompson], *Investigation of the Lincoln Way*

(Springfield, 1915); "The Lincoln Way," *Indiana Magazine of History*, v. 11 (1915), 176–77; Rexford Newcomb, *In the Lincoln Country* (Philadelphia, 1928).

14. T. G. Onstot, *Pioneers of Menard and Mason Counties* (Peoria, 1902), Ch. 6; Reep, *Lincoln and New Salem* (Petersburg, 1927); Joseph F. Booton, *The Record of the Restoration of New Salem* (Springfield, 1934); Benjamin P. Thomas, *Lincoln's New Salem* (Springfield, 1934); Richard Taylor, "The New Salem Tradition," typescript (n.p., n.d.), Special Collections, University of Illinois; Barton, "Abraham Lincoln and New Salem," *Journal of Illinois State Historical Society*, v. 19 (1927), 81; for a brief appraisal, Charles B. Hosmer, *Preservation Comes of Age* (Charlottesville, 1981), I, 398–401.

15. See, in general, Bess V. Ehrmann, *The Missing Chapter in the Life of Abraham Lincoln* (Chicago, 1938), 25 and *passim*, as well as "The Lincoln Inquiry," *Indiana Magazine of History*, v. 21 (1925), 3–17; also, Edward Murr, "Lincoln in Indiana," *ibid.*, v. 13 (1917), 307–48, v. 14 (1918), 13–75, 148–82; and Charles G. Vannest, *Lincoln the Hoosier* (Chicago, 1928).

16. On "dunghill" and "chin fly," see the booklet by Mark E. Neely, Jr., *Escape from the Frontier: Lincoln's Peculiar Relationship with Indiana* (Fort Wayne, n.d.); Swett, in *The Reminiscences of Abraham Lincoln*, Allen T. Rice, ed. (New York, 1889), 459; Lincoln to Andrew Johnston, April 18, 1846, *Collected Works*, I, 378–79.

17. Warren to Otto L. Schmidt, February 22, June 6, 1928, Lincoln Museum; Warren's own account of the foundation, *Lincoln Lore*, no. 1610 (1972).

18. *Ibid.*; Tarbell speech, newspaper clipping, September 18, 1932, and clipping from *Fairfield* (Iowa) *Daily Ledger*, Lincoln Museum. See also the Manship correspondence in the Lincoln Museum.

19. *Lincoln's Youth: Indiana Years* (Indianapolis, 1959), *passim*. On Thomas Lincoln, see John Y. Simon, *House Divided: Lincoln and His Father* (Fort Wayne, 1987); and Rice, *Reminiscences*, 458.

20. Lancaster, *For Us the Living* (New York, 1940), 141, 134, 154.

21. *Los Angeles Times*, February 12, 1984; R. Gerald McMurtry, *My Lifelong Pursuit of Lincoln* (Fort Wayne, 1981).

22. Stephenson also authored *Abraham Lincoln and the Union* in 1918 and made several other contributions, which are surveyed by Mark E. Neely Jr., in "Nathaniel W. Stephenson and the Progressive Lincoln," *Lincoln Lore*, no. 1677 (1977); Nevins to Ida Tarbell, May 30, 1939, Tarbell Papers, Allegheny College.

23. See Penelope Niven, *Carl Sandburg: A Biography* (New York, 1991), and Sandburg's memoir of his early life, *Always the Young Stranger* (New York, 1953).

24. *The Complete Poems of Carl Sandburg* (New York, 1970), 71; Hecht and White quoted in Niven, *Sandburg*, 336, 360.

25. *Complete Poems*, 20; *Carl Sandburg Remembered*, William A. Sutton, ed. (Metuchen, N.J., 1979), 84.

26. Niven, *Sandburg*, 418; Lerner, *Ideas for the Ice Age* (New York, 1941),

392; Edmund Wilson, *Patriotic Gore* (New York, 1962), 117; Williams, *Selected Essays* (New York, 1954), 278–79.

27. *Complete Poems*, 85, 71; Hansen quoted in Niven, *Sandburg*, 435.

28. *Prairie Years*, I, 16, 26 (ellipsis in original), 52, 27; Thomas, *Portrait*, 293.

29. *Prairie Years*, I, 164, 181–84, 414, 13, 40.

30. *Ibid.*, I, 47, 54, II, 197, 199, 307, 285; Van Doren quoted in Niven, *Sandburg*, 434.

31. *Ibid.*, 431; *Letters of Carl Sandburg*, Herbert Mitgang, ed. (New York, 1968), 230.

32. *Independent*, v. 116 (1926), 193; Tarbell to Sandburg, November 7, 1925, Sandburg Papers, University of Illinois; Quaife quoted in Thomas, *Portrait*, 285; Mencken quoted in Angle, *Shelf*, 52.

33. Sandburg's unsent letter, September 2, 1925, in Mitgang, *Letters*, 232–33, where no mention is made of the penciled note on the letter in the Sandburg Papers. See, in general, Claude G. Bowers, *Beveridge and the Progressive Era* (New York, 1932).

34. *Life of John Marshall* (Boston, 1919), IV, 91–93; Bowers, *Beveridge*, 561.

35. Beveridge to Lincoln, December 27, 1922, and reply, January 23, 1923, to Angle, February 23, 1926, to Charles A. Beard, July 20, 1924, Beveridge Papers, Library of Congress; to Theodore C. Pease, April 25, 1928, Beveridge Papers, University of Illinois; Thomas, *Portrait*, 249–51.

36. To Angle, February 23, 1926, Beard to Beveridge, April 23, 1925, Beveridge to Sedgwick, July 8, 1925, Beveridge Papers, Library of Congress; Benjamin P. Thomas, *Abraham Lincoln* (New York, 1952), 527.

37. Beveridge, *Abraham Lincoln* (Boston, 1928), I, 209; Beard to Beveridge, April 23, 1925, Beveridge Papers, Library of Congress.

38. Beveridge, *Lincoln*, I, Ch. 7, p. 487.

39. *Ibid.*, II, 51n, and Ch. 1 *passim*; Owsley to Beveridge, March 4, 1926, Beveridge Papers, Library of Congress. For the Fort Stevens story (footnote), see: John Henry Cramer, *Lincoln Under Fire* (Baton Rouge, 1948), 124 and *passim*; Alexander Woollcott, "Get Down, You Fool!" *Atlantic Monthly*, v. 164 (1938), 167–73; Frederick C. Hicks, "Lincoln, Wright, and Holmes at Fort Stevens," *Journal Illinois State Historical Society*, v. 39 (1946), 323–32.

40. Holmes to Beveridge, November 17, 1926, February 2, 1927, March 26, 1927, and Beveridge to Holmes, November 13, 1926, Holmes Papers, Harvard University.

41. To Hodder, May 5, 1925, Beveridge Papers, Library of Congress; Jameson to Beveridge, September 20, 1926, in "Senator Beveridge, J. Franklin Jameson, and Abraham Lincoln," Elizabeth Donnan and L. F. Stock, eds., *Mississippi Valley Historical Review*, v. 35 (1949), 667; Bowers, *Beveridge*, 575–78.

42. Morison, in *New Republic*, v. 57 (December 12, 1928), 117; Bowers, "The Lincoln of Fact," *Virginia Quarterly Review*, v. 5 (1929), 265–74.

43. Barbee, *An Excursion in Southern History* (Richmond, 1928), 60 and *passim*; Townsend to Pease, September 7, 1932, and reply, September 9, 1932,

Beveridge Papers, University of Illinois; Townsend to Rufus R. Wilson, April 27, 1933, Bullard Papers, Boston University.

44. Ford to Barrett, March 20, 1939, Sandburg Papers, University of Illinois; Sandburg to White, October 20, 1928, in *Letters*, 262; Beveridge to Jameson, November 22, 1926, *Mississippi Valley Historical Review*, v. 35 (1949), 668.

45. *New York Times*, May 23, 24, June 4, 6, 1931; Clark, *Lincoln, A Psychobiography* (New York, 1933); Shutes, *Lincoln and the Doctors* (New York, 1933) and *Lincoln's Emotional Life* (New York, 1957).

46. Masters, *Lincoln the Man* (New York, 1931), 82, 115. See Masters's *Across Spoon River: An Autobiography* (New York, 1936), and "Days in the Lincoln Country," *Journal Illinois State Historical Society*, v. 18 (1925), 779–92, and interview, *New York Times*, February 6, 1931.

47. Hertz to Warren, February 7, 1931, Lincoln Museum; *Literary Digest*, v. 108 (February 28, 1931), 34; *Congressional Record*, 71 Cong., 3 Sess., House, 1703; Lindsay to Elizabeth C. Lindsay, February 7, 1931, *Letters*, Marc Chenetier, ed. (New York, 1979), 450–51; Lytle, in *Virginia Quarterly Review*, v. 7 (1931), 620–26; Bowers, in *Saturday Review of Literature*, v. 7 (1931), 609–10.

48. Joseph Fort Newton, *River of Years* (Philadelphia, 1946), 120; Bradford, "The Wife of Abraham Lincoln," *Harper's Magazine*, v. 151 (1925), 489–98.

49. Helm, *The True Story of Mary, Wife of Lincoln* (New York, 1928); Morrow, *Mary Todd Lincoln* (New York, 1928), Sandburg and Angle, *Mary Lincoln, Wife and Widow* (New York, 1932), and William A. Evans, *Mrs. Abraham Lincoln* (New York, 1932), 334, 352, and *passim*.

50. Carnegie, *Lincoln the Unknown* (New York, 1932), 236; Richard M. Huber, *The American Idea of Success* (New York, 1971), 244; Carnegie, *How to Win Friends and Influence People* (New York, 1931), 30–34, 143, 166, 173, 273–75.

51. Bachellor, *A Man for the Ages* (Indianapolis, 1919), and *Father Abraham* (Indianapolis, 1925); Brown, *The Father* (New York, 1928); and, in general, Don E. Fehrenbacher, *Lincoln in Text and Context* (Stanford, 1987), Ch. 17. On Morrow, see her article "Lincoln—The Most Lied About Man in the World," *American*, v. 105 (February 1928), 7–9; *Mary Lincoln*, 208–48; and Grant Overton, "She Pioneered in Novels," *Mentor*, v. 15 (June 1927), 58–59. In addition to *Great Captain* (New York, 1930), Morrow wrote several short stories collected in *The Lincoln Stories* (New York, 1938).

52. *Great Captain*, 456.

53. Benét, *John Brown's Body* (New York, 1958), 72, 213, 219; *Selected Letters*, Charles Fenton, ed. (New Haven, 1960), 143; Parry Stroud, *Stephen Vincent Benét* (New York, 1962), 54–56. Edna Davis Romig's long poem in sonnet form, *Lincoln Remembers* (Philadelphia, 1930), should not be overlooked.

54. Minor to [Ellery Sedgwick], June 27, 1928, and replies, August 6, November 10, 1928, Sedgwick Papers, Massachusetts Historical Society. See also Edward Weeks, *My Green Age* (Boston, 1973), and Fehrenbacher's essay in *Text and Context*.

55. Tarbell, *Footsteps of Lincoln* (New York, 1924), 211; Snider, *Lincoln and Ann Rutledge* (St. Louis, 1912), 293–94, 301, and *Abraham Lincoln* (St. Louis, 1908), 185–86; Charles Phillips, "Abraham Lincoln," *Catholic World*, v. 128 (1929), 513, and v. 129 (1929), 48; Little, *Better Angels* (New York, 1928); Harold W. Gammens, *Spirit of Ann Rutledge: A Drama* (New York, 1927).

56. *Spoon River Anthology* (New York, 1916), 220; Carnegie, *Lincoln the Unknown*, Ch. 1; Ross Lockridge to William H. Townsend, October 11, 1926, and reply, October 25, 1926; Eleanor Atkinson, *Lincoln's Love Story* (New York, 1909), 39–31.

57. Ford, "Forged Lincoln Letters," *Proceedings of Massachusetts Historical Society*, v. 61 (1928), 183–95; Angle is quoted in *Atlantic Monthly*, v. 143 (1929), 283, also in *New York Herald Tribune*, December 3, 1928; Ford to Sedgwick, November 11, 1928, Fitzpatrick to Minor, December 31, 1928, Barton to Sedgwick, November 15, December 5, 1928, Sedgwick Papers; Barton to Minor, December 11, 1928, copy in Lincoln Museum; Angle to Sandburg, December 6, 1928, Sandburg Papers, University of Illinois; Sandburg quoted in Weeks, *Green Age*, 255.

58. *Atlantic Monthly*, v. 143 (1928), 834–37.

59. *Ibid.*, v. 143 (1929), 1–14, 215–25.

60. *New York Times*, January 22, 1929; Angle, "The Minor Collection: A Criticism," *Atlantic Monthly*, v. 143 (1929), 516–25; Barton to Sedgwick, May 6, 1929, and *Atlantic Monthly* to *New York Herald Tribune*, April 6, 1929, with attachment, Sedgwick Papers; see also, Angle Papers, Chicago Historical Society.

61. Copy of statement by Wilma Minor to Fitzpatrick, July 3, 1929, Sedgwick Papers. The statement is reprinted in full in Fehrenbacher, *Text and Context*, 267–68.

62. *Atlantic Monthly*, v. 143 (1929), 516–25; Quaife, in *Mississippi Valley Historical Review*, v. 15 (1929), 578–83; Angle to Paul Dubois, April 4, 1966, Angle Papers.

63. Barton, *Mississippi Valley Historical Review*, v. 15 (1929), 510; Angle, *Shelf*, 46.

64. On Randall, see *Dictionary of American Biography*, Supp. V (New York, 1957), 556–57, and Ruth P. Randall, *I Ruth* (Boston, 1968).

65. *Constitutional Problems Under Lincoln* (New York, 1926), 45 and *passim*; "Lincoln's Task and Wilson's," *South Atlantic Quarterly*, v. 29 (1930), 349–68, and "Lincoln's Peace and Wilson's," *ibid.*, v. 42 (1943), 225–42, also included in *Lincoln the Liberal Statesman* (London, 1947), Ch. 7.

66. On the genesis of the Lincoln biography, see Randall to Allan Nevins, July 9, 1943, Randall Papers, Library of Congress.

67. *American Historical Review*, v. 51 (1945), 576.

68. The address is included in *Liberal Statesman*, Ch. 2; see also Randall's *Lincoln and the South* (Baton Rouge, 1946).

69. *Lincoln the President* (New York, 1945), I, viii, 27–29, 36–37, 53–54;

Lincoln Lore, nos. 465 (1938), 830 (1935), 879 (1946); Randall to Henry S. Commager, January 31, 1938, Robert L. Schuyler to Hertz, February 17, 1939, and to Randall, February 27, 1939, Randall Papers.

70. *Lincoln the President*, II, Appendix, 328, 341, *passim*; Douglas L. Wilson, "Abraham Lincoln, Ann Rutledge, and the Evidence of Herndon's Informants," *Civil War History*, v. 36 (1990), 301–22.

71. Nevins, in *Saturday Review of Literature*, v. 21 (December 2, 1939), 3.

72. Thomas, *Portrait*, 294–95; Niven, *Sandburg*, 471; Randall, in *American Historical Review*, v. 45 (1940), 917–22; *Abraham Lincoln: The War Years* (New York, 1939), II, 198–200, III, 521, IV, 190–91, and Freeman, *Robert E. Lee: A Biography* (New York, 1935), III, 264.

73. *Complete Poems*, 521, 523; *War Years*, II, 124, 332–33, IV, 216–17.

74. *Time*, v. 34 (December 4, 1939); Benét, in *Atlantic Monthly*, v. 164 (1939), "Bookshelf"; Nevins, see n. 71 above; *Sandburg Remembered*, 85; Kazin, *On Native Grounds* (New York, 1942), 508; Edmund Wilson, *Letters on Literature and Politics, 1912–1972*, Elena Wilson, ed. (New York, 1977), 610. See also, Charles W. Ramsdell, "Carl Sandburg's Lincoln," *Southern Review*, v. 6 (1941), 439–53.

75. *American Historical Review*, v. 45 (1940), 917–22; Randall to F. Lauriston Bullard, March 28, 1940, Sandburg Papers, University of Illinois; Thomas J. Pressly, *Americans Interpret Their Civil War* (Princeton, 1954), 316; Randall, *Lincoln the President*, I, 107–9, 123.

76. *Ibid.*, II, 1, 120, 68–73, 141–43, 181, 189; Woodson, in *Journal of Negro History*, v. 31 (1946), 107–110, and see his "Lincoln as a Southern Man," *Negro History Bulletin*, v. 10 (1947), 105–6.

77. *Lincoln the President*, III, 421 and Ch. 17.

78. *Ibid.*, IV, 134, 262–63.

7 *Zenith*

1. In addition to Mead's *Heroic Statues in Bronze of Abraham Lincoln* (Fort Wayne, 1932), see F. Lauriston Bullard, *Lincoln in Marble and Bronze* (New Brunswick, 1952), Donald C. Durman, *He Belongs to the Ages: Sculptures of Abraham Lincoln* (Ann Arbor, 1956), and Rex A. Smith, *The Carving of Mount Rushmore* (New York, 1985).

2. *Collier's*, v. 109 (February 14, 1942), 12; *New York Times*, February 12, 13, 1955.

3. Jay Monaghan, "The Man and Memory Still Abide," *New York Times Book Review*, February 11, 1951; "Abraham Lincoln: Everybody's Business," *Reader's Digest*, v. 66 (February 1955), 94–96.

4. "The Man Who Spoke to Lincoln," *Ebony*, v. 18 (1963), 79; see also, *New York Herald Tribune*, editorial, February 12, 1949.

5. Richard Lowitt, *George W. Norris: The Triumph of Progressivism* (Urbana, 1968), 336; President Ford quoted in *Commonweal*, v. 114 (1987), 173; R. L. Duffus, "Lincoln Memorial: February 1941," *New York Times Magazine*,

February 9, 1941; Carl Holliday, *Abraham Lincoln in Heaven* (New York, 1936), 25; Thomas Curtis Clark, *Abraham Lincoln: Thirty Poems* (Chicago, 1934), 26; the sonnet is taken from Ralph H. Gabriel, *The Course of American Democratic Thought* (New York, 1940), 412.

6. Hoover, *State Papers and Other Public Writings* (Garden City, N.Y. 1970), II, 500–505, 587–90; *New York Times*, February 13, 1932, January 14, 1933; Will Rogers, *Weekly Articles* (Stillwater, Okla. 1980), V, 235.

7. On the New Deal and Jefferson, see my *The Jefferson Image in the American Mind* (New York, 1960), 355–76; Vandenburg is quoted in *New York Times*, February 18, 1934; *Ohio State Journal*, February 12, September 14, 17, October 15, November 14, December 21–22, 1934; *Congressional Record*, 75 Cong., 3 Sess., 2119, 2629; *New York Times*, March 9, 1938, quoting Barton, and November 14, 1954.

8. *Chicago Tribune*, June 8, 1935, and *New York Times*, June 8–10, 1935; see also, *Memorial of William E. Dodge*, D. Stuart Dodge, ed. (New York, 1887), 81–82, and Lucius P. Chittenden, *Recollections of President Lincoln* (New York, 1891), 77–78, reporting the interview.

9. Aiken is quoted in *Congressional Record*, 75 Cong., 3 Sess., Appendix, 1655–56; *ibid.*, 77 Cong., 1 Sess., 649–50 (Martin), 75 Cong., 1 Sess. (Dirksen), and 76 Cong., 1 Sess. 520–23 (Barton). The *Record* for successive Congresses contains many Lincoln Day speeches. On Dirksen, see Louella Dirksen, *The Honorable Mr. Marigold* (Garden City, N.Y. 1972), 75–76.

10. On the Democrats, in general, see Alfred H. Jones, *Roosevelt's Image Brokers* (Port Washington, N.Y., 1974), and Peter Karsten, *Patriot Heroes in England and America* (Madison, 1978), 105–9; Browder, *Lincoln and the Communists* (New York, 1936), 58; Wagner, in *Congressional Record*, 75 Cong., 1 Sess., Appendix, 374; *Philadelphia Record*, in *ibid.*, 73 Cong., 2 Sess., 1870; La Guardia, in *New York Times*, February 17, 1936.

11. *Public Papers and Addresses of Franklin D. Roosevelt* (New York, 1938), V, 557–58; *Collected Works of Abraham Lincoln*, Roy P. Basler, ed. (New Brunswick, 1953), VII, 301–2, VIII, 578.

12. *Public Papers*, V, 557–58.

13. Sandburg to Roosevelt, March 29, 1935, in *Letters of Carl Sandburg*, Herbert Mitgang, ed. (New York, 1968), 317–19; Robert E. Sherwood, *Abraham Lincoln in Illinois* (New York, 1939), 190 and *passim*; see also, Jones, *Image Brokers*, Ch. 3, and John Mason Brown, *The Worlds of Robert E. Sherwood, 1896–1939* (London, 1965), Ch. 28; Massey is quoted in *New York Times*, Section 9, October 30, 1938.

14. *Harper's*, v. 180 (1940), 133–36; Gabriel, *Course*, 413; Lerner, *Ideas Are Weapons* (New York, 1939), 48–53, and in *Nation*, v. 101 (1940), 753–54; Sandburg, *Home Front Memo* (New York, 1943), 29–30, and Penelope Niven, *Carl Sandburg: A Biography* (New York, 1991), 539.

15. *Public Papers*, X, 329; Niven, *Sandburg*, 558; Buck is quoted in Ted Morgan, *FDR: A Biography* (New York, 1985), 766–77.

16. See Leuchtenburg's *In the Shadow of FDR* (Ithaca, 1983); *Memoirs of Harry S. Truman* (New York, 1955), I, 120, II, 443; Richard F. Haynes, *The Awesome Power: Harry S. Truman as Commander in Chief* (Baton Rouge, 1973), 261.

17. *Truman in the White House: The Diary of Eben A. Ayers*, Robert A. Ferrell, ed. (Columbia, Mo., 1991), 171–72; *New York Times*, March 7, 1947; in general, William Seale, *The President's House: A History*, 2 v. (Washington, 1986); *Sidewalks of America*, B. A. Botkin, ed. (Indianapolis, 1954), 183–87.

18. Sherman Adams, *Firsthand Report: The Story of the Eisenhower Administration* (New York, 1961), 299; Arthur Larson, *Eisenhower: The President Nobody Knew* (New York, 1968), 134–36; *Public Papers*, II, 220; III, 422; Emmet Hughes, *The Ordeal of Power* (New York, 1963), 348.

19. *Papers of Adlai Stevenson*, Walter Johnson, ed. (Boston, 1972–79), III, 515–20, 468–71, IV, 187–88, VI, 55, 263, 766; in general, Kenneth S. Davis, *A Prophet in His Own Country . . . Stevenson* (Garden City, N.Y., 1957), 276, 506, *passim*; John B. Martin, *Adlai Stevenson of Illinois* (Garden City, N.Y., 1976), 479–99; and Porter McKeever, *Adlai Stevenson, His Life and Legacy* (New York, 1989), 250–51. For Zane's and Hay's recollections (footnote) see: Charles S. Zane, "Lincoln as I Knew Him," *Journal Illinois State Historical Society*, v. 14 (1921), 75; William Roscoe Thayer, *Life and Letters of John Hay* (Boston, 1915), I, 215; Thomas, *Lincoln*, 193.

20. Sorensen, *Kennedy* (London, 1965), 234, 241; *New York Times*, December 2, 23, 1963; *Boston Transcript*, November 24, 1963. On the cartoon which first appeared in the *Chicago Sun-Times*, see Donald R. Wrane, "Lincoln: Democracy's Touchstone," *Papers of Abraham Lincoln Association* (1979), 81.

21. Randall, *Lincoln the President: Midstream* (New York, 1952), Appendix; Ruth Painter Randall, *I Ruth* (Boston, 1968), Ch. 1; Sandburg's report, *New York Times*, July 27, 1947; the broadcast transcript, July 26, 1947, in David C. Mearns Papers, Library of Congress.

22. Randall to Adams, December 31, 1947, Randall Papers, Library of Congress; Lorant, in *Life*, v. 23 (August 25, 1947), 45–51; Merle Miller, *Plain Speaking: An Oral History of President Truman* (New York, 1973), 31; Sandburg, in *New York Times*, July 27, 1947.

23. Randall, *Midstream*, 427, 436, 446; see also, Randall, "The Great Dignity of the Rail Splitter," *New York Times Magazine*, February 8, 1948.

24. See especially, Thomas F. Schwartz, "Lincoln's Published Writings: A History," *Journal of the Abraham Lincoln Association*, v. 9 (1987), 19–34; and Roy P. Basler, *A Touchstone for Greatness* (Westport, Conn., 1973), 3–52.

25. Robert J. Donovan, *Eisenhower: The Inside Story* (New York, 1956), 207; T. Harry Williams, review article in *Mississippi Valley Historical Review*, v. 40 (1953), 89–106.

26. *New York Herald Tribune Book Review*, November 9, 1952.

27. Current, *The Lincoln Nobody Knows* (New York, 1958), 19 and *passim*; Donald, *Lincoln Reconsidered* (New York, 1956), 65, Ch. 7, *passim*.

28. *Life*, v. 35 (November 1, 1948), 66–74; *New York Times*, July 29, 1962; for convenient summaries, see *Chicago Tribune Magazine*, January 10, 1982, and *Parade*, December 12, 1982.

29. Rossiter, *Constitutional Dictatorship* (Princeton, 1948), 3, 239, Ch. 15 *passim;* Corwin, *Total War and the Constitution* (New York, 1957), 16; Dietze, *America's Political Dilemma: From Limited to Unlimited Democracy* (Baltimore, 1968), 53.

30. Meyer is quoted in George R. Nash, *The Conservative Intellectual Movement in America Since 1945* (New York, 1956), 212; Beard, *The Republic* (New York, 1943), 65, 71, quoting address of November 10, 1864; Herman Belz, "Lincoln and the Constitution: The Dictatorship Question Reconsidered," *Congress and the Presidency*, v. 15 (1988), 147–64, which may be compared to the Seventh Annual R. Gerald McMurtry Lecture (Fort Wayne, 1984).

31. Beale's essay is Ch. 3, in *Theory and Practice of Historical Study* (New York, 1946); DeVoto, in *Harper's*, v. 192 (1946), 123–26.

32. Schelsinger, "The Causes of the Civil War: A Note on Historical Sentimentalism," *Partisan Review*, v. 16 (1949), 969–81; Pressly, *Americans Interpret Their Civil War* (Princeton, 1954), Ch. 7.

33. Jaffa, *Crisis of the House Divided* (New York, 1959), 23, 302, 227–28, and "In Defense of Political Philosophy," *National Review*, v. 34 (1982), 41; Kendall and George W. Carey, *The Basic Symbols of the American Political Tradition* (Baton Rouge, 1970), 88–95, 155–56.

34. Archibald Rutledge, "Lincoln and the Theory of Secession," *South Atlantic Quarterly*, v. 41 (1942), 370–83, and see Roy Basler's response, *ibid.*, v. 42 (1943), 45–53; *Washington Post*, June 3, 1947, and *Lincoln Lore*, no. 948 (1947); Ramsdell, "Lincoln and Fort Sumter," *Journal of Southern History*, v. 3 (1937), 259–89; Tilley, *Lincoln Takes Command* (Chapel Hill, 1941), 248–49, 312; and Richard H. Current, *Lincoln and the First Shot* (Philadelphia, 1963), 189.

35. Randall, "When War Came in 1861," *Abraham Lincoln Quarterly*, v. 1 (1940), 41; Stampp, in *Mississippi Valley Historical Review*, v. 29 (1942), 438–39.

36. Williams, *Lincoln and His Generals* (New York, 1952), 291, 304–5, and *passim*.

37. Donald, *Lincoln Reconsidered*, Ch. 6, and Williams's response, *Selected Essays* (Baton Rouge, 1983); Cox and Cox, *Politics, Principles, and Prejudice* (Glencoe, Ill., 1963), 44; Trefousse, *The Radical Republicans* (New York, 1969), 121.

38. On Eisenschiml see particularly, William Hanchett, "The Historian as Gamesman: Otto Eisenschiml, 1880–1963," *Civil War History*, v. 36 (1990), 5–16; "Abraham Lincoln and the Rothschilds," *Social Justice*, February 1940, 8–9; in addition to Bates's book, see Francis Wilson, *John Wilkes Booth: Fact and Fiction in Lincoln's Assassination* (Boston, 1929), and Bernie Babcock, *Booth and the Spirit of Lincoln: The Story of a Living Dead Man* (Philadelphia, 1925); *Life*, v. 5 (July 11, 1938), 56.

39. Hill, *Lincoln the Lawyer* (New York, 1906), 194; Duff, *A. Lincoln, Prairie Lawyer* (New York, 1960), 172, 43, 149, Chs. 9, 22; Weik, "Lincoln and the Matson Negroes," *Arena*, v. 17 (1897), 752–58; Frank, *Lincoln as a Lawyer* (Urbana, 1961), 172; see also, Albert A. Woldman, *Lawyer Lincoln* (Boston, 1936).

40. Simon, *Lincoln's Preparation for Greatness: The Illinois Legislative Years* (Norman, 1965), Ch. 4, p. 292; Fehrenbacher, *Prelude to Greatness* (Stanford, 1962), 18, 161, and Ch. 4.

41. Shaw, *The Lincoln Encyclopedia* (New York, 1950), 20, 40, 285. For the "cannots" (footnote), see: *Congressional Record*, 81 Cong., 1 Sess., 534, and 88 Cong., 2 Sess., 2774; *Look*, v. 14 (January 17, 1950), 88; *Time*, v. 55 (January 30, 1950), 59; *Lincoln Lore*, no. 1085 (1950), *Washington Post*, August 19, 1992.

42. Hamilton and Ostendorf, *Lincoln in Photography* (Norman, 1963), 294, 234–35; on Lorant, see *Life*, v. 12 (January 5, 1942), 50–55, v. 38 (February 14, 1955), 22–24, and *Saturday Evening Post*, v. 220 (July 19, 1947), 25; Lorant, *Lincoln, a Picture Story of His Life*, rev. ed. (New York, 1957); Robert S. Harper, "As You Never Saw Him Before," *Collier's*, v. 131 (February 14, 1953), 20–21; Wilson, *Lincoln in Portraiture* (New York, 1935).

43. Schlesinger quoted in Michael Kammen, *A Season of Youth* (New York, 1978), 152; *Time*, v. 64 (August 30, 1954), 83.

44. Williams, *House-Divided* (New York, 1947), Prologue, 480, 491–92, 767–69, 1060, 1273, 1375, and *passim*.

45. *New York Times*, May 3, 1925; James F. Evans, *Prairie Farmer and WLS* (Urbana, 1969), 196–97; "The Lonesome Train," in *Radio Dramas in Action: Twenty-five Plays*, Erik Barnouw, ed. (New York, 1945); *Prologue to Glory*, in *Federal Theatre Plays* (New York, 1938).

46. A. M. R. Wright, *The Dramatic Life of Abraham Lincoln* (New York, 1925), and *Literary Digest*, v. 80 (February 23, 1924), 30–31; *ibid.*, v. 108 (September 20, 1930), 15–16, and Richard Schickel, *D. W. Griffith: An American Life* (New York, 1984), 555–57.

47. The two films, like Griffith's and others discussed herein, may be viewed on videocassette.

48. Levin, "Abraham Lincoln Through the Picture Tube," *Reporter*, v. 8 (April 14, 1953), 31–33; Laurence Balgreen, *James Agee: A Life* (New York, 1984), 370, 368–74; Jack Gould, in *New York Times*, February 26, 1956; Van Doren, *The Last Days of Lincoln* (New York, 1959), 141, and *passim*, and *Autobiography* (New York, 1958), 336–37; *New York Times*, October 19, 1961.

49. See, in general, Kenneth A. Bernard, *Lincoln and the Music of the Civil War* (Caldwell, Idaho, 1966); Aaron Copland and Vivian Perles, *Copland: 1900 Through 1942* (New York, 1984), 342–45.

50. *The Life and Writings of Frederick Douglass*, Philip S. Foner, ed. (New York, 1955), IV, 360; Maynard P. Turner, Jr., "The Emancipation in Retrospect and Prospect," *Negro History Bulletin*, v. 22 (1959), 161–63; John Haley, *Charles H. Hunter and Race Relations in North Carolina* (Chapel Hill, 1987),

272; DuBois quoted in John David Smith, "Black Images in the Age of Jim Crow," *Lincoln Lore*, no. 1681 (1978), 3, and in *Crisis*, v. 24 (122), 103, 199; clipping, *Chicago Daily News*, December 24, 1986, Lincoln Museum.

51. Quarles, *Lincoln and the Negro* (New York, 1962), Foreword, 18, 36, 93–97, 123, 299, *passim*.

52. Hofstadter, *The American Political Tradition* (New York, 1948), 110, 115, 131; Jaffa, *Crisis*, Ch. 17. For two different views, see George M. Fredrickson, "A Man but Not a Brother," *Journal of Southern History*, v. 41 (1975), 39–48, and Don E. Fehrenbacher, "Only His Stepchildren," in *Lincoln in Text and Context* (Stanford, 1987), 95–112; on the Wadsworth letter, see Ludwell E. Johnson, "Lincoln and Equal Rights: The Authenticity of the Wadsworth Letter," *Journal of Southern History*, v. 32 (1966), 83–87, and Harold M. Hyman, "Lincoln and Equal Rights for Negroes: The Irrelevancy of the 'Wadsworth Letter,' " *Civil War History*, v. 12 (1966), 258–61.

53. For Lincoln and modern "repatriation," see Ernest S. Cox, *White America* (Richmond, 1923) and *Lincoln's Negro Policy* (Richmond, 1938); expanded ed. 1972); *Congressional Record*, 76 Cong., 1 Sess., 4660–71; Taylor Branch, *Parting the Waters: America in the King Years, 1954–63* (New York, 1988), 170–71.

54. Marian Anderson, *My Lord, What a Morning* (New York, 1956), Ch. 17; T. H. Watkins, *Righteous Pilgrim: The Life and Times of Harold L. Ickes* (New York, 1990), 249–53; *Washington Post*, April 19, 1939.

55. For King, see Branch, *Parting the Waters*, 213, 215–218, and *passim*; *Chicago Defender*, May 11, 25, 1957; *Cooper v. Aaron*, 358 U.S. 1 (1958).

56. Branch, *Parting the Waters*, 518, 555, 572.

57. "An Appeal to the Honorable John F. Kennedy . . . for National Rededication to the Principles of the Emancipation Proclamation," May 17, 1962, King Library and Archives; Branch, *Parting the Waters*, 589–90, 685, 687, 694.

58. *Ibid.*, Ch. 22; Thomas Gentile, *March on Washington: August 28, 1963* (Washington, 1985); *Atlanta Constitution*, August 29, 1963; A *Testament of Hope: The Essential Writings of Martin Luther King*, James M. Washington, ed. (New York, 1986), 216–20; David Howard-Pitney, *The Afro-American Jeremiad* (Philadelphia, 1990), 4–5.

59. *Testament of Hope*, 169 (my italics); *Negro History Bulletin*, v. 28 (1965), 119.

60. Julius Lester, *Look Out, Whitey!* (New York, 1968) 58; Bennett, in *Ebony*, v. 23 (1968), 35–42, and see reply by Herbert Mitgang, "Was Lincoln Just a Honkie?" *New York Times Magazine*, February 11, 1968; Vincent Harding, *There Is a River: The Black Struggle for Freedom in America* (New York, 1981), 236; and, in general, Stephen B. Oates, *Abraham Lincoln: The Man Behind the Myth* (New York, 1984), 21–30.

61. Charles Preston, Communication, *Freedomways*, v. 5 (1965) 426; *Wall Street Journal*, advertisement, February 11, 1964; Neil L. McMillen, *The Citizens' Councils* (Urbana, 1971), 181–83; Putnam, *Race and Reason: A Yankee View* (Washington, 1961), 60–62; Willmore Kendall, *The Conservative Affir-*

mation (Chicago, 1963), 249–53, and with Carey, *Basic Symbols*, 88–95, 155–56.

62. On Chesterton's observation and its explication, see Sidney Mead, *The Nation with the Soul of a Church* (New York, 1975); *New York Times*, February 8, 1954, February 13, 1962; Ward, "That All Have an Equal Chance," *New York Times Magazine*, February 15, 1956.

63. Wolf, *The Religion of Abraham Lincoln* (New York, 1963); *Collected Works*, IV, 236 (my italics), I, 382; Wilson, *Patriotic Gore* (New York, 1962), 101; see also, Elton Trueblood, *Abraham Lincoln, Theologian* (New York, 1972), and Glen E. Thurow, *Abraham Lincoln and American Political Religion* (New York, 1976).

64. Editorial, *Christian Century*, v. 60 (1943), 160; Niebuhr, *The Irony of American History* (New York, 1952), 171–74, and "The Religion of Abraham Lincoln," *Christian Century*, v. 82 (1965), 172–75, which is based on an address included in *Lincoln and the Gettysburg Address*, Allan Nevins, ed. (Urbana, 1964).

65. The seminal article, originally published in *Daedalus*, is in *American Civil Religion*, Russell E. Richey and Donald G. Jones, eds. (New York, 1974), 31–32; Richard V. Pierard and Robert Linder, *Civil Religion and the Presidency* (Grand Rapids, 1988), Ch. 4.

66. Wilson, "A Historian's Approach to Civil Religion," in Richey and Jones, *Civil Religion*, 122; for the five meanings, Introduction and Ch. 7 (Martin Marty, "Two Kinds of Two Kinds of Civil Religion"), in *ibid.*, also, William J. Wolf, *Freedom's Holy Light: American Identity and the Future of Theology* (Wakefield, Mass., 1977), Ch. 3; Warner, *The Living and the Dead: A Study of the Symbolic Life of Americans* (New Haven, 1959), 270; Wilson, *Patriotic Gore*, 99–125, 95.

67. See *Abraham Lincoln Sesquicentennial* 1959–60; *Final Report* . . . (Washington, 1960).

68. On Lincoln's birthday, see *The American Book of Days*, 3d ed., James Hatch, ed. (New York, 1978), 169–74.

69. *New York Herald Tribune*, February 12, 1949; Schwengel quoted in *Final Report*, 48; advertisement in *Life*, v. 32 (February 11, 1952), 7; *New York Times*, February 12, 1963, February 12, 1962.

70. On Springfield, see especially, A. J. Liebling, "Abe Lincoln in Springfield," *New Yorker*, v. 26 (June 24, 1950), 29–48; *New York Times*, February 13, 1955, February 12, 1961; Ralph Gray, "Vacation Tour Through Lincoln Land," *National Geographic Magazine*, v. 101 (1952), 141–84; *New Yorker*, v. 40 (April 4, 1964), 84–87; Calvin M. Logue, *Ralph McGill Speaks* (Durham, 1969), 154–67; *Lincoln Group of Boston*, 1938–1988, in Robert Barton Papers, University of Illinois.

71. *Final Report*, 69–99. The General Files of the Lincoln Sesquicentennial Commission are in the National Archives.

72. Craven, "Southern Attitudes Toward Abraham Lincoln," *Papers in Illinois History* (1942), 10; *New York Times*, February 13, 14, 1954.

73. *Final Report*, 100–126.

74. Bruno Lasker, "Lincoln and World Human Rights," *Survey*, v. 88 (1952), 82–83; *Jawaharlal Nehru's Speeches* (Calcutta, 1958), III, 46, and M. J. Akbar, *Nehru: The Making of India* (London, 1988), 565–66.

75. *Final Report*, 61–88; Randle Bond Truett, *Lincoln in Philately* (Washington, 1959) and *Supplement* (Washington, 1961); *New York Times*, January 25, 1959.

76. *Final Report*, 19–27, 24; *Abraham Lincoln Through the Eyes of High School Youth*, Jean D. Grambs, ed. (Washington, 1959), 6 and *passim: Abraham Lincoln: An Exhibit at the Library of Congress* (Washington, 1959); *Chicago Defender*, February 24, 1959.

77. *Final Report*, 31–36; Introduction, *Lincoln Day by Day: A Chronology 1809–1865*, Earl S. Miers, ed., 3 v. (Washington, 1960).

78. *Final Report*, 127–31; Victor Searcher, *Lincoln Today: An Introduction to Modern Lincolniana* (New York, 1969), surveys films, music, and miscellaneous Lincolniana.

79. Quoted in Sherman Paul, *Edmund Wilson* (Urbana, 1965), 203; Wilson, *Patriotic Gore*, xvii, 99–130, 122, 106; Foote, *The Civil War: A Narrative* (New York, 1958), I. Ch. 1.

80. *Final Report*, 38–46; Niven, *Sandburg*, 679–81; *New York Times*, February 13, 1959; the speech and proceedings are in *Congressional Record*, 86 Cong., 1 Sess., 2265–66.

81. *Newspaper Columns by William E. B. Du Bois*, Herbert Aptheker, ed. (White Plains, N.Y., 1986), II, 1023–24; Michael Kammen, *Mystic Chords of Memory* (New York, 1991), 590–610; *Washington Post*, September 22, 23, 1962; *Afro-American* (Baltimore), January 5, 19, 1963.

82. *New York Times*, November 20, 1963; *Lincoln and the Gettysburg Address*, 89; *A Portion of That Field: The Centennial of the Burial of Lincoln* (Urbana, 1967).

83. *Ibid.*, 27, 15; Warren, *The Legacy of the Civil War: Meditations on the Centennial* (New York, 1961), 4, 54–66, 107; Catton is quoted in *Portion of That Field*, 84; *New York Times*, February 12, 1966.

8 *Lincoln Everlasting*

1. Williams, in statement to the author, 1991; see, in general, Stephen B. Oates, *The Man Behind the Myth* (New York, 1984).

2. Johannsen, "In Search of the Real Lincoln: Or Lincoln at the Crossroads," *Journal of Illinois State Historical Society*, v. 61 (1968), 231, 229–47.

3. On the ALA, see *Lincoln Lore*, no. 1697 (1979). Since 1969 Frank J. Williams has authored an annual article, "Lincolniana," in the *ALA Journal*, which tracks scholarship and related activities in the field; see also the quarterly "Lincoln News Digest" in the *Lincoln Herald*, published at Lincoln Memorial University. On history and heritage, see Patrick Wright, *On Living in an Old Country* (Thetford, 1985) and Peter Burke, "History as Social Memory," in

Memory: History, Culture, and the Mind, Thomas Butler, ed. (Oxford, 1989); on Springfield, *Chicago Tribune,* February 12, 1988; on presenters, *Washington Post,* July 4, 1992; on Sanders, *New York Times,* August 16, 1973.

4. *Ibid.,* May 18, 1984, March 23, August 9, 1966; Williams, "Lincolniana in 1991," *ALA Journal,* v. 13 (1992), 93; *Washington Post,* November 21, December 17, 1992.

5. On Marfan's, see Gabor S. Boritt and Adam Boritt, "Lincoln and the Marfan Syndrome: The Medical Diagnosis of a Historical Figure," *Civil War History,* v. 29 (1983), 212–29, and Gabor Boritt, *How Big Was Lincoln's Toe? or, Finding a Footnote* (Redlands, Calif. 1989); and Williams, "Lincolniana in 1991," *ALA Journal,* v. 13 (1992), 95–96; *Time,* v. 111 (May 22, 1978), 83; *New York Times,* February 10, May 3, 1991.

6. Current, in *The Historian's Lincoln,* Gabor S. Boritt and Norman O. Forness, eds. (Urbana, 1988), 384. In addition to the biography, see Oates's *Man Behind the Myth.*

7. *Abraham Lincoln and the American Political Tradition,* John L. Thomas, ed. (Amherst, 1986); see also, the "Lincoln and Lincolniana" issue of *Books at Brown,* v. 31–32 (1984–85), *Lincoln and the American Political Tradition: An Exhibition . . .* Jennifer B. Lee, curator (Providence, 1984), and Williams, "Lincolniana in 1984," *Papers of the ALA,* v. 6 (1984), 29–35.

8. Oates, *Man Behind the Myth,* 30; Alexander, in *Vital Speeches,* v. 52 (1986), 326–29; Tice, *Lincoln* (New Brunswick, 1984); Meehan, in *New Yorker,* v. 47 (August 28, 1971), 35; for Newhart, *New York Times,* February 19, 1986; *Los Angeles Times,* February 12, 1988; on the Boston monument, *Chicago Tribune,* February 28, 1982; on Disneyland, see AP dispatches, August 5, 19, 1990, and Williams, in *ALA Journal,* v. 13 (1992), 94.

9. *Washington Post,* August 3, 1977, for both the Anderson-Whitten column and one by Haynes Johnson; *ibid.,* April 14, 1985, for feature on Booth; for Kirstein's play, see *New York Times,* August 9, 27, 1967; for Shyre, *ibid.,* August 7, 1969; Hewitt, in *Washington Post,* April 16, 1984, and Whitmore, *Ibid.,* February 3, 1993. Neely, in "The Lincoln Theme Since Randall's Call," *ALA Papers,* v. 1 (1979), 44–45; Hanchett, *Lincoln Murder Conspiracies* (Urbana, 1983), 186.

10. On McPherson, see especially, *Abraham Lincoln and the Second American Revolution* (New York, 1990), Chs. 1 and 2; Jaffa, article on Lincoln in *Encyclopedia of the Constitution,* Leonard Levy, ed. (New York, 1987), III, 1162–63.

11. Wilson's article in *The New Yorker* was collected in *Eight Essays* (New York, 1954), 182–202; Lyceum Address, in *Collected Works,* I, 114, and *passim;* Jaffa, *Crisis of the House Divided* (New York, 1959), Ch. 9. For another reading: Thomas F. Schwartz, "The Springfield Lyceum and Lincoln's 1838 Speech," *Illinois History Journal,* v. 83 (1990), 45–49.

12. For a statement of Nevins's view, see *The Emergence of Lincoln* (New York, 1950), II, 468; Arendt, *Crisis of the Republic* (New York, 1972), 61; see also, David Lighter, "Abraham Lincoln and the Ideal of Equality," *Journal Il-*

linois State Historical Society v. 75 (1982), 289–309, and my own *"This Grand Pertinacity": Abraham Lincoln and the Declaration of Independence* (Fort Wayne, 1992); for Bradford, see his "The Heresy of Equality: A Reply to Harry Jaffa," in *A Better Guide than Reason* (Lasalle, Ill. 1979), 42–50, "Against Lincoln, An Address at Gettysburg," *Historian's Lincoln*, 107–15, and "The Lincoln Legacy: A Long View," in *Remembering Who We Are: Observations of a Southern Conservative* (Athens, 1985); Will, in *Washington Post*, November 11, 1981.

13. Franklin, "The Two Worlds of Race," in *Race and History: Selected Essays, 1938–1988* (Baton Rouge, 1989), 138; Robinson, in *Historian's Lincoln*, 206; Cox, *Lincoln and Black Freedom: A Study of Presidential Leadership* (Columbia, 1981), and McCrary, *Abraham Lincoln and Reconstruction* (Princeton, 1987); Lightner, "Ideal of Equality," 305, and Boritt, "The Voyage to the Colony of Lincolnia: The Sixteenth President, Black Colonization, and the Defense Mechanism of Avoidance," *Historian*, v. 37 (1975), 619–32.

14. *The Papers of Woodrow Wilson*, Arthur Link, ed. (Princeton, 1966), XXXVIII, 142–45; Croly, "The Paradox of Lincoln," *New Republic*, v. 21 (1920), 352.

15. *Collected Works*, IV, 438; the diary passage, May 7, 1861, is in Tyler Dennett, *Lincoln and the Civil War* (New York, 1939), 19–20; and in Nicolay and Hay's *History*, IV, 258; Safire, *Freedom* (New York, 1987), 123, and see his column, *New York Times*, February 24, 1986.

16. "Abraham Lincoln and Our Unfinished Business," in *Building the Myth: Selected Speeches*, Waldo W. Braden, ed. (Urbana, 1990), 232–43; Will, in *Washington Post*, January 4, 1990; Holzer, in *American Heritage*, v. 34 (1983), 56–64; Auth, in *Boston Globe*, November 10, 1986.

17. "Commemoration Ode," *Works* (Boston, 1890), X, 22–24; Donald Capps, "Lincoln's Martyrdom: A Study of Exemplary Mythic Patterns," in *The Biographical Process: Studies in the History and Psychology of Religion*, Frank E. Reynolds and Donald Capps, eds. (The Hague, 1976) discusses the conflict between the vulgar and the spiritual Lincoln; Adin Baber, *The Hanks Family of Virginia and Westward* (Kansas, Ill. 1965). For children's literature, see Henry Mayer, "Abe, Honestly and Otherwise," *New York Times Book Review*, February 12, 1989.

18. *Collected Works*, IV, 438; Cuomo, in Braden, *Building the Myth*, 240–41; Hofstadter, *The American Political Tradition and the Men Who Made It* (New York, 1948), 104–5, and Boritt, *Lincoln and the Economics of the American Dream* (Memphis, 1978); Thomas L. Purvis, "The Making of a Myth," *Journal Illinois State Historical Society*, v. 75 (1982), 149–60.

19. Hofstadter, *American Political Tradition*, 105; Kincaid, "The Ageless Lincoln," in *Vital Speeches*, v. 16 (1950), 305; Steven Wilson, orally to the author in 1990.

20. *New York Times*, February 13, 1976, March 29, 1987; *Washington Post*, June 8, 1976.

21. "The Lincoln Relics," in *The Poems of Stanley Kunitz, 1928–1978* (Boston, 1979), 22–25.

22. *U.S. News and World Report*, cited in *ALA Journal*, v. 11 (1990), 61; on the Wolper production, *New York Times*, September 6, 1974; Mitgang, *Mister Lincoln: A Drama in Two Acts* (Carbondale, Ill., 1982), and John O'Connor's review, *New York Times*, February 9, 1981; Lawrence Shyer, "Robert Wilson: Current Projects," *Theater*, v. 14 (1983), 83–97; on Burns and film, see Frank J. Williams, "Lincolniana in 1991," *ALA Journal*, v. 13 (1992), 97–98.

23. See, in general, Don E. Fehrenbacher, *Lincoln in Text and Context* (Stanford, 1987), Ch. 16; Langer's address is included in *Our Selves/Our Past: Psychological Approaches to American History*, Robert Brugger, ed. (Baltimore, 1981).

24. In addition to Forgie's book, see his op-ed article in *New York Times*, February 12, 1979, and his presentation in *Historian's Lincoln*, 285–301, which offer minor variations; Richard O. Curry, "Conscious or Subconscious Caesarism," *Journal of Illinois State Historical Society*, v. 77 (1984), 70.

25. In addition to the book, see the presentation in *Historian's Lincoln*, together with Robert Bruce's comments, 253–78.

26. *Ibid.*, 211–52, for presentation and comments by Jean Baker and Herman Belz; Fehrenbacher's review in *Reviews in American History*, v. 11 (1983), 14–16, was later expanded into the essay in *Text and Context*; Strozier, *Quest for Union*, 233.

27. Simon, "Abraham Lincoln and Ann Rutledge," *ALA Journal*, v. 11 (1990), 13–33; Wilson, "Abraham Lincoln, Ann Rutledge, and the Evidence of Herndon's Informants," *Civil War History*, v. 36 (1990), 301–23.

28. Vidal, in *Los Angeles Times*, February 8, 1981, and letters to the editor, February 12, 21, 1981; "Gore Vidal's Lincoln: An Exchange," *New York Review of Books*, v. 35 (April 29, 1988), 58.

29. *Lincoln* (New York, 1984) 657; see also, Vidal's contribution to *Paths to Resistance: The Art and Craft of the Political Novel*, William Zinsser, ed (Boston, 1989); among critiques by historians: Richard N. Current, "Fiction and History," *Journal of Southern History*, v. 52 (1986), 77–90; Roy P. Balser, "Lincoln and American Writers," *Papers of the ALA*, v. 7 (1985), 7–17, and Don E. Fehrenbacher, in *Historian's Lincoln*, 387–91, and *Text and Context*, 240–43; Safire, *Freedom*, 977, 1110.

30. Vidal, "Of History and Hagiography," *Los Angeles Times*, March 20, 1988; "Gore Vidal's Lincoln," *New York Review of Books*, v. 35 (April 29, 1988), 58, and (August 18, 1988), 66–69; Current, in *Christian Science Monitor*, June 28, 1988.

31. "Communications," *American Historical Review*, v. 96 (1991), 324–28.

32. James, *Memories and Studies* (New York, 1912), 268; Sandburg, *Congressional Record*, 86 Cong., 1 Sess., 2265.

33. *Lincoln on Democracy* (New York, 1990), Preface; Williams, "Lincolniana in 1991," *ALA Journal*, v. 13 (1992), 68; McPherson, *Lincoln and the Second American Revolution* (New York, 1991), Ch. 5; *Safire's Political Dictionary* (New York, 1978), 256; Adler and Gorman, *The American Testament* (New York, 1975), and Lowell, in *Lincoln and the Gettysburg Address*, Allan Nevins,

ed. (Urbana, 1964), 89; Wills, *Lincoln at Gettysburg* (New York, 1992); see also, Peterson, *"This Grand Pertinacity,"* Herman Belz, "The 'Philosophical Cause' of 'Our Free Government and Consequent Prosperity': The Problem of Lincoln's Political Thought," *ALA Journal*, v. 10 (1988–89), 47–71, and Cushing Strout, "American Dilemma: Lincoln's Jefferson and the Irony of History," in *Making American Traditions* (New Brunswick, 1990), 133–51.

34. *Collected Works*, V, 537; Bruce, *Lincoln and the Riddle of Death* (Fort Wayne, 1981), 20–23.

ACKNOWLEDGMENTS

I wish to thank my colleagues at the University of Virginia, especially William W. Abbot, Norman Graebner, and William H. Harbaugh, for encouragement to undertake this work. The support of distinguished historians elsewhere, David Donald, Don E. Fehrenbacher, and Michael Kammen, was also important to me. In the course of my research I received valued assistance from Thomas F. Schwartz of the Illinois State Historical Library; Mark E. Neely, Jr., formerly director of the Lincoln Museum, in Fort Wayne, and his associate, Ruth Cook; Robert W. Johannsen of the University of Illinois; Jennifer B. Lee, Curator of the Lincoln Collection, Brown University; Brenda M. Lawson of the Massachusetts Historical Society, and others too numerous to mention. Special thanks go to my friend Knight Aldrich for the end papers. The interlibrary loan office of Alderman Library, University of Virginia, proved indispensable to the work; and as in the past I am indebted to the library and its former director, Ray W. Frantz, Jr. for many aids. Frank J. Williams, president of the Abraham Lincoln Association, himself a Lincoln scholar, collector, and bibliophile, rendered me the inestimable service of reading the manuscript, chapter by chapter, as it came from my typewriter. Expert word processing was provided by Donna Packard.

I thank the following for permission to quote from certain copyrighted works: Harcourt Brace & Company for *Chicago Poems* and *The People, Yes*, by Carl Sandburg; Henry Holt and Company for *John Brown's Body* by Stephen Vincent Benét and *A Book of Americans* by Rosemary and Stephen Vincent Benét; Little, Brown and Company for *The Poems of Stanley Kunitz 1928–1978*; Penguin USA for *Selected Poems of Alfred Kreymbourg 1912–1944*; University of California Press for *Collected Poems* by Charles Olson; Estate of James Oppenheim for "Memories of Whitman and Lincoln" by James Oppenheim; Shawnee Press, Inc. for *The Lonesome Train* by Earl Robinson and Millard Lampell; Charles Scribner's Sons, an imprint of Macmillan Publishing Company, for *R. E. Lee: A Biography*, vol. III, by Douglas Southall Freeman.

INDEX

NOTE: AL is the abbreviation of Abraham Lincoln. Italicized page numbers indicate photographs.